Elements of
the Mechanical Behavior
of Solids

Elements of
the Mechanical Behavior
of Solids

NAM P. SUH
Massachusetts Institute of Technology
Cambridge, Massachusetts

and

ARTHUR P. L. TURNER
Argonne National Laboratory
Argonne, Illinois

SCRIPTA BOOK COMPANY
Washington, D.C.

McGRAW-HILL BOOK COMPANY
New York   St. Louis   San Francisco   Auckland   Düsseldorf   Johannesburg
Kuala Lumpur   London   Mexico   Montreal   New Delhi   Panama
Paris   São Paulo   Singapore   Sydney   Tokyo   Toronto

Library of Congress Cataloging in Publication Data

Suh, Nam P.,   date.
    Elements of the mechanical behavior of solids.

    1. Strength of materials.   I. Turner, Arthur P. L.,
joint author.   II. Title.
TA405.S824   1975   620.1'12   74-10949
ISBN 0-07-061765-1

This book was set in Baskerville by Scripta Graphica.
The editor was B. J. Clark; the designer was Victor Enfield;
the compositor was Rebekah McKinney; and the
production supervisor was Keith Wilkinson.
The printer and binder was The Kingsport Press.

To
Young and Marianne

Contents

III   ELASTIC BEHAVIOR 68

IV   PLASTIC RESPONSE—CONTINUUM TREATMENT 125

V MICROSCOPIC BASIS OF PLASTIC BEHAVIOR 237

VI VISCO-ELASTIC-PLASTIC DEFORMATION OF POLYMERS . . . 293

Preface

The mechanical behavior of materials refers to a subject matter which deals with the response of materials under the influence of external loads and displacements. The materials responses treated in this book are deformation, fracture, friction, and wear. The subject matter is founded on the disciplines of applied mechanics and materials science and engineering, and bridges the gap between them. Understanding of the subject is essential to work in such fields as materials processing and machine design. However, it has been one of the most difficult subjects to teach and to learn because of its diverse nature and its heavy dependence on empiricism. This book attempts to give the subject a unified, logical treatment based on the mechanical engineering students' prior background in elementary applied mechanics. The emphasis of the book is on the physical understanding and the quantitative analysis of the macroscopic behavior of solids rather than qualitative treatment of the microscopic mechanisms. Examples are used extensively throughout the book to clarify the statements made in the main text.

Preliminary versions of the book have been used at MIT and two other universities during the past three years. It was used as the text in a junior-level one-semester required course at MIT. The prerequisite for the MIT course was the introductory applied mechanics course based on "An Introduction to the Mechanics of Solids" by Crandall, Dahl and Lardner, McGraw-Hill, New York, 1973. In lectures the physical basis of the macroscopic behavior and its significance were discussed extensively so that even the student who did not have a background in the elementary aspects of materials science could successfully follow the topics covered in the book.

We are indebted to our students for their patience and valuable criticisms during the development of the book. During the early stages of our effort we were very fortunate to have the able assistance of Scott C. Holden who, among other things, solved many of the homework problems and prepared tables and figures. Michael Shakespear and Lance Antrim also provided similar assistance during the latter stages of the development. It is our great pleasure to acknowledge the unreserving support given to this undertaking by Professor Ascher H. Shapiro. Our special thanks are also due to Professors Nathan H. Cook and E. Rabinowicz for their valuable comments, criticisms, and assistance. We have also benefited enormously through our association with Professors Ali S. Argon and Frank A. McClintock. Our secretaries deserve special thanks for their patience and help.

N. P. Suh and A. P. L. Turner

Elements of
the Mechanical Behavior
of Solids

CHAPTER ONE

Introduction

I-1 SCOPE OF THE SUBJECT MATTER

It is difficult, if not impossible, for a mechanical engineer to
seriously ignore the realm of engineering materials, regardless of his
particular field of interest. One of the first things a new engineer
learns in his profession is that lack of proper engineering materials
and lack of knowledge of those that are available can impose sharp
limitations on engineering design. The purpose of this book is to
provide mechanical engineers with a basic knowledge of the
mechanical behavior of common structural materials. Most mechanical engineers deal with one or more of the following aspects of
materials:

   a) *Selection* of materials for structures, machine components, etc.

   b) *Processing* of these materials into useful parts

   c) *Diagnosis* and *cure* of material failures resulting from fracture,
wear, corrosion, etc.

Unlike the courses in thermodynamics and applied mechanics, the subject of the mechanical behavior of materials cannot be arranged and presented as a logical unified scheme, since our modern understanding of the theory is as yet incomplete. As a consequence, when one comes to deal with engineering materials, he is introduced to what at first appear to be a multitude of diverse subjects. However, after a period of struggle, the bewildered student will find that there is after all an underlying order that ties all materials engineering together. It is our hope that this hidden order can be clearly revealed by relating the fundamentals of the subject matter to real problems and applications. This will be done by discussing the selection, processing, and failure of materials whenever appropriate.

The central unifying scheme for engineering materials would by logic seem to be the arrangement of the constituent atoms, the bond strength between them, and their interactions with other atomic species. Indeed, broad categorization of materials can actually be made on this basis. For example, it is known that metals with high melting points generally also have high yield strengths, that metals with a body-centered cubic structure are more difficult to deform than face-centered cubic metals, and that metallic elements with similar atomic radii can be combined to form alloys. However, a simple atomic treatment of metals cannot explain all microscopic behavior in metals, mainly because all engineering materials have certain defects. The existence of these defects forces us to look at engineering materials macroscopically, i.e., in terms of their bulk properties rather than their atomic properties. In contrast to the atomic regime, materials in the macroscopic regime are often treated as *homogeneous continua*. Between the atomic and macroscopic regimes there lies yet a third regime, which may be called the microscopic regime. On this level, the interest is in the behavior of groups of atoms and in characteristics such as line defects or dislocations, voids, and impurities.

The unfortunate part of the whole story is that these regimes are essentially three separate islands connected by only weak bridges. During the past several decades great progress has been made in strengthening the intercommunication between these islands, but, at the present time, they still stand joined in a fairly elementary, qualitative linkage. Perhaps the treatment of engineering materials may never attain the ultimate in rationality, but still the challenge of quantitatively deriving the behavior of one regime from knowledge of another will always remain.

The subject matter to be covered in this book is concerned with the response of materials to an externally imposed load and/or displacement. It seeks to provide a framework which can be used for the analysis of problems in the fabrication and design of structural members which must meet certain strength and performance requirements. The approach is primarily from a macroscopic point of view with the emphasis on explaining the engineering significance of various experimentally measured quantities such as yield stress, creep rupture life, or relaxation modulus rather than on the relationship between microstructure and mechanical properties. Descriptions of the microscopic processes which give rise to the macroscopic behavior are included to the extent that they give insight into the exceptions to and limitations of the relationships presented. For more extensive treatments of the individual subjects discussed, the reader is referred to specialized books on the specific topics.

I-2 MATERIALS PROBLEMS AS AN OPTIMIZATION PROBLEM

The purpose of the subject matter described at length in Sec. I-1 can be put into a much simpler form. When a material is subjected to external loads, it either deforms or fractures. The purpose of the text is to study how and why materials deform or fracture so as to provide the optimum solution to a given engineering problem. This statement is shown pictorially in Fig. I.1.

In selecting materials for a given application the choice has to be optimized by satisfying constraining parameters, the most important of which is often *cost*. There are also physical constraining parameters such as load, dimension, frequency of loading, loading rate, service temperature, and maximum allowable deformation and wear.

I-3 DEFORMATION

Here, deformation is taken to mean the change in the geometric shape of a body. Depending on the nature of the engineering materials and the magnitude of the applied force, the material can

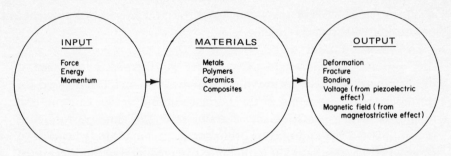

Fig. I.1 Schematic representation of the response of materials to various inputs. In some cases, the input-output roles may be reversed. For example, a magnetic field can produce a stress in a constrained body.

deform in several different ways. It can deform *elastically*, *plastically*, *viscoelastically*, or in a *viscous* manner. Each of these deformations may be briefly defined and characterized as follows:

a) *Elastic deformation* may be defined as a reversible deformation, i.e., after the applied load is removed, the body returns to its original shape and all stored energy can be recovered. The physical process involved is a simple stretching of atomic bonds by changing the distance between atoms. Any time a load is applied, there is a corresponding elastic deformation regardless of whether the stresses are shear, normal or hydrostatic. Therefore, if the magnitude of the applied stress is given, the corresponding deformation can be determined *uniquely*. The material properties which characterize the elastic nature of the isotropic material are Young's modulus E, shear modulus G, bulk modulus K, and Poisson's ratio v. For isotropic materials only two of these properties are independent. Most metals at low stresses and ceramics are elastic. Elastomers such as rubber may be considered elastic at suitable strain rates.

b) *Plastic deformation* is a *permanent* deformation, i.e., after the applied load is removed, the body remains deformed and net work has been done. The physical process involves the sliding of atoms past each other, permanently changing their relative positions. In the applications normally encountered with metals, the normal stress does not significantly affect the sliding process. Therefore, it may be stated that only the shear stress can and does induce plastic deformation in metals. The degree of deformation depends on the loading history and therefore cannot be uniquely specified in terms of stress. Some of the important parameters involved are the critical shear stress for plastic deformation, and work-hardening rate. For most quasi-static loading conditions, plastic deformation is time-independent, i.e., it does not depend on the duration of loading. At high temperatures, metals undergo a time-dependent type of plastic deformation called creep. Most metals at high stresses exhibit plastic properties.

c) *Viscous deformation* is a permanent deformation which is characterized by a time rate of deformation which is proportional to the applied load. Therefore, a viscous material can be deformed substantially by a small load if the duration of loading is long. Net work has to be done to deform viscous materials, and viscous deformation is normally induced by shear stresses only. However, the rate of deformation is influenced by the hydrostatic stress. Viscous materials can be characterized by specifying the viscosity. Most fluids behave in a viscous manner.

d) *Viscoelastic deformation* occurs in certain materials such as plastics. It has aspects similar to both the elastic and viscous behavior. Therefore, the deformation of viscoelastic materials is partially recoverable, but is also time and rate dependent. The viscous work done during deformation is not recoverable. The elastic portion of deformation is induced by any stress components, whereas the viscous portion is governed by only the shear stresses.* Viscoelastic materials are characterized by the relaxation modulus (or creep compliance) and viscosity.

I-4 FRACTURE

After a certain amount of deformation, most materials eventually fracture, i.e., if the applied load is continuously increased, solids separate into two or more pieces unless the material is viscous. Under certain loading conditions some materials may rupture by a continuous decrease of the cross-sectional area.

Fracture may be classified into different types on the basis of the behavior of the material prior to fracture and the loading conditions which cause the fracture,

a) *Brittle fracture* occurs when the deformation which precedes fracture is predominantly elastic, i.e., there is little or no prior

---

*Volumetric viscous deformation due to crazing may also occur under normal stress, but will not be considered in this book.

deformation. Glass and ceramics fracture in a brittle manner with essentially no preceding plastic deformation. Many metals also fracture in a brittle manner under suitable conditions. However, in the case of metals, some plastic strain does occur in thin layers of the material at the fracture surface. Since the region which deforms plastically is of limited extent and forms only a small percentage of the volume of the body, the average strain and overall deflection of the body are still predominantly elastic and no appreciable permanent shape change occurs before fracture. Many glassy polymers also fracture in a brittle manner, but as in the case of metals, permanent deformation occurs near the fracture surface. Since the energy which must be supplied to propagate a brittle crack is usually less than the elastic strain energy released by the growth of the crack, brittle cracks can grow without additional work being done by the applied loads. For this reason, brittle cracks are often able to grow at very high velocities, comparable to the speed of sound in the material.

b) *Ductile fracture* occurs when fracture is preceded by plastic deformation of a large percentage of the volume of the body. Ductile fracture differs from brittle fracture in that a certain amount of plastic deformation of the material is required to produce conditions favorable for fracture. Because of the plastic deformation required to produce the fracture, ductile fracture is preceded by considerable permanent change in the shape of the body. Therefore, the work which must be done by the external loads in order to produce fracture is much larger than the elastic energy stored in the body. Most structural metals fracture in a ductile manner at temperatures at or above room temperature.

c) When a body is subjected repeatedly to a load which is less than that required to cause fracture in a single application, it may fracture. This type of fracture is defined as *fatigue fracture*. As in the case of brittle fracture, fatigue fracture is often not preceded by general plastic deformation of the body. Fatigue fracture is in many ways similar to brittle fracture of metals except that the crack grows only a small amount on each cycle of loading. Metals which are exceedingly ductile under monotonic loading will fracture by fatigue without ever undergoing any general plastic deformation.

d) Fracture in materials which are used at high temperatures where deformation occurs by creep is often called *creep fracture*. On a macroscopic basis, creep fracture is similar to ductile fracture in that it is preceded by general permanent deformation of the body. On a microscopic basis the processes which give rise to creep fracture are often different from those which occur at low temperatures.

I-5 MATERIALS PROCESSING

Raw materials in the form of ingots and powders must be transformed into end products through proper processing techniques. In processing these materials their basic deformation and fracture characteristics must be considered. The choice of a particular processing technique can be influenced by manufacturing cost, its effect on the physical properties of end products, dimensional accuracy, and esthetic qualities.

I-5-a Metals

Processing of metals involves either *shear* deformation or *fracture* or both. Chip forming processes such as metal cutting involve both shear deformation and fracture processes. Chipless forming processes are accomplished utilizing only shear deformation and any fracture is detrimental to the process. In metal processing, interfacial problems such as friction and wear of tools are as important as the mechanical behavior of work materials.

Metals are processed at both high and low temperatures. Hot working is normally done at temperatures higher than the recrystallization temperature (about half of the absolute melting point) of the metal. Therefore, hot-worked metals have minimal residual stresses. Hot working requires less force and energy although the coefficient of friction between the tool and work is generally much higher (greater than 0.6) in hot working than in cold working (about 0.1 to 0.3). The advantage of cold working is that it generally improves the physical properties of metals such as toughness and strength. For example it minimizes the brittleness of steel by breaking up interconnecting carbides into small fragments. Cold working also increases the flow strength of metals through work hardening. Cold-worked metal parts have better surface finishes. Most structural metals are hot worked, whereas most metals used in making machine parts are cold worked. In making sheet metals, billets are hot rolled first and cold rolled afterwards.

There are other processing methods. Metals can be removed by melting the surface with electric arcing (electric discharge machining), by chipping the surface by impact of abrasive particles (ultrasonic grinding), by electrochemical means (electrochemical machining), and by melting the surface with electrons and electro-magnetic beams (electron beam machining and laser beam machining).

I-5-b Plastics

Thermoplastics, which deform in a viscous manner, are usually processed by raising their temperature to either the melting point or the glass transition temperature. At these temperatures thermoplas-tics behave as non-Newtonian viscous fluids. They are usually molded or extruded. Thermosetting plastics, which do not soften with increase in temperature, are usually molded from their liquid components, which react and solidify, either under pressure or in open molds. These plastics can be machined, but machining is usually confined to high molecular weight thermoplastics and thermosetting plastics. Some of the high molecular weight thermoplastics such as Teflon and certain polyethylenes are either sintered from compacted powder or cast from the molten state.

I-5-c Ceramics

Ceramic parts are normally manufactured from micropowder. Ceramic powders are either processed cold or hot into final shapes. Cold-pressed powders must be sintered in a kiln for agglomeration. During pressing, binders such as liquid wax are sometimes added to obtain "green" strength. Ceramic powders are also processed by first wetting them with liquid (usually water) and then casting the slurry into a mold or by extruding it through a die. The final step of the process involves firing the molded piece in a kiln. Ceramic parts are also made from the molten liquid state. Sputtering and vapor deposition are some of the newer techniques of processing ceramics.

I-6 SELECTION OF ENGINEERING MATERIALS

The single most important factor in choosing engineering materials is their *cost*. The cost of materials is influenced by the availability of natural resources, their usefulness as engineering materials, and the ease of processing them. This fact is illustrated by Table I.1, which shows the availability of natural resources within 10 miles of the

TABLE I.1 Availability of Natural Resources
(within 10 miles of the earth's surface)

Oxygen (O)	46.68%
Silicon (Si)	27.60%
Aluminum (Al)	8.05%
Iron (Fe)	5.03%
Calcium (Ca)	3.63%
Sodium (Na)	2.72%
Potassium (K)	2.56%
Magnesium (Mg)	2.07%

earth's surface. The costs of some engineering materials are shown in Table I.2. It can be seen from the two tables that readily available materials are generally cheaper. Glass (silicon oxide) is cheap because oxygen and silicon are available in almost limitless quantities. On the

TABLE I.2 Cost of Some Engineering Materials*

Metals	Ingot form Hopper car loads per pound	Extruded 30,000-lb lots per pound	1/2" Round extruded 1,000-lb lots per foot
Steel 1020	$.04	$.08	$.19
tool	.25–.60	—	.50
Aluminum 1100F (pure)	.24	.48	.30
6061	.25	.51	.32
2024	.28	.53	.32
Copper	.31	.87	1.14
Brass 7030	.48	.78	.53
free cutting	—	.53	.38
Beryllium	—	—	.56

Plastics	Pellets Hopper car loads per pound	1/2" Round extruded 2,000-lb lots per foot
PVC (rigid)	$.36	$.12
Plexiglas	.45	.27
Polystyrene	.15	.10
Teflon	3.25	1.37
Nylon	.75	.26
Polyethylene	.14	.15
Polypropylene	.21	

*1971 prices, except for plastics as pellets in hopper car loads, which are 1973 prices.

other hand, although aluminum is more abundant than steel, it is more expensive, because of its processing cost. In fact, in 1886, at the time of Hall's invention of electrolysis of aluminum in the fused condition, the price of aluminum was \$8/lb.

Plastics have been becoming increasingly important engineering materials during the past two decades. The main reasons are that the price of plastics has been decreasing with the increase in consumption rate* and that the physical properties of various plastics have improved to meet engineering requirements.

There are other factors that determine the usefulness of a particular material as an engineering material. The functionality is largely determined by the bulk and the surface properties. The important bulk properties are modulus (i.e., stiffness), hardness, yield strength, tensile strength, fracture strength, ductility, fatigue strength, creep rate, and melting point. The important surface properties are coefficient of friction, corrosion resistance, and wear resistance. In structural materials bulk properties are more important, whereas surface properties dictate choice of materials for bearings, cams, sliders, etc. In some applications both bulk and surface properties are equally important.

I-7 TYPICAL ENGINEERING MATERIALS

I-7-a Metals

1 Steel

Steel is an alloy based on iron and small amounts of carbon, usually less than 1%. Various other alloying elements are often added to steel for specific end uses. It is the most commonly used engineering material due to its high modulus as well as its low cost. The Young's modulus of steel is about 30×10^6 psi, whereas aluminum has a modulus of only 10×10^6 psi. The modulus is one of the most important parameters in structural applications. Other attractive features of steel are high strength and ease of processing.

There are many grades of steel for various applications. They are classified by their chemical composition. The AISI[†] classification

---

*Price ($/lb) = 1.5×10^3 (annual production rate in lbs)$^{-0.43}$. (From Rodriguez, *Principles of Polymer Systems*, McGraw-Hill, 1970.)

[†]AISI = American Iron and Steel Institute.

TABLE I.3 AISI-SAE Steel Classification

CARBON STEELS
    10XX   Nonsulfurized carbon steel (plain carbon)
    11XX   Resulfurized carbon steel (free machining)
    12XX   Resulfurized and rephosphurized carbon steel

LOW-ALLOY STEELS
    13XX   Manganese 1.75
    23XX   Nickel 3.50
    25XX   Nickel 5.00
    31XX   Nickel 1.25, chromium 0.65
    33XX   Nickel 3.50, chromium 1.55
    40XX   Molybdenum 0.25
    41XX   Chromium 0.50 or 0.95, molybdenum 0.12 or 0.20
    43XX   Nickel 1.80, chromium 0.50 or 0.80, molybdenum 0.25
    46XX   Nickel 1.55 or 1.80, molybdenum 0.20 or 0.125
    47XX   Nickel 1.05, chromium 0.45, molybdenum 0.20
    48XX   Nickel 3.50, molybdenum 0.25
    50XX   Chromium 0.28 or 0.40
    51XX   Chromium 0.80 to 1.05
  5XXXX   Chromium 0.50 to 1.45, carbon 1.00
    61XX   Chromium 0.80 or 0.95, vanadium 0.10 or 0.15 min
    86XX   Nickel 0.55, chromium 0.50 or 0.65, molybdenum 0.20
    87XX   Nickel 0.55, chromium 0.50, molybdenum 0.25
    92XX   Manganese 0.85, silicon 2.00
    93XX   Nickel 3.25, chromium 1.20, molybdenum 0.12
    98XX   Nickel 1.00, chromium 0.80, molybdenum 0.25

HEAT- AND CORROSION-RESISTANT STEELS
    2XX   Chromium-nickel-manganese (nonhardenable, austenitic, nonmagnetic)
    3XX   Chromium-nickel (nonhardenable, austenitic, nonmagnetic)
    4XX   Chromium (hardenable, martensitic, magnetic)
    4XX   Chromium (hardenable, ferritic, magnetic)
    5XX   Chromium (low-chromium, heat-resisting)

system, shown in Table I.3, is most commonly used. The first numeral specifies the type of steel, i.e., 2 for nickel alloy steel (as in 23XX) and 4 for nickel, chromium, and molybdenum alloy steel (as in 43XX). For simple alloys the second numeral specifies the percentage of principal alloying element (e.g., 23XX contains approximately 3% nickel). For complex alloys the meaning of the second digit is a matter of definition. (See, for example, the 4XXX series in Table I.3.) The last two digits indicate the approximate content of carbon, e.g., AISI 1020 is plain carbon steel with 0.20% carbon. Sometimes prefixes such as C (for the basic open hearth steel making process), B (for the acid Bessemer process) and O (for the basic oxygen process) are used to specify the particular steel making process used. The prefixes are useful since the impurity contents depend upon the process used. There are also ASTM* specifications for steel which specify the requirement of steel for various applications, e.g., A36 is plain carbon steel for bridges and buildings.

Commercially available steels and their uses may be divided into the following groups:

i) Plain carbon steel is the cheapest type of steel. It is primarily used for structural purposes. It is available in commercial form as bars, plates, channels, etc.

ii) High-strength steels are low-alloy steels which are used in structures where the strength requirement cannot be met by plain carbon steel.

iii) Alloy steels have better hardenability than plain carbon steels. They are generally used for machine parts but rarely for structural purposes because of the increased cost. The main hardening agent is still carbon.

iv) Ultrahigh-strength steels are alloy steels with exceptionally high strength. Typical examples are modified AISI 4330 steel with maximum tensile strength of 300,000 psi, maraging steel (high nickel content) with 400,000 psi strength and high ductility, and Ausformed steel (made by deformation of untransformed austenite at 1000°F prior to cooling). These steels are used in applications requiring high strengths, such as aircraft landing gear, turbine shafts, etc.

v) Stainless steels are high-alloy steels which have more than 12% chromium content. There are three grades: ferritic (AISI 400 series), austenitic (AISI 300 series), and heat treatable. Austenitic stainless steel is most commonly used, but ferritic stainless may be used for economy or in case of a shortage of nickel. Stainless steel is used where corrosion is a problem.

vi) Tool steels are high-alloy steels with tungsten or molybdenum as the main alloying elements. They are used for molds, dies and cutting tools.

2 Aluminum

After steel, aluminum is the most commonly used metal. It is available in many forms, and is much easier to machine and form

---

*ASTM = American Society for Testing and Materials.

than steel. Its advantages are light weight (its density is 1/3 that of steel), good thermal conductivity, high corrosion resistance in air and water, and high reflectivity. Table I.4 shows the classification system of aluminum. The first numeral designates the alloy group, the second digit specifies the impurity limit or modification of the original alloy, and the last two numerals specify aluminum purity. These numbers are sometimes followed by temper designations such as F (as fabricated), H (strain hardened), T (heat treated) and O (annealed and recrystallized).

3 Magnesium

Magnesium is the lightest structural material, its density being only 2/3 that of aluminum. Magnesium alloys have high strength-to-weight ratios. It is used to make office machines and luggage.

4 Copper

Copper is mainly used where its high electrical and thermal conductivities are useful, as in electrical transmission lines and refrigerators. It is also extensively used in plumbing (because of antiquated building codes), and originally its ease of joining was a factor.

5 Brass

Brass is an alloy of copper and zinc. It is easily machined, formed, and soldered, and has better corrosion resistance than copper and steel. It has a good combination of strength and ductility giving it excellent cold working properties. It is used to make bearings, valves, plumbing fittings, and shell casings.

6 Bronze

Bronze is basically an alloy of copper and tin. Other elements are sometimes added for various applications. Aluminum bronze is used

TABLE I.4 Classification of Aluminum

1XXX	Aluminum 99% pure
2XXX	Copper 2.5-5%, magnesium 0.3-1.5%
3XXX	Manganese 1.0-1.2%, magnesium 0.3-1.0%
4XXX	Silicon 5-10%, copper 0-4.0%
5XXX	Magnesium 1-6%, manganese <1.0%, chrome 0.1-.25%
6XXX	Magnesium 0.5-1.5%, silicon 0.3-1.5%
7XXX	Zinc 4-8%, magnesium 1.5-3.0%, copper 0.5-2.0%
8XXX	Other elements (nickel, iron, lead, tin)

to make gears and cams. Manganese bronze is used to make bearings, pump parts, and ship propellers, as a result of its excellent corrosion resistance. Beryllium bronze is hard and strong and is used to make springs, gears and bearings, especially if sparking is to be avoided.

I-7-b Thermoplastics*

1 Polyethylene

Polyethylene is one of the most commonly used plastics due to its many desirable properties. It is lighter than water, translucent, highly pliable, nontoxic, odorless, tough, quite stable chemically, and easy to process. It is used to make bottles, thin films for laundry bags, toys, chemical containers, electrical insulation, pipes, and linings in corrosive applications. It is one of the cheapest plastics.

2 Polystyrene

Polystyrene is also a commonly used thermoplastic because of its low cost. It is transparent in solid form, quite brittle, rigid, nontoxic, odorless, and easy to process. It is also widely used in the form of foam. It is used to make cups, panels, imitation wood furniture, ornaments, and toys. To increase its toughness, polystyrene can be copolymerized with rubber, e.g., impact grade polystyrene. ABS (acrylonitrile-butadiene-styrene) is a copolymer which is very tough. It is used to make furniture parts, automobile crash pads and dashboards, pipes, and helmets.

3 Polyvinylchloride (PVC)

Another commonly used low-cost thermoplastic is PVC. PVC is normally rigid and brittle but becomes tough and highly pliable when it is plasticized by the addition of such substances as mineral oil. Since it is self-extinguishing, it is used in applications where fire resistance is important. It is used to make tiles, house siding, pipes, upholstery, phonograph records, garments, shoes, and shoe soles. Foamed PVC is the strongest of the foam plastics, and is used to make various garment items.

---

*Within a given thermoplastic there is a wide variation in properties. One should always consult the manufacturer's data if they are available.

4 Polymethylmethacrylate (PMMA)

This plastic is known for its transparency and yet is not as brittle as polystyrene. It is also very stable under light and outdoor exposure and is easy to process. Its uses include outdoor signs, light pipe, airplane canopies, and decorative items. Its trade names are Lucite, Plexiglas, and Perspex.

5 Polytetrafluoroethylene (PTFE)

PTFE is mainly known for its low coefficient of friction and chemical stability. It is translucent, tough even at low temperatures, waxy or nonwetting (does not adhere easily to any other surface), and difficult to process. It is molded into different shapes by compacting and sintering powder. It is stable to temperatures as high as $500°F$. It has low creep strength and therefore lacks dimensional stability. Its abrasion resistance is very low. Gaskets, O-rings, fillers for bearings, and inner linings for chemical chambers and cookware are made of this polymer, whose trade name is Teflon.

6 Cellulose Plastics

Cellulose is distinguished by the fact that it is a natural polymer produced by plants. Cellulose itself is not a thermoplastic, but its derivatives, such as cellulose nitrate and cellulose acetate, are. Cellulose nitrate is tough, dimensionally stable and has low water absorption, but is highly inflammable and unstable in heat and sunlight. Cellulose nitrate with low nitrogen content is widely used as lacquers and decorative products. Cellulose nitrate with high nitrogen content ($>11\%$) is used in solid propellants. Cellulose acetate is not highly flammable and is processed easily by conventional means. Rayon is derived from cellulose.

7 Polyamides (nylon)

Nylon is one of the well-known synthetic plastics. Nylon possesses outstanding resistance to impact and fatigue, low coefficient of friction, excellent abrasion resistance, chemical stability, and good dielectric properties. It is self-extinguishing and can be processed by conventional means. Nylon is normally formed by extrusion or injection molding, but is also available in a special grade for casting.

Drawn nylon fibers have exceptionally high tensile strength but cost more than other polymer fibers. It is used to make gears, cams, fibers for brush bristles and fabrics, machine parts, tubing, appliance parts, rope and fishing line.

8 Polypropylene

Polypropylene is similar to polyethylene in many ways, but it has lower density, higher strength and rigidity, and better dielectric properties. It is used to make automotive parts, films, appliance parts, wire coatings and fibers.

9 Acetal

Acetal, whose trade name is Delrin, is a strong and stiff thermoplastic. At room temperature it has a tensile strength of 10,000 psi. It also has good creep resistance, good fatigue properties, strong impact resistance over a wide temperature range (from $-40°F$ to $212°F$), strong solvent resistance, low coefficient of friction, acceptable abrasion resistance, and good dielectric properties. It is used to make cams, gears, automotive parts, and bearings.

10 Phenylene Oxide

Phenylene oxide is a tough and strong thermoplastic which has a broad temperature usage (from $-275°F$ to $375°F$), outstanding dimensional stability, low creep rate, high modulus, and low moisture absorption. It also has an excellent dielectric strength, good impact resistance, and resistance to moisture and chemicals. It is used to make hospital utensils, computer housings, and pump components. Its trade name is Noryl.

11 Ionomer

Ionomer plastic is similar to polyethylene, except that it has ionic bonding between its polymer chains, making it stronger and tougher. It is transparent, flexible, abrasion resistant, and reasonably resistant to organic solvents. It is used to make molded housewares, toys, containers, films, sheets, tubing, and coatings.

12 Polycarbonate

Polycarbonate, which is still relatively expensive, is transparent and distinguished by high heat resistance, dimensional stability, high impact strength and good optical and electrical properties. Its creep resistance is especially outstanding. It is also very easy to process by any conventional means. Polycarbonates are used in appliances, electrical and electronics equipment, and in the food handling and sporting goods industry. Its trade names are Lexan and Merlon. It should be noted that polycarbonate is a polyester which is thermoplastic.

13 Phenoxy

This plastic, which was introduced in 1962, is also transparent, highly rigid, strong, very tough and dimensionally stable, with a good resistance to various chemicals. It meets FDA (Food and Drug Administration) requirements for food contact uses and therefore can be used as food packaging materials.

14 Polysulfone

This is a relatively new thermoplastic which has very good dimensional stability at high temperatures ($300°$F) and is self-extinguishing. Polysulfone also has very good dielectric properties even at high temperatures, making it very suitable for use in electrical parts. Commercial applications include a variety of electronic, appliance, and automotive parts.

15 Polyimide

This is one of the recently developed high-temperature thermo-plastics. It can withstand a continuous exposure at $500°$F and short-duration exposure up to $900°$F. It has a low coefficient of friction and good wear resistance. Its impact resistance is very good. It also has a low coefficient of thermal expansion. Because of its good physical and thermal properties, it can be used as machine parts such as in internal combustion engines. It is very expensive, costing three times as much as Teflon. It is so difficult to process that it is normally sintered or solution cast.

I-7-c   Thermosetting Plastics*

1 Epoxy Resins

Due to its uses as protective coatings, potting material, adhesives, and laminating material epoxy is one of the best known thermosetting

---

*It should again be emphasized that the variations in physical properties within a given resin family are great. Consult the manufacturers' data before using in large quantities.

plastics although it is much more expensive than polyester. Epoxies offer many good physical properties such as good bonding characteristics, high temperature stability, high dielectric strength, high resistivity, good chemical resistance, and reasonably good mechanical properties. Mechanical properties can be improved a great deal by the addition of fillers such as glass or cloth fibers and hollow glass beads. The final polymerization process during molding does not yield any condensates, enabling forming with no external pressure.

2 Polyester Resins

Polyester resins may be made as thermoplastics or as thermosets. Thermosetting polyester does not have desirable mechanical strength by itself, but in combination with reinforcements, such as glass fibers, it possesses good physical properties. It is the cheapest thermosetting plastic and has good high-temperature resistance. It is used to make corrugated and flat sheets, boats, tanks, and large structural lay-ups. The final polymerization step does not generate any condensates and therefore it can be molded without applying any external pressure.

3 Polyurethanes

Polyurethanes are distinguished by their diverse nature: they can be produced in the form of foams, elastomers, rigid thermosets, adhesives, and fibers. Flexible urethane foams are used to make mattresses and seat cushioning. Rigid urethane foams are used in appliances, e.g., insulators in "thin-walled" refrigerators, in construction of cabinets and buildings, in transportation as insulators for refrigerated tank cars and as crash pads in automobiles. Polyurethane is also widely used as a coating material. Rigid polyurethanes are much tougher than epoxy and polyester and can also be molded without external pressure.

4 Amino Resins

Among the amino resins, urea-formaldehyde and melamine-formaldehyde are well known. The consumption rate of these plastics is less than that of other thermosetting plastics. They are used in laminates and as adhesives, coatings, and molding compounds. They are transparent, resistant to chemical solvents and moisture, tasteless, odorless, and self-extinguishing. They are also resistant to abrasion and have excellent electrical properties. Amino resins produce condensates during the cross-linking reaction, and therefore must be molded under pressure.

5 Phenolic Resins

Phenolic resin is more commonly known as Bakelite, is the oldest of the synthetic plastics. It is relatively cheap and widely used. Fillers such as wood flour and chopped cotton fibers are usually incorporated in order to improve the physical properties. Typical applications are utensil and appliance handles and parts, electrical switch parts, and laminates. Phenolic must also be molded under pressure since it produces condensates.

6 Silicone Plastics

Silicones are produced in the form of resins, elastomers, and fluids. They have very good thermal stability (even at 500°F), and good surface and electrical properties. Since silicone oils are relatively inert, they are used as mold release, parting agents, and heat transfer fluids. Silicone resins are used as impregnating varnishes, molding compounds, and in laminates. Silicone elastomers are used as potting and encapsulating compounds, and as sealants. Silicone rubber is weaker and more expensive than other rubber. Silicone plastics are easy to form and do not require external pressure to mold.

I-7-d Other Engineering Materials

1 Glass

Glass is a ceramic*, belonging to the general class of amorphous silicon oxides with very high viscosity. It is very brittle (showing little plasticity), and does not conduct heat or electricity well. It is relatively cheap because of the abundance of its raw materials and due to ease of processing. There are many special grades of glass, such as Pyrex and Vycor, for various purposes. Pyrex is resistant to thermal shock, due to its low coefficient of thermal expansion, while Vycor is used as the material for vessels where diffusion of elements from the vessel is not desirable. Quartz is silicon oxide in a crystalline form and thus has a high melting point. It is also very shock resistant. Vycor is synthetic quartz. Glass is used for containers, windows,

---

*A ceramic is defined as a combination of one or more metals with a nonmetallic element, usually oxygen. The atoms are held together by ionic bonds, with some covalent bonding.

optical devices, insulators, and in the form of fibers for reinforcing materials and fabrics.

2 Other Ceramics

Most of the ceramics are brittle but have high melting points and high hardness. Ceramics are widely used to make high-temperature containers such as crucibles, bricks, structural slabs for buildings, heat shields, semiconductors, various insulators, abrasives, and pigments. Some of the important raw materials are aluminum oxide, magnesium oxide, calcium oxide, silicon oxide, silicon carbide, tungsten carbide, and titanium dioxide. Ceramics are usually processed by compacting and sintering a powder.

3 Cements and Concrete

Concrete is an aggregate of sand, gravel, and cement which bonds together when the cement undergoes hydration. Concrete is one of the major structural materials used in construction of buildings, roads, dams and other large fixed structures. Recently ferrocement, which is a composite of steel wires and cement, has been used to build ship hulls. As is the case with most ceramics, concrete is chiefly used for applications where it is under a primarily compressive load.

4 Composite Materials

Several different materials can be combined to obtain optimum physical and chemical properties. Thermosetting plastics and thermoplastics can be reinforced with glass fibers, glass beads, paper, graphite fibers and metal powders to increase the mechanical strength, fracture resistance, and thermal and electrical conductivities. Other commonly used composites are prestressed concrete, ferrocement, printed-circuit board and composite solid propellants for rockets (this normally consists of a rubber matrix and ammonium perchlorate). Reinforced plastic parts are used as insulators, structural parts replacing many nonferrous metals, and housings for small machine parts. Prestressed concrete is finding extensive use in buildings and bridges.

5 Wood

Wood is one of the oldest and most commonly used engineering composite materials in the construction of houses and other simple structures. There are some modified woods with better physical properties than natural wood. Its modulus is relatively low, being on the order of 1.5 x 10⁶ psi. Its main advantages are ease of processing, relatively good overall physical properties, and its relatively low cost.

REFERENCES

1. VanHorn, K. R. (ed.): "Properties, Physical Metallurgy and Phase Diagrams" and "Design and Applications," Aluminum, vols. 1 and 3, ASM, Metals Park, Ohio, 1967.
2. Lyman, Taylor (ed.): "Properties and Selection of Metals" and "Heat Treating, Cleaning and Finishing," Metals Handbook, vols. 1 and 2, ASM, Metals Park, Ohio, 1964.
3. Hanson, A. and J. G. Parr: "The Engineer's Guide to Steel," Addison-Wesley, Reading, Mass., 1965.
4. Kinney, G. F.: "Engineering Properties and Applications of Plastics," John Wiley, New York, 1957.
5. "Modern Plastics Encyclopedia," McGraw-Hill, New York, latest edition.
6. Parker, E. R.: "Materials Data Book," McGraw-Hill, New York, 1967.
7. "Metal Progress Databook," Metals Park, Ohio, 1970.

PROBLEMS

Mechanical engineers are often confronted with many interesting but puzzling questions which cannot be answered without understanding the material taught in this book. For each one of the problems listed below, discuss why and how the phenomenon occurs. If the question deals with certain material failures, give the most important physical properties that characterize its resistance to such failures.

I.1   Some single crystal whiskers are about 100 times stronger than the commercial grade structural metal of the same kind. Why? (Just imagine what a breakthrough it would be if we could make all structural metal as strong as the whiskers.)

I.2   Glass, which is normally quite brittle, can be successfully used as a structural material in the construction of deep diving oceanographic research vessels. How? Why?

I.3   A very pure copper deforms plastically if you apply a uniaxial load of about 5 gm/cm². Now, if you pound the same copper with a hammer, after a while it will not deform permanently unless you apply a much greater load. Why?

I.4   A steel rod, which can be stretched plastically at a given stress, cannot be indented locally by an indenter under the same stress. Why?

I.5   Sometimes an automobile axle breaks suddenly without any warning. Why?

I.6   If you hang a picture frame on a wall using a hook which is attached to the wall with epoxy cement, it will probably stay there. However, if you do the same with the adhesive used in making Scotch tape, the picture will eventually drop. Why?

I.7   The rivets used in airplane frames are always bigger than the holes. Why?

I.8   If you shoot a bullet against a rigid wall, it requires much greater stress to deform than when it is loaded slowly. Why?

I.9   Why does a material behave elastically? Why plastically?

I.10   Is there any way of making glass panels flexible so that they can be used in convertible cars?

I.11   Sometimes, large ships break in half (mostly during winter) in the middle of the Atlantic Ocean. Why?

I.12   Sometimes, an oil pipeline several miles long cracks in half, especially when the weather is cold. Why?

I.13   Why is there friction between sliding surfaces?

I.14   A tensile specimen deforms plastically at 10,000 psi. If you put this specimen in a pressurized chamber filled with fluid, is it going to yield at the same uniaxial stress?

I.15   After manufacturing the cylindrical housing for missiles out of steel, the load capacity was determined by filling the chamber with water and pressurizing the fluid. Later, when a load less than the maximum test load was applied to the chamber in actual service, the chamber fractured. Why?

I.16   Silver soldered joints can withstand much higher normal stress than a tensile specimen made of the same solder material can. Why?

I.17   In order to observe the combustion phenomenon of solid propellants at 2,000 psi, a 1/8"-thick Lucite tube was used as a test chamber. It was found that the tube fractures when the internal pressure rises to 400 psi. It was suggested that the thickness of the tube be increased to 1", maintaining the internal diameter constant, so as to prevent fracture until the desired pressure is reached. Is the suggested idea sound? What would you do?

I.18   Which plastic do you expect to creep more, thermoplastic or thermosetting plastic? Why?

I.19   A rubber band is stretched between two rigid posts. If heat is applied, what will happen to the force required to keep the rubber band stretched?

I.20   Comet jet airplanes operated for over a year with no crashes. Then three crashed within a period of three months. What type of failure would you guess to be responsible for the crashes? Why?

CHAPTER TWO

Continuum Mechanics

II-1 INTRODUCTION

Before proceeding with the discussion of specific types of material behavior, the more general aspects of the response of bodies to applied loads will be discussed. When a body is subjected to loads, either from forces acting on its external surface or from the influence of gravity or similar "forces" which act throughout the body, stresses and deformation are produced within the body. The study of the effect of forces on continuous bodies forms the subject of continuum mechanics. An essential part of the formulation of problems in continuum mechanics is a description of the relationship between the stresses acting on an element of the body and the deformations which occur in that element. This relationship can take many forms, depending on the material and the magnitude of the loads. It is with the various forms of this relationship that this book will be primarily concerned. However, first, those aspects of continuum mechanics which are common to all materials (i.e., equilibrium and geometric compatibility) will be reviewed in this chapter.

II-2 STRESS

In considering the transmission of force through a body, one must be concerned not only with resultant forces but also with the distribution of the force. In order to do this it is necessary to define a quantity which describes the intensity of a distributed force at a point. This quantity is called the stress. Consider first the stress acting on a surface. If a resultant force $\mathbf{F}_1$ is applied as shown in Fig. II.1 to a surface of area A_1, then the average stress on the surface by definition has magnitude F_1/A_1. This stress acts in the same direction as the resultant force $\mathbf{F}_1$. By defining a coordinate system such as the one indicated by the unit vectors $\mathbf{e}_1$, $\mathbf{e}_2$, and $\mathbf{e}_3$, the average stress acting on the surface can be expressed in terms of its components in the direction of the three unit vectors as

$$\sigma_{11} = \frac{f_{11}}{A_1} \; ; \; \sigma_{12} = \frac{f_{12}}{A_1} \; ; \; \sigma_{13} = \frac{f_{13}}{A_1} \qquad (II.1)$$

In these expressions* the constant subscript indicates that the surface on which the stress acts is perpendicular to the positive

*Notation for coordinates, components of stress, and vectors:
Coordinate axes are labeled x_1, x_2, x_3. Unit vectors along the coordinate axes are indicated by $\mathbf{e}_1$, $\mathbf{e}_2$, $\mathbf{e}_3$. Components of stress in a Cartesian coordinate system are denoted by σ_{11}, $\sigma_{12}, \ldots, \sigma_{23}, \ldots$ where the significance of the subscripts is defined above and in the following discussion. Subscripts i, j, etc. can be taken to have any value from 1 to 3. Hence the statement "the x_i-axis" is to be read as shorthand for the statement "the x_1-axis or the x_2-axis, or the x_3-axis." Similarly the symbol σ_{ij} can stand for any component of stress. Many other books use x, y, z for coordinates and σ_x, $\sigma_y, \ldots, \tau_{xy}, \tau_{yz}, \ldots$ for stresses. Note that vectors are indicated by bold types in the text and with arrow heads in figures.

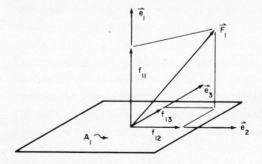

Fig. II.1 Components of a force acting on a surface.

x_1-axis. The second subscript indicates the component of the resultant force from which the stress is derived (i.e., σ_{13} is the average force per unit area in the e_3 direction). Note that the stress component σ_{11} is associated with the resultant force component which is perpendicular to the surface while σ_{12} and σ_{13} are associated with the force components parallel to the surface. σ_{11} is called a normal stress component while σ_{12} and σ_{13} are called shear stress components.

The stress components on the surface defined above are average stress components and would represent the stress at a point on the surface only if the stress were uniformly distributed. However, if A_1 is very small, or in the limiting case when A_1 goes to zero, the stress above represents the stress at the center of area A_1. Therefore, the rigorously defined stresses are given by the following limits:

$$\sigma_{11} = \lim_{A_1 \to 0} \frac{f_{11}}{A_1}$$

$$\sigma_{12} = \lim_{A_1 \to 0} \frac{f_{12}}{A_1} \qquad (\text{II}.2)$$

$$\sigma_{13} = \lim_{A_1 \to 0} \frac{f_{13}}{A_1}$$

where f_{11}, f_{12}, and f_{13} are the components of the resultant force $\mathbf{F}_1$ on the area A_1 which is perpendicular to the direction e_1.

The definition of stress on a surface involved the direction of the surface in space and the area of the surface. In order to define stress components at a point in a solid, one must also make reference to a surface through the point. The direction chosen for this reference surface is arbitrary and in fact it is necessary to give the stress components on at least three surfaces in order to completely describe the state of stress at the point. By convention the three surfaces chosen as reference surfaces are taken perpendicular to the three coordinate axes. This can be illustrated by using the infinitesimal stress element shown in Fig. II.2. The state of stress at the center of the rectangular parallelepiped element with surfaces perpendicular to the coordinate axes is defined in terms of the resultant forces, $\mathbf{F}_i$, acting on the surfaces of the element. If the forces on the faces are given by

$$\mathbf{F}_1 = f_{11}\mathbf{e}_1 + f_{12}\mathbf{e}_2 + f_{13}\mathbf{e}_3$$
$$\mathbf{F}_2 = f_{21}\mathbf{e}_1 + f_{22}\mathbf{e}_2 + f_{23}\mathbf{e}_3 \qquad \text{(II.3)}$$
$$\mathbf{F}_3 = \cdots$$

then the stress components σ_{ij} are defined by

$$\sigma_{11} = \lim_{A_1 \to 0} \frac{f_{11}}{A_1}$$

$$\sigma_{12} = \lim_{A_1 \to 0} \frac{f_{12}}{A_1} \qquad \text{(II.4)}$$

$$\sigma_{13} = \cdots$$
$$\vdots$$
$$\sigma_{ij} = \lim_{A_i \to 0} \frac{f_{ij}}{A_i}$$

where A_i is the area of the surface on which the stress component acts. Since there are three stress components on each of the three faces of the element, there are nine stress components in all.

In the preceding discussion, the stress components were always taken to be acting on surfaces whose outward normal vectors were in the positive direction of the coordinate unit vectors. Such surfaces will henceforth be called positive surfaces. From the definition of the

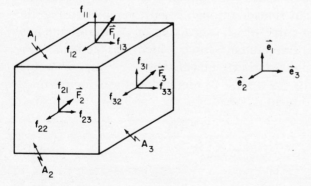

Fig. II.2 Components of forces acting on the surfaces of a three-dimensional stressed element.

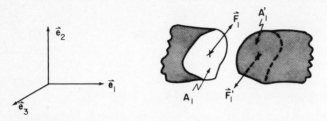

Fig. II.3 Making an imaginary cut through a body perpendicular to the e_1-direction creates two surfaces, A_1 and A'_1. By definition, A_1 is a positive surface, and A'_1 is a negative surface. F_1 and F'_1 are the resultant forces acting on A_1 and A'_1, respectively. Equilibrium requires that $\mathbf{F}'_1 = -\mathbf{F}_1$.

stress components in Eq. (II.4), one can see that a stress component σ_{ij} on a positive surface is positive if the resultant force component f_{ij} is positive, and negative if the resultant force component is negative. It is convenient to establish a sign convention for stress components on surfaces which have normal vectors in the negative-axis directions (i.e., negative surfaces). The sign convention for negative surfaces is that a stress component σ_{ij} on a negative surface is positive if the resultant force component f_{ij} is negative and negative if f_{ij} is positive. To reiterate, the sign convention to be used for stress components is: A stress component is positive if the resulting force component is positive on a positive surface or negative on a negative surface. A stress component is negative if the resultant force component is negative on a positive surface or positive on a negative surface.

The reason for adopting this sign convention can be illustrated by considering the stresses acting on both of the two surfaces formed when an imaginary cut is made through a body as shown in Fig. II.3. If $\mathbf{F}_1$ is the resultant force which was acting on A_1 before the body was cut and $\mathbf{F}'_1$ is the resultant force which was acting on A'_1, equality of action and reaction requires that $\mathbf{F}_1 = -\mathbf{F}'_1$. Therefore, the force components $f_{1j} = -f'_{1j}$. By definition the components of the average stress acting on A_1 are

$$\sigma_{1j} = \frac{f_{1j}}{A_1} \tag{II.5}$$

and because A'_1 is a negative surface, the average stress components on A'_1 are

$$\sigma'_{1j} = -\frac{f'_{ij}}{A'_1} = \frac{f_{ij}}{A_1} = \sigma_{1j} \tag{II.6}$$

Thus, the stress acting on A_1 is equal to the stress acting on A'_1 when the sign convention described above is used. This is a convenient definition since the stress should be the same on a surface, independent of which side of the surface is considered. One should note that under this sign convention a normal stress component directed out of the element (tension) is positive, while a normal stress component which is directed into the element (compression) is negative.

EXAMPLE II.1—Dependence of Stresses on the Choice of Plane at a Point

Show that the magnitudes of the normal and the resultant shear stresses at a given point in a body depends on the orientation of the plane on which the stresses act by considering a uniaxially loaded tensile specimen.

Solution:

Consider a tensile specimen with a uniform cross-sectional area A_0 shown in Fig. II.4. The normal stress acting on a plane perpendicular to the x_1-axis is

$$\sigma_{11} = \frac{F_1}{A_0} \tag{a}$$

All other stresses on that plane are zero, i.e., $\sigma_{12} = 0$ and $\sigma_{13} = 0$.

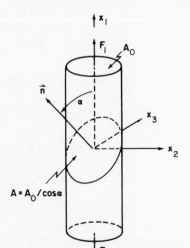

Fig. II.4 Geometry of a uni-axially loaded specimen.

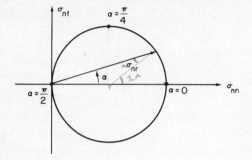

Fig. II.5 Mohr's circle representation of the state of stress in a
uniaxially loaded tensile specimen.

Now if the plane is rotated about the x_2-axis so that the normal to the plane makes an angle α with the x_1-axis, the normal force component F_n and the tangential force component F_t acting on the plane are obtained (from the equilibrium consideration) as

$$F_n = F_1 \cos \alpha$$
$$F_t = F_1 \sin \alpha \qquad \text{(b)}$$

The area of the plane A is related to the cross-sectional area A_0 of the specimen and the inclination angle α as

$$A = \frac{A_0}{\cos \alpha} \qquad \text{(c)}$$

Then the normal stress σ_{nn} and the resultant shear stress σ_{nt} are given by

$$\sigma_{nn} = \frac{F_1}{A_0} \cos^2 \alpha$$

$$\sigma_{nt} = \frac{F_1}{A_0} \cos \alpha \sin \alpha \qquad \text{(d)}$$

The resultant stress acting on the plane is

$$\sigma_{nr} = \frac{F_1}{A_0} \cos \alpha \qquad \text{(e)}$$

The state of stress given by Eqs. (d) can be represented graphically as shown in Fig. II.5. In the σ_{nn}-σ_{nt} space the state of stress of a uniaxially loaded tensile specimen is represented by a circle. A more general case is treated in Sec. II-4.

II-3 EQUILIBRIUM OF STRESSES

Since stresses are related to the resultant forces acting on surfaces within bodies, the laws of mechanics can be used to derive certain useful properties of stress. One such property of stress distributions is the condition of equilibrium. The equations which express equilibrium of stress distributions are derived by considering the dynamic equilibrium of a small element of a body such as the one shown in Fig. II.6. Just as $\mathbf{F} = \mathbf{ma}$ must hold for the entire body, it must also hold for this small element of the body. The resultant forces and moments acting on this body can be expressed in terms of the stress components acting on each of the faces and the area of the face. If the element is sufficiently small, it can be assumed that the magnitude of the stress does not change significantly over the area of the face. Then the resultant force from a stress component is simply

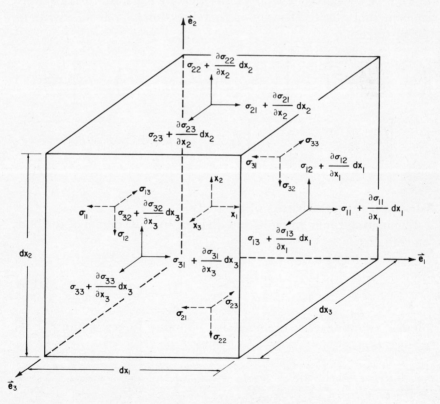

Fig. II.6 The stresses acting on an infinitesimal element of a stressed body.

the value of the stress at the center of the face times the area of the face. Furthermore, the difference in stress on two surfaces separated by a small distance can be approximated by the product of the rate of change of the stress in the direction perpendicular to the surface and the distance separating the surfaces. For example, if the normal stress on the negative x_1-face is σ_{11}, then the normal stress on the positive x_1-face will be $\sigma_{11} + \partial\sigma_{11}/\partial x_1 \cdot dx_1$ as indicated in Fig. II.6. The values of all of the other stress components on positive faces are determined in a similar manner.

If in addition to the stresses there is also a force per unit volume which acts on the body, such as gravitational, electrostatic, or magnetic attraction, there can be an additional resultant force acting on the body. This force is indicated in Fig. II.6 by the components X_i acting at the center of the element.

Since it is necessary that the resultant force acting on the element in each of the coordinate directions must be equal to the mass of the element times its acceleration in that direction, a_i, one may write for the e_1-direction

$$\left[\left(\sigma_{11} + \frac{\partial\sigma_{11}}{\partial x_1}\,dx_1\right) - \sigma_{11}\right]dx_2dx_3 + \left[\left(\sigma_{21} + \frac{\partial\sigma_{21}}{\partial x_2}\,dx_2\right) - \sigma_{21}\right]dx_1dx_3$$

$$+ \left[\left(\sigma_{31} + \frac{\partial\sigma_{31}}{\partial x_3}\,dx_3\right) - \sigma_{31}\right]dx_2dx_1 + X_1dx_1dx_2dx_3 = \rho a_1dx_1dx_2dx_3$$

$$(\text{II.7})$$

where ρ is the mass density of the body.

Eliminating common terms and dividing by the volume of the body one obtains

$$\frac{\partial\sigma_{11}}{\partial x_1} + \frac{\partial\sigma_{21}}{\partial x_2} + \frac{\partial\sigma_{31}}{\partial x_3} + X_1 = \rho a_1 \qquad (\text{II.8a})$$

Similar equations are obtained by summing forces in the e_2- and e_3-directions.

$$\frac{\partial\sigma_{12}}{\partial x_1} + \frac{\partial\sigma_{22}}{\partial x_2} + \frac{\partial\sigma_{32}}{\partial x_3} + X_2 = \rho a_2 \qquad (\text{II.8b})$$

$$\frac{\partial \sigma_{13}}{\partial x_1} + \frac{\partial \sigma_{23}}{\partial x_2} + \frac{\partial \sigma_{33}}{\partial x_3} + X_3 = \rho a_3 \qquad \text{(II.8c)}$$

It should be noted that all of the quantities in these expressions are generally functions of position.

The element must also obey the condition that the resultant torque **T** must be equal to the rate of change of angular momentum $I\alpha$. Since the moment of inertia becomes zero in the limit as the dimensions of the body go to zero, this condition implies that the resultant torques must vanish. Consider the torque about the x_3-axis taken about the center of the element:

$$\left[\left(\sigma_{12} + \frac{\partial \sigma_{12}}{\partial x_1} dx_1 \right) + \sigma_{12} \right] dx_2 dx_3 \frac{dx_1}{2}$$

$$- \left[\left(\sigma_{21} + \frac{\partial \sigma_{21}}{\partial x_2} dx_2 \right) + \sigma_{21} \right] dx_1 dx_3 \frac{dx_2}{2} = 0$$

$$\sigma_{12} + \frac{1}{2} \frac{\partial \sigma_{12}}{\partial x_1} dx_1 - \sigma_{21} - \frac{1}{2} \frac{\partial \sigma_{21}}{\partial x_2} dx_2 = 0 \quad \text{(II.9)}$$

In the limit as dx_1 and dx_2 go to zero, this becomes

$$\sigma_{12} = \sigma_{21}$$
$$\sigma_{23} = \sigma_{32} \quad \text{(Similarly)} \qquad \text{(II.10)}$$
$$\sigma_{31} = \sigma_{13}$$

It should be emphasized that these equations of equilibrium are simply a specific expression of a general physical principle, Newton's Second Law. No properties of the material have to be defined in order to derive the expressions given above. These expressions therefore hold for all materials including both solids and fluids regardless of the type of material behavior involved.*

*See Appendix II-A for the form of the equilibrium equations in non-Cartesian coordinate systems.

EXAMPLE II.2—Stress Distribution in a
        Cantilever Beam

Consider an elastic cantilever beam loaded at its end by a concentrated shear load. Find the stress distribution in the beam.

Solution:
    The bending moment in the beam is determined by static equilibrium to be

$$M_b = -Fx_1 \tag{a}$$

and the shear force is $F = \int_A \sigma_{12} \, dA$. From the strength-of-materials solution for the bending stress distribution in the beam,

$$\sigma_{11} = -\frac{M_b x_2}{I_{22}} = \frac{Fx_1 x_2}{I_{22}} \tag{b}$$

Assume σ_{13} to be zero since there is no shear force in the x_3-direction and because σ_{13} must be zero on the side surfaces of the beam.
    The first equilibrium equation is

$$\frac{\partial \sigma_{11}}{\partial x_1} + \frac{\partial \sigma_{12}}{\partial x_2} + \frac{\partial \sigma_{13}}{\partial x_3} = 0 \tag{c}$$

which gives

$$\frac{Fx_2}{I_{22}} + \frac{\partial \sigma_{12}}{\partial x_2} = 0 \tag{d}$$

By integrating

$$\sigma_{12} = -\frac{Fx_2{}^2}{2I_{22}} + f(x_1, x_3) \tag{e}$$

The second equilibrium equation is

$$\frac{\partial \sigma_{12}}{\partial x_1} + \frac{\partial \sigma_{22}}{\partial x_2} + \frac{\partial \sigma_{32}}{\partial x_3} = 0 \tag{f}$$

If σ_{22} and σ_{32} are zero everywhere, which would be expected if the beam were slender, then $\partial \sigma_{12}/\partial x_1$ must be zero, implying that f is a function of x_3 only. This is also consistent with the fact that the shear force F is independent of x_1. Therefore, σ_{12} can be rewritten

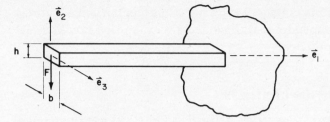

Fig. II.7 Cantilever beam carrying a concentrated end load.

$$\sigma_{12} = -\frac{Fx_2^2}{2I_{22}} + C + f'(x_3) \tag{g}$$

where C is a constant.

The stress must satisfy the boundary condition $\sigma_{12} = 0$ at $x_2 = \pm h/2$. So

$$-\frac{Fh^2}{8I_{22}} + C + f'(x_3) = 0 \tag{h}$$

Since this must be satisfied for all x_3 it is convenient to assume $f' = 0$, and therefore,

$$C = \frac{Fh^2}{8I_{22}} \tag{i}$$

As a check, the shear force can be calculated

$$F_s = \int_A \sigma_{12} \, dA = \int_A \frac{F}{2I_{22}} \left(\frac{h^2}{4} - x_2^2 \right) dA \tag{j}$$

Since $\int_A x_2^2 \, dA = I_{22}$ and $\int_A dA = A$, this becomes

$$F_s = \frac{F}{2I_{22}} \left(\frac{h^2}{4} A - I_{22} \right) = \frac{3F}{2} - \frac{F}{2} = F \tag{k}$$

because $I_{22} = bh^3/12$. This is the same as the shear force determined by static analysis of the beam and is therefore consistent. In order to arrive at this solution it was necessary to assume that several terms were zero. These assumptions cannot be checked at this time because the equilibrium equations alone are not a complete set of equations. These assumptions can therefore only be checked after the remaining equations which describe elastic behavior are presented.

Additional examples of the use of the stress equilibrium equations are contained in Appendix II-B. Included are examples for cylindrical and spherical coordinates.

II-4 TRANSFORMATION OF STRESS

If the nine stress components σ_{ij} defined in Eq. (II.2) are known, the stress on any surface passed through a point in the body can be determined. This can be shown by considering the equilibrium of the element shown in Fig. II.8. If the stress components σ_{ij} are known, it will be shown that the stress components on an inclined surface with normal vector $e_{1'}$ can be determined. Since the element is in static equilibrium, the vectorial sum of forces on the element vanishes.

By definition, the force on the surface with normal $e_{1'}$ is

$$F_{1'} = (\sigma_{1'1'}e_{1'} + \sigma_{1'2'}e_{2'} + \sigma_{1'3'}e_{3'})A_{1'} \qquad (II.11)$$

on the other faces

$$F_i = -(\sigma_{i1}e_1 + \sigma_{i2}e_2 + \sigma_{i3}e_3)A_i \qquad (II.12)$$

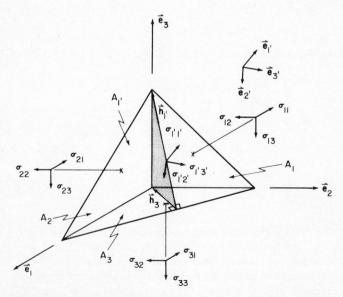

Fig. II.8 Element for determining the stresses on a plane inclined relative to the coordinate axes.

Since the component of a vector **v** in the direction of the unit vector $\mathbf{e}_{1'}$ is $\mathbf{v} \cdot \mathbf{e}_{1'}$, the force $\mathbf{F}_i$ can be resolved into its components in the $\mathbf{e}_{1'}$, $\mathbf{e}_{2'}$, and $\mathbf{e}_{3'}$ directions by taking the dot products $\mathbf{F}_i \cdot \mathbf{e}_{1'}$, $\mathbf{F}_i \cdot \mathbf{e}_{2'}$ and $\mathbf{F}_i \cdot \mathbf{e}_{3'}$ to give

$$
\begin{aligned}
\mathbf{F}_i = \; & -(\sigma_{i1}\mathbf{e}_1 \cdot \mathbf{e}_{1'} + \sigma_{i2}\mathbf{e}_2 \cdot \mathbf{e}_{1'} + \sigma_{i3}\mathbf{e}_3 \cdot \mathbf{e}_{1'})\mathbf{e}_{1'} A_i \\
& - (\sigma_{i1}\mathbf{e}_1 \cdot \mathbf{e}_{2'} + \sigma_{i2}\mathbf{e}_2 \cdot \mathbf{e}_{2'} + \sigma_{i3}\mathbf{e}_3 \cdot \mathbf{e}_{2'})\mathbf{e}_{2'} A_i \qquad \text{(II.13)} \\
& - (\sigma_{i1}\mathbf{e}_1 \cdot \mathbf{e}_{3'} + \sigma_{i2}\mathbf{e}_2 \cdot \mathbf{e}_{3'} + \sigma_{i3}\mathbf{e}_3 \cdot \mathbf{e}_{3'})\mathbf{e}_{3'} A_i
\end{aligned}
$$

where i can be 1, 2, or 3. The negative sign arises because all of the faces A_i have negative unit normal vectors, so that positive stress components produce negative resultant forces.

For simplification, we define the direction cosines as follows:

$$
\begin{aligned}
l_{1'1} &= \mathbf{e}_{1'} \cdot \mathbf{e}_1 = \cos\theta_{1'1} \\
l_{1'2} &= \mathbf{e}_{1'} \cdot \mathbf{e}_2 = \cos\theta_{1'2} \\
& \cdots\cdots\cdots\cdots \qquad\qquad\qquad \text{(II.14)} \\
l_{i'j} &= \mathbf{e}_{i'} \cdot \mathbf{e}_j = \cos\theta_{i'j}
\end{aligned}
$$

where $\theta_{i'j}$ is the angle between the unit vector $\mathbf{e}_{i'}$ and the unit vector $\mathbf{e}_j$. Equation (II.13) can then be written as

$$
\mathbf{F}_i = -\left(\sum_{j=1}^{3} \sigma_{ij} l_{1'j}\mathbf{e}_{1'} + \sum_{j=1}^{3} \sigma_{ij} l_{2'j}\mathbf{e}_{2'} + \sum_{j=1}^{3} \sigma_{ij} l_{3'j}\mathbf{e}_{3'} \right) A_i
$$

or

$$
\mathbf{F}_i = -\sum_{k'=1}^{3} \sum_{j=1}^{3} \sigma_{ij} l_{k'j}\mathbf{e}_{k'} A_i \qquad \text{(II.15)}
$$

From $\mathbf{F} = 0$

$$
\mathbf{F}_{1'} + \mathbf{F}_1 + \mathbf{F}_2 + \mathbf{F}_3 = 0 \qquad \text{(II.16)}
$$

or for the component in the $\mathbf{e}_{1'}$ direction

$$
\sigma_{1'1'} A_{1'} - \sum_{j=1}^{3} \sigma_{1j} l_{1'j} A_1 - \sum_{j=1}^{3} \sigma_{2j} l_{1'j} A_2 - \sum_{j=1}^{3} \sigma_{3j} l_{1'j} A_3 = 0 \quad \text{(II.17)}
$$

plus two other equations for the force in the $\mathbf{e}_{2'}$- and $\mathbf{e}_{3'}$-directions.

The areas A_i and $A_{1'}$ will now be eliminated from the expression so that the limit as the element shrinks to a point can be obtained. This can be done by relating the areas A_i to $A_{1'}$. Note that each triangle A_i has a common side with the triangle $A_{1'}$. If this is taken to be the base of the triangle in each case, the ratio of the areas are then the same as the ratios of the altitudes. Looking along the line of intersection of $A_{1'}$ and A_3, such that the plane of the page in Fig. II.8 is the plane defined by the x_3-axis and the altitudes $h_{1'}$ and h_3 of Fig. II.8, it is seen that

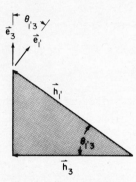

$$\frac{A_3}{A_{1'}} = \frac{h_3}{h_{1'}} = \frac{h_{1'}\cos\theta_{1'3}}{h_{1'}} = l_{1'3} \quad (II.18)$$

Fig. II.9 Shaded plane in Fig. II.8, viewed in the normal direction.

Similarly,

$$\frac{A_2}{A_{1'}} = l_{1'2} \quad \text{and} \quad \frac{A_1}{A_{1'}} = l_{1'1} \qquad (II.19)$$

Using these relationships in Eq. (II.17) and dividing by $A_{1'}$ we are left with

$$\sigma_{1'1'} - \sum_{j=1}^{3} \sigma_{1j}l_{1'j}l_{1'1} - \sum_{j=1}^{3} \sigma_{2j}l_{1'j}l_{1'2} - \sum_{j=1}^{3} \sigma_{3j}l_{1'j}l_{1'3} = 0 \quad (II.20)$$

or

$$\sigma_{1'1'} = \sum_{i=1}^{3} \sum_{j=1}^{3} \sigma_{ij}l_{1'i}l_{1'j} \qquad (II.21)$$

Similar expressions are obtained by summing forces in the $e_{2'}$- and $e_{3'}$-directions, so that the new stress components in general can be written as

$$\sigma_{i'j'} = \sum_{m=1}^{3} \sum_{n=1}^{3} \sigma_{mn}l_{i'm}l_{j'n} \qquad (II.22)$$

This rule allows one to define the stress on any surface when the nine stress components σ_{ij} are known and the orientation of the new surface is defined by the direction cosines $l_{i'j}$. This rule is also known as the tensor transformation rule. Any quantity which is a second-order Cartesian tensor will transform according to Eq. (II.22) when the quantity, expressed in terms of the coordinate system x_1, x_2, x_3 is re-expressed in terms of the coordinate system $x_{1'}$, $x_{2'}$, $x_{3'}$.

This expression for the transformation of stress components from one coordinate system to another is quite tedious in practice, because of the large number of terms in the summation. For this reason, simplified expressions which apply to special cases, such as plane stress and plane strain, are often used. An example of the simplification possible when a state of plane stress exists is given in the following example. It may be noted in passing that, although the task of summing up the terms of Eq. (II.22) by hand is difficult, it is quite easily done using a computer.

EXAMPLE II.3 – Transformation of Boundary
Stresses

The transformation of stress is often used to express the conditions of stress on a boundary. A trivial example of boundary conditions is a surface perpendicular to the e_1 unit vector which is stress free. This implies that the three stress components on this surface, σ_{11}, σ_{12} and σ_{13}, must be zero. The situation is more complicated when the surface is inclined to the coordinate system at an angle θ as in Fig. II.10(b). Given a stress-free surface making an angle with the coordinate direction e_2, find the stress components σ_{11}, σ_{22} and σ_{12}.

Solution:
If the surface is stress free, $\sigma_{1'1'} = 0$, $\sigma_{1'2'} = 0$, and $\sigma_{1'3'} = 0$. The direction cosines are

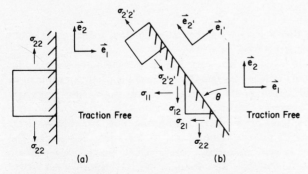

Fig. II.10 Stresses adjacent to stress-free boundaries.

$$\begin{pmatrix} l_{1'1} = \cos\theta & l_{1'2} = \sin\theta & l_{1'3} = 0 \\ l_{2'1} = -\sin\theta & l_{2'2} = \cos\theta & l_{2'3} = 0 \\ l_{3'1} = 0 & l_{3'2} = 0 & l_{3'3} = 1 \end{pmatrix} \qquad \text{(a)}$$

$$\sigma_{11} = \sum_{i'=1}^{3} \sum_{j'=1}^{3} \sigma_{i'j'} l_{1i'} l_{1j'}$$

$$= \sum_{i'=1}^{3} \sum_{j'=1}^{3} \sigma_{i'j'} l_{i'1} l_{j'1} \qquad \text{(b)}$$

$$\sigma_{11} = \sigma_{2'1'}(-\sin\theta\cos\theta) + \sigma_{2'2'}\sin^2\theta$$

since $\sigma_{1'1'} = 0$ and $l_{i'3} = l_{3'i} = 0$ for $i \neq 3$. From equilibrium, $\sigma_{2'1'} = \sigma_{1'2'}$. Thus,

$$\begin{aligned} \sigma_{11} &= \sigma_{2'2'}\sin^2\theta \\ \sigma_{22} &= \sigma_{2'2'}\cos^2\theta \\ \sigma_{33} &= \sigma_{3'3'} \\ \sigma_{12} &= -\sigma_{2'2'}\cos\theta\sin\theta \\ \sigma_{13} &= -\sigma_{2'3'}\sin\theta = \sigma_{31} \\ \sigma_{23} &= \sigma_{2'3'}\cos\theta = \sigma_{32} \end{aligned} \qquad \text{(c)}$$

Therefore, the conditions

$$\frac{\sigma_{11}}{\sigma_{22}} = \tan^2\theta$$

$$-\frac{2\sigma_{12}}{\sigma_{22} - \sigma_{11}} = \tan 2\theta \qquad \text{(d)}$$

$$\frac{\sigma_{13}}{\sigma_{23}} = -\tan\theta$$

express the condition that a surface inclined at an angle θ to the x_1-axis and parallel to the x_3-axis is stress free.

If $\sigma_{2'3'} = \sigma_{3'3'} = 0$, as would be the case if the stressed body were a thin sheet, the problem would reduce to a two-dimensional plane stress problem. In this case the transformation could have been made using the geometric construction of Mohr's circle.

Mohr's circle is a graphical means of evaluating the transformation of stress for rotation about one of the coordinate axes, for example the x_3-axis. The direction cosines for this type of rotation are given in Eq. (a). Using these direction cosines in Eq. II.22, we obtain for the transformed stress components

$$\sigma_{1'1'} = \cos^2\theta\,\sigma_{11} + \sin^2\theta\,\sigma_{22} + 2\cos\theta\,\sin\theta\,\sigma_{12}$$

$$\sigma_{2'2'} = \sin^2\theta\,\sigma_{11} + \cos^2\theta\,\sigma_{22} - 2\cos\theta\,\sin\theta\,\sigma_{12} \tag{e}$$

$$\sigma_{1'2'} = -\cos\theta\,\sin\theta\,\sigma_{11} + \cos\theta\,\sin\theta\,\sigma_{22} + (\cos^2\theta - \sin^2\theta)\sigma_{12}$$

After suitable manipulation, this can be rewritten as

$$\sigma_{1'1'} = \frac{\sigma_{11} + \sigma_{22}}{2} + \frac{\sigma_{11} - \sigma_{22}}{2}\cos 2\theta + \sigma_{12}\,\sin 2\theta$$

$$\sigma_{2'2'} = \frac{\sigma_{11} + \sigma_{22}}{2} - \frac{\sigma_{11} - \sigma_{22}}{2}\cos 2\theta - \sigma_{12}\,\sin 2\theta \tag{f}$$

$$\sigma_{1'2'} = -\frac{\sigma_{11} - \sigma_{22}}{2}\sin 2\theta + \sigma_{12}\,\cos 2\theta$$

As recognized by Mohr, these equations can be represented by a circle. The circle is constructed by using a coordinate system where normal stresses are plotted along the horizontal axis and shear stress along the vertical axis. Positive normal stresses are plotted to the right of the origin along the horizontal axis. The positive shear stress acting on the plane perpendicular to the clockwise axis (i.e., the x_1 axis in the x_1-x_2 coordinate system) is plotted below the horizontal axis, and conversely the positive shear stress acting on the plane perpendicular to the counterclockwise axis (i.e., the x_2-axis in the x_1-x_2 coordinate space) is plotted above the horizontal axis. This is illustrated in Fig. II.11. Note that the points for σ_{11} and σ_{22} are at opposite ends of a diameter of a circle centered at $(\sigma_{11} + \sigma_{22})/2$. Note also that the shear stresses which are 90° apart in physical space are 180° apart in the Mohr's circle. The equations

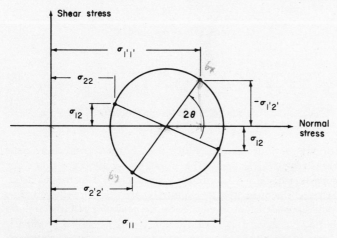

Fig. II.11 Mohr's circle representation of a general two-dimensional state of stress.

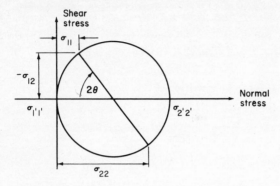

Fig. II.12 Transformation of stress at a traction-free boundary using Mohr's circle.

above state that the stress components in a new coordinate system rotated by an angle θ from the x_1, x_2 coordinates are represented by opposite ends of a diameter of the circle rotated by an angle 2θ from the original diameter.

Returning now to the specific case of the stresses at a traction-free boundary, Mohr's circle for that state of stress can be constructed as shown in Fig. II.12.

The circle must pass through the origin since $\sigma_{1'1'}$ and $\sigma_{1'2'}$ are both zero. A normal stress and a shear stress can only be simultaneously zero if Mohr's circle passes through the origin. The circle must also pass through $\sigma_{2'2'}$. Therefore, the horizontal diameter of Mohr's circle represents the stresses in the primed coordinate system. The unprimed coordinate system is rotated by an angle θ in the clockwise direction from the unprimed system. This means that a diameter of Mohr's circle rotated 2θ from horizontal gives the stresses in the unprimed system as shown. This would give the stresses as

$$\sigma_{11} = \frac{\sigma_{2'2'}}{2}(1 - \cos 2\theta) = \sigma_{2'2'} \sin^2 \theta$$

$$\sigma_{22} = \frac{\sigma_{2'2'}}{2}(1 + \cos 2\theta) = \sigma_{2'2'} \cos^2 \theta \qquad \text{(g)}$$

$$\sigma_{12} = -\frac{\sigma_{2'2'}}{2} \sin 2\theta = -\sigma_{2'2'} \sin \theta \cos \theta$$

which agrees with our previous answer. Notice that, in the primed coordinate system, the shear stress $\sigma_{1'2'}$ is zero. The principal stresses at a point are defined as the normal stresses in a coordinate system directed such as to make the shear stresses zero. The quantities $\sigma_{1'1'}$ and $\sigma_{2'2'}$ are therefore the principal stresses of the two-dimensional problem. Since the center of Mohr's circle is always on the horizontal axis, it is clear that, in two-dimensional problems, principal coordinates in which the shear stress is zero must exist. The student may want to use Eq. (II.22) to show that principal axes for which

$\sigma_{1'2'} = \sigma_{1'3'} = \sigma_{2'3'} = 0$ also exist for all three-dimensional stress states. To do this you will need to use the orthogonality condition for the direction cosines. (See Appendix II-C for a more detailed discussion of this condition.) This condition is given by

$$\sum_{i'=1}^{3} \sum_{n'=1}^{3} l_{i'j} l_{n'm} = \delta_{jm}$$

$$\delta_{jm} = 0 \quad j \neq m$$

$$= 1 \quad j = m$$

(h)

Another method of specifying boundary conditions at a surface makes use of the concept of boundary tractions. The force per unit area applied to a surface can be described by a vector $\mathbf{T}$, which gives the direction and the intensity of the applied traction, as shown in Fig. II.13. The orientation of the surface can be described by the direction of its unit normal vector $\mathbf{n}$. The boundary tractions $\mathbf{T}$ can be related to the stress components σ_{ij} in the material adjacent to the surface by considering the equilibrium of an element of the body adjacent to the surface. For the two-dimensional case shown in Fig. II.13, the equilibrium of the element requires that

$$T_1 ds = \sigma_{11} dx_2 + \sigma_{21} dx_1$$
$$T_2 ds = \sigma_{12} dx_2 + \sigma_{22} dx_1$$

(II.23)

Elements dx_1 and dx_2 are related to ds through the components of the unit formal vector:

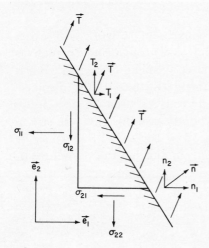

Fig. II.13 Relationship between the stresses adjacent to a surface and the vector traction applied to the surface.

$$dx_1 = n_2\,ds$$
$$dx_2 = n_1\,ds \qquad (II.24)$$

so that Eq. (II.23) can be rewritten as

$$T_1 = \sigma_{11} n_1 + \sigma_{21} n_2$$
$$T_2 = \sigma_{12} n_1 + \sigma_{22} n_2 \qquad (II.25)$$

Generalizing this to the three-dimensional case, the boundary tractions and the stresses are related by the equation

$$T_i = \sum_{j=1}^{3} \sigma_{ji} n_j \qquad (II.26)$$

II-5 DEFINITION OF STRAIN COMPONENTS AND COMPATIBILITY

In addition to describing the stresses in a body, problems in continuum mechanics also require the description of the deformations of the body. The deformation of the body can be completely described by giving the displacement of each point in the body from its undeformed position to its deformed position. Thus, one can completely specify the deformation of a body by giving the three components of the displacement vector **u** for every point in the body.

For many applications it is more convenient to work with other functions of the displacements. For *small displacements* such as occur in many elastic problems, it is convenient to use the linear elastic strains, defined in terms of the derivatives of the displacements. The combinations of the derivatives used in the strain definitions are chosen so that the rigid body motions give zero strain, thus reflecting the fact that the position of the entire body may be changed without any change occurring in the internal structure of the body. The defining equations for the linear strains are

$$\epsilon_{11} = \frac{\partial u_1}{\partial x_1} \qquad \epsilon_{12} = \frac{1}{2}\left(\frac{\partial u_1}{\partial x_2} + \frac{\partial u_2}{\partial x_1}\right) = \epsilon_{21} \qquad (II.27)$$

$$\epsilon_{22} = \frac{\partial u_2}{\partial x_2} \qquad \epsilon_{23} = \frac{1}{2}\left(\frac{\partial u_2}{\partial x_3} + \frac{\partial u_3}{\partial x_2}\right) = \epsilon_{32}$$

(II.27)
(Cont.)

$$\epsilon_{33} = \frac{\partial u_3}{\partial x_3} \qquad \epsilon_{31} = \frac{1}{2}\left(\frac{\partial u_3}{\partial x_1} + \frac{\partial u_1}{\partial x_3}\right) = \epsilon_{13}$$

or, in general,*

$$\epsilon_{ij} = \frac{1}{2}\left(\frac{\partial u_i}{\partial x_j} + \frac{\partial u_j}{\partial x_i}\right)$$

(II.28)

where u_i is the displacement component in the x_i-direction.

Physically the above expressions can be related to the changes in shape of the body. The *normal strains* ϵ_{11}, ϵ_{22}, ϵ_{33} give the *relative changes in length of lines parallel to the coordinate axes.* The *shear strains* ϵ_{12}, etc., are equal to *half of the change in angle between two orthogonal lines initially parallel to two of the coordinate axes.* (See Fig. II.14.)

The linear expressions used above are descriptions of the internal changes in the structure of the body only when the displacements u_i

---

*The strains defined in this way are known as tensor strains because, using this definition of strain, they are components of a second-order tensor. The shear strains ϵ_{12}, ϵ_{23}, ... differ by a factor of $1/2$ from the engineering shear strains given by

$$\gamma_{12} = \frac{\partial u_1}{\partial x_2} + \frac{\partial u_2}{\partial x_1}$$

or

$$\gamma_{xy} = \frac{\partial u}{\partial y} + \frac{\partial v}{\partial x}$$

(See Appendix II-A for the defining equations of strain in non-Cartesian coordinates.) Because strain is a second-order tensor, the tensor transformation rule derived for stresses holds for strains. That is,

$$\epsilon_{i'j'} = \sum_{m=1}^{3} \sum_{n=1}^{3} l_{i'm} l_{j'n} \epsilon_{mn}$$

The proof of this is an exercise in vector calculus which may be found in Wang (Ref. 5). All of the properties of second-order tensors (i.e., existence of principal axes, invariants, etc.) also apply to strains.

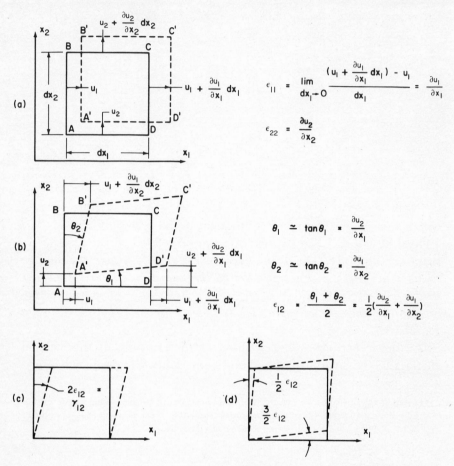

Fig. II.14 Geometric interpretations of normal and shear strains. All three states of shear—(b), (c), and (d)—are identical except for a rigid-body rotation.

are sufficiently small that second-order terms can be neglected. For larger displacements no simple linear expressions are suitable for describing the total deformation of the body. One means of avoiding these problems with nonlinear terms is to consider the deformation incrementally, i.e., to divide the displacement **u** into small increments **du**. Since each increment can be taken infinitesimally small, the linear definitions can be used to define increments of strain $d\epsilon_{ij}$. Another means of treating large-deformation problems is to use the time rate of change of strain, or the strain rate $(d\dot{\epsilon}_{ij})$. The final shape of the body and the total strains can then be found by integrating over the increments. In this process, the coordinate system is fixed in space and the coordinates are always taken to refer to the current position of a point in the body, not to its original position. This procedure will be considered more thoroughly when we take up plastic deformation. Unlike elastic deformation, plastic deformation can involve large displacements.

EXAMPLE II.4–Shear Strain Approximations

The meaning of the strain components and the implications of the approximations involved can be understood by considering the two states of strain represented by the Figs. II.15(a) and (b) with displacement vectors shown. In both cases the lengths of the sides remain constant, but the sides rotate. Find the strains from the given displacements and show that for $\theta \ll 1$ the two cases are equivalent if $\theta = 2\phi$.

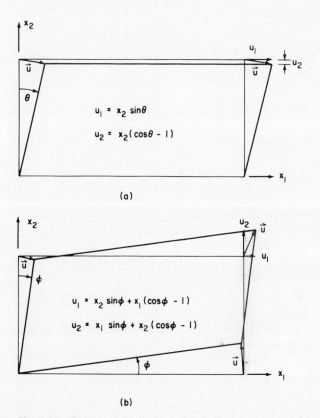

(a)

(b)

Fig. II.15 Two possible displacement fields for a shear deformation.

Solution:

A) The strains for (a) can be represented by

$$\epsilon_{12} = \frac{1}{2}\left(\frac{\partial u_1}{\partial x_2} + \frac{\partial u_2}{\partial x_1}\right) = \frac{1}{2}\sin\theta \simeq \frac{\theta}{2}$$

$$\epsilon_{11} = 0 \tag{a}$$

$$\epsilon_{22} = \cos\theta - 1 \simeq -\frac{\theta^2}{2}$$

Since we are concerned with small strains in the elastic region, the quantity $\theta^2/2$ is small compared with $\theta/2$.

The area of the deformed element is

$$A' = L_1 L_2 (\cos\theta) = L_1 L_2 \left(1 - \frac{\theta^2}{2}\right) \simeq L_1 L_2 \tag{b}$$

The change in area is therefore zero to the first order for small θ.

B) For case (b) the strains are given by

$$\epsilon_{12} = \sin\phi \simeq \phi = \frac{\theta}{2}$$

$$\epsilon_{11} = \cos\phi - 1 \simeq -\frac{\phi^2}{2} \ll \phi \tag{c}$$

$$\epsilon_{22} = \cos\phi - 1 \simeq -\frac{\phi^2}{2} \ll \phi$$

Again a small angle of rotation has been assumed. The change in area can be seen from

$$A' = L_1 L_2 \cos 2\phi = L_1 L_2 \left[1 - \frac{(2\phi)^2}{2}\right] \tag{d}$$

which is equivalent to (b) if $\theta = 2\phi$. When $\theta = 2\phi$, case (b) differs from (a) by only a rigid counterclockwise rotation of ϕ; thus, the strains should be the same in the linear approximation. If θ and ϕ become large, the linear definitions no longer give the same values of strain in the two elements which differ from each other by only rotation.

EXAMPLE II.5—Strains in a Sheet Bent to a Constant Radius of Curvature

As a further example of the relationship between strains and displacements, find the strains which occur when an initially flat sheet is bent around a cylinder of constant radius.

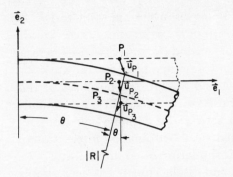

Fig. II.16 Displacements in a sheet bent to a constant radius of curvature.

Solution:

From a knowledge of the strength of materials solution to bending it can be guessed that the centerplane of the sheet remains the same length during the deformation. This will be assumed. Therefore the displacements will be as shown in Fig. II.16 and are given by

$$\mathbf{u} = x_2 \sin\theta\, \mathbf{e}_1 - R(1 - \cos\theta)\mathbf{e}_2 \qquad \text{(a)}$$

where $\mathbf{e}_1$ and $\mathbf{e}_2$ are unit vectors along the x_1 and x_2 axis. Since the defining equations for strain are valid only for small displacements, θ must be small. Therefore, $\sin\theta \simeq \theta$ and $\cos\theta \simeq 1 - \theta^2/2$. Thus, substituting $\theta = -x_1/R$,*

$$\mathbf{u} = -\frac{x_2 x_1}{R}\, \mathbf{e}_1 + \frac{x_1^2}{2R}\, \mathbf{e}_2 \qquad \text{(b)}$$

Using Eq. (II.23), the strains are

$$\epsilon_{11} = \frac{\partial u_1}{\partial x_1} - \frac{x_2}{R} \qquad \epsilon_{22} = \epsilon_{12} = \epsilon_{21} = 0 \qquad \text{(c)}$$

This is the same strain distribution derived by the normal methods of strength of materials.

Since six strain components are derived from three displacement components, it is clear that the strain components are not all independent quantities. That is, one may not say that any arbitrary set of six functions are the components of strain in a body. Additional restrictions must be placed on the functions to ensure that the strains can be integrated to give a displacement field which is single valued and independent of the path of integration. These conditions are called *compatibility* conditions. Physically these conditions of compatibility ensure that if one were to divide the body up into small elements and deform each element according to the local value of the strain, he could then reassemble the elements

---

*R is equal to $\partial^2 u_2/\partial x_1^2$ and is negative for the case shown here.

into a complete body with no holes or overlapping material. In mathematical formulations of continuum mechanics, these conditions are expressed as a series of second-order partial differential equations. These equations may be found in any of several works on the theory of elasticity (Refs. 3, 4 and 6). For the purposes of this book it will be easier to demonstrate compatibility by actually deriving the displacements which satisfy the defining equations for the strains. If allowable displacements can be found by integration of the strains, they must satisfy any formulation of compatibility.

This entire discussion of strains and displacements has been simply a discussion of geometry and definition. As in the case of stress equilibrium, it has not been necessary to state anything about the material in question. It is therefore clear that the definitions and principles of this section are once again universal and must be satisfied for all bodies independent of the nature of mechanical behavior of the material involved.

EXAMPLE II.6–Incompatible Strain (The Screw Dislocation)

Consider the strain given by

$$\epsilon_{z\theta} = \frac{b}{4\pi} \frac{1}{r}$$

$$\epsilon_{rr} = \epsilon_{\theta\theta} = \epsilon_{zz} = \epsilon_{zr} = \epsilon_{r\theta} = 0$$

(a)

in reference to the solid cylinder shown in Fig. II.17. Find the displacements which these strains represent and show that the displacement field is not admissible for the solid cylinder. The equations for the strain components in cylindrical coordinates can be found in Appendix II-A.

Solution:

Direct integration of these strains is possible but difficult because of the extra terms in the defining equations for strain in cylindrical coordinates.

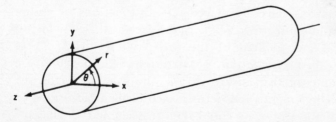

Fig. II.17 The strains given in Eq. (a) cannot be produced in the solid cylinder without cutting it.

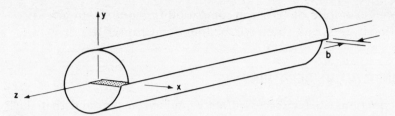

Fig. II.18 If the cylinder is split along the positive x-z-plane and given the displacements prescribed by Eq. (c), the strains given in Eq. (a) are produced. (This set of displacements corresponds to the screw dislocation which will be discussed in Chapter V.)

However, if one can guess a set of displacements which work, it can be shown that they are unique except for possible rigid-body motions.

Notice that a strain $\epsilon_{z\theta}$ would imply either a displacement $u_z = f_1(\theta)$ or a displacement $u_\theta = f_2(z)$. Therefore, try

$$u_r = 0 \qquad u_\theta = f_2(z) \qquad u_z = f_1(\theta)$$

$$\epsilon_{rr} = 0 \qquad \epsilon_{\theta\theta} = 0 \qquad \epsilon_{zr} = 0$$

$$\epsilon_{r\theta} = -\frac{1}{2}\frac{f_2(z)}{r} = 0 \text{ (only if } f_2 = 0)$$

$$\epsilon_{z\theta} = \frac{1}{2r}\frac{\partial f_1}{\partial \theta} = \frac{b}{4\pi r}; \quad f_1(\theta) = \frac{b\theta}{2\pi}$$

(b)

The function $f_1(\theta)$ is defined to within a rigid-body translation. The displacements are given by

$$u_r = u_\theta = 0$$

$$u_z = \frac{b\theta}{2\pi}$$

(c)

Consider a point on the body with corrdinates r, θ, z. This point can also be defined as having coordinates r, $2n\pi + \theta$, z. However, the displacement depends linearly on θ, so that different displacements are calculated for the body, depending on how its position is described. This is not an allowable displacement field for the solid cylinder. In fact, the strains given in Eq. (a) cannot be produced in the solid cylinder by any conceivable loading of its outside surface (i.e., the strain distribution is incompatible with the solid cylinder).

If, however, the cylinder is slit along the positive x-z-plane, the displacements given by Eq. (c) can be produced by sliding the surfaces of the cut relative to each other by an amount b parallel to the z-axis. Compatibility is satisfied because no path of integration for the strains can encircle the z-axis. (See Fig. II.18).

Additional examples of how the compatibility requirement gives rise to additional stress and strain will be found in Chapter III.

II-6 CONSTITUTIVE RELATIONS

If the equations so far discussed are examined, it is found that the equations which involve stresses involve only stresses, body forces, and accelerations, but do not involve strains or displacements. Furthermore, those equations which involve strains and displacements do not contain stresses. Since it was initially stated that the application of forces to a body produces both stress and deformation, it is expected that the stresses on an element can be related to the deformation which these stresses produce. Such a set of relationships will then complete the description of the action of loads on the body. Experience has demonstrated that these relationships will depend on the material in question.

The general subject of continuum mechanics is divided into separate disciplines on the basis of the form of the stress-strain relationships which are to be used. Thus, we have the subjects of elasticity, plasticity, viscoelasticity, and fluid mechanics as manifestations of continuum mechanics for bodies made from materials with varying types of stress-strain behavior.

REFERENCES

1. Crandall, S. H., N. C. Dahl, and T. J. Lardner (eds): "Introduction to the Mechanics of Solids," 2d ed., McGraw-Hill, New York, 1972.
2. Flügge, W. (ed.): "Handbook of Engineering Mechanics," McGraw-Hill, New York, 1962.
3. Sokolnikoff, I. S.: "Mathematical Theory of Elasticity," McGraw-Hill, New York, 1956.
4. Timoshenko, S., and J. N. Goodier: "Theory of Elasticity," 2d ed., McGraw-Hill, New York, 1951.
5. Wang, C.: "Applied Elasticity," McGraw-Hill, New York, 1953.
6. Housner, G. W., and T. Vreeland, Jr.: "The Analysis of Stress and Deformation," Macmillan, New York, 1966.

PROBLEMS

II.1   A circular hole is drilled in a member which carries stress. What conditions are required on the stress at the edge of the hole if the surface of the hole is stress free? Express your answer in terms of Cartesian coordinates.

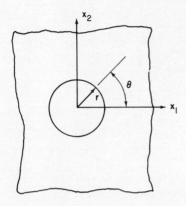

II.2   The state of strain is given as

$$\epsilon_{ij} = \begin{bmatrix} \epsilon_{11} \epsilon_{12} \epsilon_{13} \\ \epsilon_{21} \epsilon_{22} \epsilon_{23} \\ \epsilon_{31} \epsilon_{32} \epsilon_{33} \end{bmatrix} = \begin{bmatrix} 0.002 & 0 & 0.003 \\ 0 & 0.0006 & 0 \\ 0.003 & 0 & 0.004 \end{bmatrix}$$

Calculate: 1) principal strains; 2) maximum shear strain; 3) dilatation; and 4) equivalent strain.

II.3   Show that the strains given by:

$$\epsilon_{\theta\theta} = \frac{\alpha_0}{2\pi} \left(1 - \frac{r_0}{r} \right)$$

$$\epsilon_{rr} = \epsilon_{zz} \qquad \epsilon_{r\theta} = \epsilon_{z\theta} = \epsilon_{zr} = 0$$

correspond to displacements which would occur if the thin-walled section shown below were closed. This corresponds to the strains associated with forming of welded seam tubing.

Would these strains be possible in an initially complete tube? Explain.

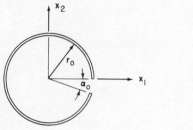 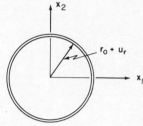

Note: $(2\pi - \alpha_0)r_0 = 2\pi(r_0 + u_r)$

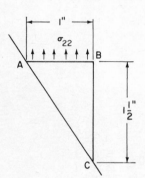

II.4 The bonding at the boundary AC of the bracket shown can withstand a maximum shear stress of 5,000 psi, a maximum tension of 7,000 psi, and essentially unlimited compression perpendicular to the joint. Boundary BC is stress free. If the stress is uniform in ABC, what are the maximum and minimum (negative) allowable values of σ_{22}?

II.5 A trapezoidal body $ABCD$ is uniformly deformed such that B and D have moved to B' and D' as shown. Determine the new

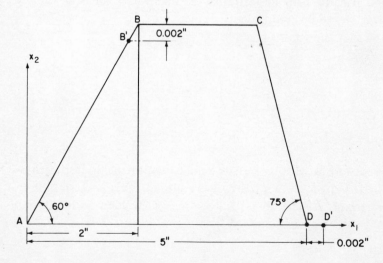

position of C and the strains ϵ_{11}, ϵ_{22}, and ϵ_{12}. Also determine the maximum principle strains and their directions.

II.6   A flexible cable (assumed to have no bending stiffness) passes through a guide in the form of an elastic pipe built in at one end as shown. The friction between the cable and guide is kept very low by lubrication. The cable carries a tension T and leaves the guide at an angle θ relative to the horizontal. Write an equilibrium equation for a section of the guide and cable and determine the bending moment in the guide as a function of position. Compare this with the bending moment in a beam where the cable is attached at the end of the beam (as in (c)).

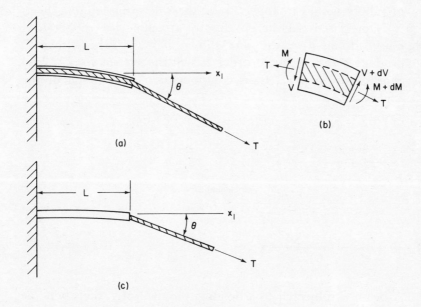

II.7   A thin-walled tube with radius R, length L, and wall thickness t, is bonded to two rigid plates as shown in (a). The tube is then stretched an amount ΔL, and the top plate is twisted an amount θ_0 relative to the bottom plate. Assume that Poisson's ratio ν is zero for the tube.

(a) Find the directions of the principal axes relative to the cylindrical coordinate unit vectors, e_θ, e_τ, and e_z.

(b) Find the maximum tensile and maximum shear strains.

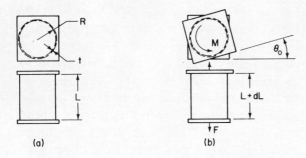

(a)   (b)

II.8 An investigator who is studying the deformation characteristics of rubber wishes to apply a pure shear deformation to a sheet. It is proposed that he do this by attaching the rubber to a hinged frame like the one shown below. If he wishes to produce shear strains of order 1, will the system work? Write expressions for the strain components as a function of the angle ϕ.

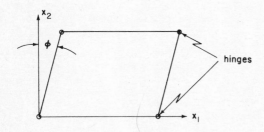

II.9 A beam is subjected to a bending moment M_{13}. This bending moment is transmitted through the beam as a distribution of stress on a cross section. Show that any stress distribution of the form $\sigma_{11} = Af(x_2)$ satisfies the equilibrium equations and the boundary conditions on the lateral surfaces. (i.e., $x_2 = \pm h$). Give an expression for A in terms of the moment M_{13} and appropriate integrals of $f(x_2)$. What other conditions must be satisfied by f?

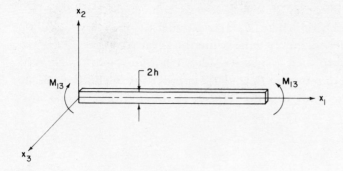

II.10 A parallelogram-shaped body $ABCD$ is deformed to the shape $AB'C'D'$. The strain is uniform in the body. What are the strains?

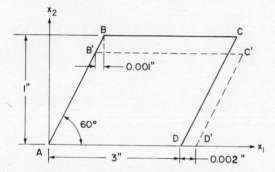

II.11 The parallelogram-shaped body shown below is subjected to the surface tractions shown. What is the state of stress in the body?

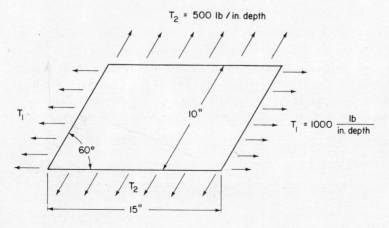

II.12 The parallelogram shown below is subjected to the surface tractions shown where T_2 acts parallel to the x_2-axis and T_1 acts parallel to the x_1-axis, while the shear tractions S_1 and S_2 act parallel to their respective sides. What must be the values of S_1 and S_2 in terms of T_1 and T_2 in order for the body to be in equilibrium and for the stress to be uniform inside the body?

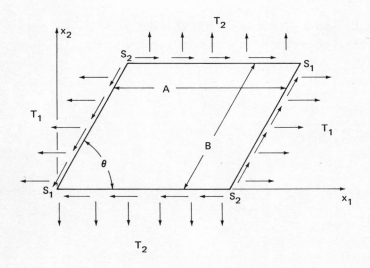

II.13 A composite beam is subjected to an external bending moment M_{13} as shown. The outer layers are made of a material A which always deforms at a constant stress σ_0 (i.e., a rigid plastic material). The inner core is made of an elastic material B which has a stress-strain relationship given by $\sigma_{11} = E\epsilon_{11} = (E/\rho)x_2$, where ρ is the radius of curvature, and E is the time-independent modulus. Determine the equilibrium conditions

by relating M_{13} to the stress distribution in the beam and also determine the location of the neutral axis. If the modulus of the inner core is time dependent [i.e., viscoelastic, $E = E_r(t)$], what are the equilibrium conditions? Does the neutral axis remain at the same location?

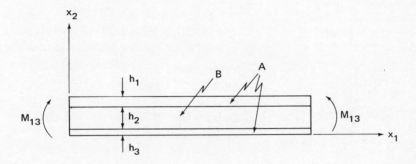

II.14 A stress tensor is given by

$$
\sigma_{ij} = \begin{bmatrix} \sigma_{11} & \sigma_{12} & \sigma_{13} \\ \sigma_{21} & \sigma_{22} & \sigma_{23} \\ \sigma_{31} & \sigma_{32} & \sigma_{33} \end{bmatrix} = \begin{bmatrix} 10{,}000 & 5{,}000 & 7{,}000 \\ 5{,}000 & 10{,}000 & 2{,}000 \\ 7{,}000 & 2{,}000 & 1{,}000 \end{bmatrix} \text{ psi}
$$

If a new coordinate system, $i'j'$, is generated by rotating the original coordinate system by $\pi/4$ about the x_1-axis, determine the stress components about the new coordinate system. Also determine the magnitudes of all invariants.

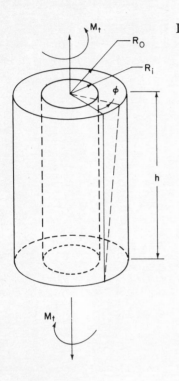

II.15 A composite rod is twisted as shown. The outer layer of the rod is viscoelastic and the inner core is a rigid plastic material which flows at a constant shear stress τ_0. The shear stress $\sigma_{\theta z}$ in the outer layer is related to shear strain $\epsilon_{\theta z}$ and circumferential displacement u_θ /

$$\sigma_{\theta z} = 2G_r(t)\epsilon_{\theta z} = 2G_r(t)(d\phi/dz)r,$$

where $G_r(t)$ is the time-dependent shear modulus, and $(d\phi/dz)$ is the angle of twist per unit length. Sketch the M_t vs. time curve for $G_r(t) = G_0 e^{-At}$, where $(d\phi/dz)$ is maintained constant for $t > 0$.

APPENDIX II-A

Continuum Mechanics in Cylindrical and Spherical Coordinate Systems

For many problems it is convenient to work in special coordinate systems such as cylindrical or spherical. The form of the expressions for stress and strain in these coordinate systems can be derived by use of the formalism of vector calculus for generalized orthogonal coordinate systems or by repeating physical developments in the new coordinate systems. These derivations can be found in detail in many works on the theory of elasticity; therefore only the results will be summarized here.

1 Stress

Since stress is a property defined at a point, the definitions of stress are unchanged by using other coordinate systems. However, one must remember that the reference surfaces on which the stress components are given change orientation from place to place in the coordinate system as the unit vectors of the coordinate system rotate.

2 Equilibrium

Since the equilibrium equations are differential equations, terms arise as a result of changing directions and sizes of elements as coordinates vary. For cylindrical coordinates equilibrium is given by

$$\sigma_{r\theta} = \sigma_{\theta r}, \quad \sigma_{rz} = \sigma_{zr}, \quad \sigma_{z\theta} = \sigma_{\theta z}$$

$$\frac{\partial \sigma_{rr}}{\partial r} + \frac{1}{r}\frac{\partial \sigma_{r\theta}}{\partial \theta} + \frac{\partial \sigma_{rz}}{\partial z} + \frac{\sigma_{rr} - \sigma_{\theta\theta}}{r} + X_r = \rho a_r$$

$$\frac{\partial \sigma_{r\theta}}{\partial r} + \frac{1}{r}\frac{\partial \sigma_{\theta\theta}}{\partial \theta} + \frac{\partial \sigma_{z\theta}}{\partial z} + \frac{2\sigma_{r\theta}}{r} + X_\theta = \rho a_\theta \qquad \text{(A1)}$$

$$\frac{\partial \sigma_{rz}}{\partial r} + \frac{1}{r}\frac{\partial \sigma_{z\theta}}{\partial \theta} + \frac{\partial \sigma_{zz}}{\partial z} + \frac{\sigma_{zr}}{r} + X_z = \rho a_z$$

In spherical coordinates (r, θ, ϕ) the equilibrium conditions are given by

$$\frac{\partial \sigma_{rr}}{\partial r} + \frac{1}{r\sin\phi}\frac{\partial \sigma_{r\theta}}{\partial \theta} + \frac{1}{r}\frac{\partial \sigma_{r\phi}}{\partial \phi} + \frac{2\sigma_{rr} - \sigma_{\theta\theta} - \sigma_{\phi\phi} + \sigma_{r\phi}\cot\phi}{r}$$

$$+ X_r = \rho a_r$$

$$\frac{\partial \sigma_{r\theta}}{\partial r} + \frac{1}{r\sin\phi}\frac{\partial \sigma_{\theta\theta}}{\partial \theta} + \frac{1}{r}\frac{\partial \sigma_{\theta\phi}}{\partial \phi} + \frac{3\sigma_{r\theta} + 2\sigma_{\theta\phi}\cot\phi}{r} + X_\theta = \rho a_\theta$$

$$\frac{\partial \sigma_{r\phi}}{\partial r} + \frac{1}{r\sin\phi}\frac{\partial \sigma_{\theta\phi}}{\partial \theta} + \frac{1}{r}\frac{\partial \sigma_{\phi\phi}}{\partial \phi} + \frac{3\sigma_{r\phi} + (\sigma_{\phi\phi} - \sigma_{\theta\theta})\cot\phi}{r}$$

$$+ X_\phi = \rho a_\phi \quad \text{(A2)}$$

3 Strains

Strains are differential quantities. Thus, the form of the definition of strain changes with the coordinate system. For cylindrical coordinates, the strains are defined as

$$\epsilon_{rr} = \frac{\partial u_r}{\partial r}, \qquad \epsilon_{\theta\theta} = \frac{1}{r}\frac{\partial u_\theta}{\partial \theta} + \frac{u_r}{r}, \qquad \epsilon_{zz} = \frac{\partial u_z}{\partial z}$$

$$\epsilon_{zr} = \frac{1}{2}\left(\frac{\partial u_r}{\partial z} + \frac{\partial u_z}{\partial r}\right)$$

$$\epsilon_{r\theta} = \frac{1}{2}\left(\frac{\partial u_\theta}{\partial r} + \frac{1}{r}\frac{\partial u_r}{\partial \theta} - \frac{u_\theta}{r}\right) \qquad\qquad \text{(A3)}$$

$$\epsilon_{z\theta} = \frac{1}{2}\left(\frac{1}{r}\frac{\partial u_z}{\partial \theta} + \frac{\partial u_\theta}{\partial z}\right)$$

Strains in spherical coordinates are defined by

$$\epsilon_{rr} = \frac{\partial u_r}{\partial r}, \quad \epsilon_{\phi\phi} = \frac{1}{r}\frac{\partial u_\phi}{\partial \phi} + \frac{u_r}{r}, \quad \epsilon_{\theta\theta} = \frac{1}{r\sin\phi}\frac{\partial u_\theta}{\partial \theta} + \frac{u_r}{r} + \frac{u_\phi}{r}\cot\phi$$

$$\epsilon_{r\theta} = \frac{1}{2}\left(\frac{1}{r\sin\phi}\frac{\partial u_r}{\partial \theta} + \frac{\partial u_\theta}{\partial r} - \frac{u_\theta}{r}\right)$$

$$\epsilon_{r\phi} = \frac{1}{2}\left(\frac{\partial u_\phi}{\partial r} - \frac{u_\phi}{r} + \frac{1}{r}\frac{\partial u_r}{\partial \phi}\right) \qquad\qquad \text{(A4)}$$

$$\epsilon_{\phi\theta} = \frac{1}{2}\left(\frac{1}{r}\frac{\partial u_\theta}{\partial \phi} - \frac{u_\theta}{r}\cos\phi + \frac{1}{r\sin\phi}\frac{\partial u_\phi}{\partial \theta}\right)$$

APPENDIX II-B

Two Examples of Common Stress Distributions

EXAMPLE B-1—Stress in a Thin-walled Tube

Derive the stress distribution in a thin-walled tube with an internal pressure.

Solution:

The boundary conditions for σ_{rr} are

$$\sigma_{rr} = \begin{cases} -p_i & \text{at } r = r_i \\ 0 & \text{at } r = r_0 \end{cases} \qquad\qquad \text{(a)}$$

The gradient of σ_{rr} across the thickness may be approximated by

$$\frac{\partial \sigma_{rr}}{\partial r} \simeq \frac{+p_i}{t} \quad \text{for} \quad t \ll r_i \qquad \text{(b)}$$

The stresses $\sigma_{r\theta}$ and σ_{rz} must be zero from the symmetry of the problem. Thus, the equilibrium equation becomes

$$\frac{+p_i}{t} + \frac{\sigma_{rr}}{r} - \frac{\sigma_{\theta\theta}}{r} = 0 \qquad \text{(c)}$$

Since $-p_i \leq \sigma_{rr} \leq 0$ and $t \ll r$

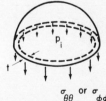

Fig. II.B1 Free-body diagram for the determination of the equilibrium condition for a cylinder containing an internal pressure p_i.

$$\frac{p_i}{t} \gg \frac{\sigma_{rr}}{r} \qquad \text{(d)}$$

Therefore,

$$\frac{p_i}{t} \simeq \frac{\sigma_{\theta\theta}}{r} \quad \text{or} \quad \sigma_{\theta\theta} = \frac{p_i r}{t} \qquad \text{(e)}$$

This result agrees with the result obtained by considering the equilibrium of half of the tube as shown in Fig. II.B1.

EXAMPLE B-2—Stress in a Thin-walled
Spherical Shell

Derive the stress distribution in a thin-walled sphere containing a pressure p_i.

Solution:
The derivation is similar to that for the cylinder.

$$\sigma_{rr} = \begin{cases} -p_i & \text{at} \quad r = r_i \\ 0 & \text{at} \quad r = r_0 \end{cases} \qquad \text{(a)}$$

$$\frac{\partial \sigma_{rr}}{\partial r} \simeq \frac{p_i}{t} \quad \text{for} \quad t \ll r_i \qquad \text{(b)}$$

$\sigma_{r\theta}$, $\sigma_{r\phi}$ are zero from symmetry so Eq. (A2) reduces to

$$\frac{p_i}{t} + \frac{2\sigma_{rr} - \sigma_{\theta\theta} - \sigma_{\phi\phi}}{r} = 0 \qquad \text{(c)}$$

Again, $\sigma_{rr}/r \ll p_i/t$. Therefore,

$$\sigma_{\theta\theta} + \sigma_{\phi\phi} = \frac{p_i r}{t} \qquad \text{(d)}$$

Fig. II.B2 Free-body diagram for the determination of the equilibrium of a spherical shell containing an internal pressure p_i.

From the third equation of Eq. (A2) we obtain

$$\frac{1}{r}\frac{\partial \sigma_{\phi\phi}}{\partial \phi} + \frac{1}{r}[(\sigma_{\phi\phi} - \sigma_{\theta\theta})\cot\phi] = 0 \tag{e}$$

since $\sigma_{r\phi}$ and $\sigma_{\theta\phi}$ must be zero by symmetry. Symmetry also requires that the stresses be independent of the angular coordinates. Thus, $\partial\sigma_{\phi\phi}/\partial\phi = 0$ and Eq. (e) becomes

$$\sigma_{\phi\phi} - \sigma_{\theta\theta} = 0$$

or $\tag{f}$

$$\sigma_{\theta\theta} = \sigma_{\phi\phi}$$

Therefore, Eq. (d) becomes

$$\sigma_{\theta\theta} = \sigma_{\phi\phi} = \frac{p_i r}{2t} \tag{g}$$

which again agrees with the stress determined by consideration of the equilibrium of the hemisphere shown in Fig. II.B2.

APPENDIX II-C

Stress and Strain Invariants

Certain combinations of the stress and strain components have numerical values which are the same in all possible coordinate systems. Because they do not change from one coordinate system to another, they are called invariants. Some of the invariants can be related to physical quantities which can be defined independently without use of a coordinate system. For example, in the case of small strains the specific volume change of a distorted body is equal to the sum of the three normal strains:

$$\frac{\Delta V}{V} = \epsilon_{11} + \epsilon_{22} + \epsilon_{33} \tag{C1}$$

Since the change in volume of a body can be measured without reference to a coordinate system, it is clear that the sum of the normal strains must be an invariant:

$$I_1 = \epsilon_{11} + \epsilon_{22} + \epsilon_{33} \tag{C2}$$

Since both stress and strain are symmetric second-order tensors, a similar combination of the stress components must be zero. Therefore,

$$J_1 = \sigma_{11} + \sigma_{22} + \sigma_{33} \tag{C3}$$

is an invariant of stress. The expression $J_1/3$ is often called the mean normal stress or the hydrostatic stress.

Two additional invariant combinations arise in the calculation of principal stress or principal strains. These combinations are

$$I_2 = \epsilon_{11}\epsilon_{22} + \epsilon_{22}\epsilon_{33} + \epsilon_{33}\epsilon_{11} - \epsilon_{12}{}^2 - \epsilon_{23}{}^2 - \epsilon_{31}{}^2$$

$$I_3 = \epsilon_{11}\epsilon_{22}\epsilon_{33} + 2\epsilon_{12}\epsilon_{23}\epsilon_{31} - \epsilon_{11}\epsilon_{23}{}^2 - \epsilon_{22}\epsilon_{31}{}^2 - \epsilon_{33}\epsilon_{12}{}^2 \tag{C4}$$

and

$$J_2 = \sigma_{11}\sigma_{22} + \sigma_{22}\sigma_{33} + \sigma_{33}\sigma_{11} - \sigma_{12}{}^2 - \sigma_{23}{}^2 - \sigma_{31}{}^2$$

$$J_3 = \sigma_{11}\sigma_{22}\sigma_{33} + 2\sigma_{12}\sigma_{23}\sigma_{31} - \sigma_{11}\sigma_{23}{}^2 - \sigma_{22}\sigma_{31}{}^2 - \sigma_{33}\sigma_{12}{}^2 \tag{C5}$$

Other invariants of stress and strain can be derived by using combinations of J_1, J_2, and J_3 or I_1, I_2, and I_3.

In later chapters it will be shown that additional physical material properties such as maximum strength can be conveniently expressed in terms of the stress and strain invariants. A particular combination of the stress invariants given by

$$\bar{\sigma} = (J_1{}^2 - 3J_2)^{1/2} \tag{C6}$$

is called the equivalent stress. A similar combination of the strain invariants

$$\bar{\epsilon} = \left[\frac{4}{9}(I_1{}^2 - 3I_2)\right]^{1/2} \tag{C7}$$

is called the equivalent strain. The physical significance of these quantities will be discussed in relation to the plastic behavior of materials in Chapter IV.

1 Proof of Invariance of J_1

Define the normal stress, and consequently the first stress invariant, relative to a new coordinate system by

$$\sigma_{1'1'} = \sum_{i=1}^{3} \sum_{j=1}^{3} l_{1'i} l_{1'j} \sigma_{ij}$$

$$\sigma_{2'2'} = \sum_{i=1}^{3} \sum_{j=1}^{3} l_{2'i} l_{2'j} \sigma_{ij}$$

$$\sigma_{3'3'} = \sum_{i=1}^{3} \sum_{j=1}^{3} l_{3'i} l_{3'j} \sigma_{ij}$$

(C8)

$$J_{1'} = \sigma_{1'1'} + \sigma_{2'2'} + \sigma_{3'3'} = \sum_{i=1}^{3} \sum_{j=1}^{3} \sigma_{ij} \left(\sum_{k'=1}^{3} l_{k'i} l_{k'j} \right)$$

However, the fact that both the primed and unprimed coordinate systems are orthogonal allows one to derive conditions on the term

$$\sum_{k'=1}^{3} l_{k'i} l_{k'j}$$

For this purpose note that

$$\mathbf{e}_{k'} = l_{k'1} \mathbf{e}_1 + l_{k'2} \mathbf{e}_2 + l_{k'3} \mathbf{e}_3$$

$$|\mathbf{e}_{k'}|^2 = \mathbf{e}_{k'} \cdot \mathbf{e}_{k'} = l_{k'1}^2 + l_{k'2}^2 + l_{k'3}^2 = 1$$

(C9)

also, for $k' \neq m'$

$$\mathbf{e}_{k'} \cdot \mathbf{e}_{m'} = 0$$

(C10)

since $\mathbf{e}_{k'}$ is perpendicular to $\mathbf{e}_{m'}$. Therefore,

$$\mathbf{e}_{k'} \cdot \mathbf{e}_{m'} = l_{k'1} l_{m'1} + l_{k'2} l_{m'2} + l_{k'3} l_{m'3}$$

(C11)

For convenience, define the symbol

$$\delta_{ij} = \begin{cases} 1 & i = j \\ 0 & i \neq j \end{cases} \qquad \text{(C12)}$$

Then, the above conditions, known as orthogonality conditions for the direction cosines, can be written as

$$\sum_{k'=1}^{3} l_{k'i} l_{k'j} = \delta_{ij}$$

$$\sum_{i=1}^{3} l_{k'i} l_{m'i} = \delta_{k'm'} \qquad \text{(C13)}$$

Using this result,

$$J_{1'} = \sum_{i=1}^{3} \sum_{j=1}^{3} \sigma_{ij} \delta_{ij} = \sum_{i=1}^{3} \sigma_{ii} = \sigma_{11} + \sigma_{22} + \sigma_{33} = J_1 \quad \text{(C14)}$$

Proofs for the invariance of I_1, I_2, I_3, J_2, and J_3 are similar.

CHAPTER THREE

Elastic Behavior

III-1 STRUCTURAL CHANGES ASSOCIATED WITH ELASTIC DEFORMATION

In discussing the mechanical behavior of materials there are several reasons to begin with the subject of elastic behavior. Elastic deformation arises from the small shifts in the equilibrium positions of atoms which occur when external forces are applied to a solid body; it is therefore a fundamental property of solids which can be related directly to the binding forces between atoms. Elastic deformation is homogeneous on a finer scale than other types of deformation, since it involves shifts in all of the atomic positions, not just in the positions of certain critically located atoms.

Because of this intrinsic nature of elastic deflections, several properties of elastic behavior can be directly deduced. Elastic deformation is completely reversible, since elastic distortions involve small-scale shifts in the equilibrium positions of the atoms, without wholesale rearrangement of the atomic positions. Removal of an applied load causes the interatomic potentials to return to their original state, with the result that the atoms return to their undisturbed positions, and the body returns to its original shape.

Because elastic deformation involves all of the atoms in the body, small changes in composition do not have a very large effect on the elastic response of the material. The exchange of a few atoms of one element for a few atoms of another in a solid changes the atomic binding forces only in the region of the atoms exchanged. The size of the effect on the elastic response of the material is roughly proportional to the number of substituted atoms and the difference in the involved binding forces. For the types of elements which are generally used in alloying, or accidentally introduced as impurities, the variation is usually only a few percent. Accordingly, the differences in elastic response among the various alloys and purities of a material are generally only a few percent. For example, nearly all of the many steel alloys have a practically identical Young's modulus of 28 to 29×10^6 psi. In contrast, the yield strengths* of these various forms of steel differ by as much as an order of magnitude.

For reasons similar to those discussed in connection with changes in composition, the elastic behavior of the material is only very slightly affected by the previous history of the material. Since elasticity is directly related to the binding between atoms, any processing which results in a material with the same crystal structure will give a material with the same elastic behavior. For example, a metal may be subjected to a number of different heat treatments which can have a profound effect on other mechanical properties. As a rule, heat treatments affect the distribution of phases in a material, but they do not alter the identity of these phases. Since the phases retain their identity, the atomic binding forces remain essentially the same, and the elastic behavior of the material undergoes no change.

Elastic behavior is similarly unaffected by prior deformation history, except when there has been a major realignment of the material. In crystalline materials, permanent deformation occurs via the mechanism of dislocation motion. It is possible through deformation to increase the number of dislocations in a body by many orders of magnitude. However, even in the most heavily

*The yield strength of a material is the stress at which a material begins to behave plastically.

cold-worked material, the number of atoms adjacent to defects such as dislocations, and thus able to have their binding energies significantly disturbed, are but a small percentage of the total. Since the elastic properties of the body are predominantly determined by the great mass of undisturbed material, they therefore remain unchanged.

Prior deformation can have a significant effect on the elastic properties of a material, if some fundamental change in the structure of the material has occurred. This would be the case, for example, in a drawn polymer which has been sufficiently stretched to align most of the molecules in a single direction. The elastic properties of the structure are then said to be anisotropic, that is, they depend upon direction. Along the drawing axis, the elasticity will be primarily a function of the binding forces between atoms of the same molecule, while in directions perpendicular to that axis they will depend on forces between atoms of different molecules. In contrast, the elastic properties of the undrawn, unoriented polymer represent averages of the properties derived from these two different interatomic forces. A similar effect can occur in metals when the deformation results in a structure which has a preferred orientation of the small crystals (crystallites) which make up the polycrystalline bulk. In both of these cases, the change in elastic properties is observed because the prior deformation has produced fundamental changes in the structure of the material over a large fraction of the total volume.

Since small shifts in atomic positions are not resisted by any rate-dependent force other than inertia, one would expect no dependence of the elastic behavior on the rate of deformation, except for the inertial effects, which have already been included in the equilibrium equations. Actually, there is a small difference between adiabatic and isothermal deformation, but this effect is usually quite small and is considered only when the internal damping of the material is important.

III-2 ELASTIC CONSTITUTIVE RELATIONS

Mathematical expressions to describe the elastic behavior of materials will now be proposed. For small deformations it can be assumed that the relationship between the stresses and strains is linear. In later chapters it will be shown that, in most engineering materials, elastic strains are always sufficiently small that the linear stress-strain relationship is valid. The most general set of linear relations which could be written is

$$\sigma_{ij} = \sum_{k=1}^{3} \sum_{l=1}^{3} C_{ijkl}\epsilon_{kl} \tag{III.1}$$

Alternatively,

$$\epsilon_{ij} = \sum_{k=1}^{3} \sum_{l=1}^{3} S_{ijkl}\sigma_{kl} \tag{III.2}$$

The C_{ijkl}'s and S_{ijkl}'s are proportionality constants, where i and j can have any value from 1 to 3. The number of constants C_{ijkl} required is therefore $3^4 = 81$. From equilibrium it is known that $\sigma_{ij} = \sigma_{ji}$ and by definition $\epsilon_{ij} = \epsilon_{ji}$. From energy arguments it can be shown that $C_{ijkl} = C_{klij}$. These conditions combined imply that all of the equalities of the following types hold for the elastic moduli:

$$C_{ijkl} = C_{jikl} = C_{ijlk} = C_{klij} \dots \text{etc.} \tag{III.3}$$

This means that the number of independent constants which must be specified in the most general linear elastic formulation is 21. Since materials generally have a great deal of symmetry, arguments based on the equivalence of various directions in a material can be used to further reduce the number of constants required.

For any material in which the three coordinate directions are equivalent (for example, in a cubic crystal with coordinates along the cube edges), it is clear that relationships of the type

$$\begin{aligned} C_{1111} &= C_{2222} \dots \text{etc.} \\ C_{1212} &= C_{1313} = C_{2323} \dots \end{aligned} \tag{III.4}$$

must hold. It is also true that a shear stress σ_{12} can produce no strain other than ϵ_{12} without violating the cubic symmetry of the body. Similarly, a normal stress cannot produce a shear strain. Therefore, constants of the type

$$C_{1123} = C_{1213} = \cdots = 0 \qquad \text{(III.5)}$$

Careful completion of these arguments reveals that, in the case of cubic symmetry, the only nonzero elastic constants are

$$C_{1111} = C_{2222} = C_{3333}$$
$$C_{1122} = C_{1133} = \cdots \qquad \text{(III.6)}$$
$$C_{1212} = C_{1313} = C_{2323} \cdots$$

For an isotropic material, one in which all directions are equivalent, it can be further shown that

$$C_{1212} = \tfrac{1}{2}(C_{1111} - C_{1122}) \qquad \text{(III.7)}$$

An isotropic material, therefore, has only two independent elastic constants and a great many zeros in the general formulation above.

For the reduced number of nonzero elastic constants that appear in the relations for an isotropic linearly elastic solid, it is convenient to rewrite the general equation above in the more familiar form of Hooke's law:

$$\epsilon_{11} = \frac{1}{E}[\sigma_{11} - \nu(\sigma_{22} + \sigma_{33})]$$

$$\epsilon_{22} = \cdots$$

$$\epsilon_{33} = \cdots$$

$$\epsilon_{12} = \frac{\sigma_{12}}{2G} \qquad \text{(III.8)}$$

$$\epsilon_{23} = \cdots$$

$$\epsilon_{31} = \cdots$$

As was stated above, only two of the three constants represented by the symbols E, ν, and G can be independent.* It can be shown that

$$G = \frac{E}{2(1 + \nu)} \qquad \text{(III.9)}$$

*Poisson's ratio ν is sometimes denoted by μ. The symbol E is Young's modulus, and G is the shear modulus.

When the constants E and G are specified, the six equations of Eq. (III.8) form the link between the stress and the strain relations of continuum mechanics. These equations, the equilibrium equations, and the definitions of strain, form a complete description of the behavior of linearly elastic materials under the action of loads.

Unlike the equilibrium equations and the strain-displacement equations (compatibility conditions), which have general applicability, Hooke's law as given above applies only to a very specific class of materials. Which, if any, materials behave in the way described? Nearly all crystalline materials at sufficiently low stresses behave elastically. Single crystals are not isotropic, but polycrystalline aggregates taken in bodies of a size sufficient to involve many individual crystal grains with random orientation can be adequately described as isotropic. This class of materials includes all of the structural metals, most ceramics, and most rocks. Glassy polymers can be described as isotropic elastic at sufficiently low temperatures and short times such that the elastic behavior predominates over the viscous behavior. Partially crystalline polymers with no preferred orientation are also essentially elastic at low temperatures. Drawn polymers such as nylon fiber can be elastic under suitable conditions, but are not isotropic.

Most materials of economic importance can therefore be described by linear elastic constitutive relationships in some circumstances of interest. Some materials such as rubber behave elastically to large strains, so that a nonlinear formulation of elasticity is necessary to describe their behavior.

III-3 DETERMINATION OF STRESS DISTRIBUTIONS IN ELASTIC BODIES

The general equations of continuum mechanics plus the elastic constitutive equations form a complete set of equations whose number equals the number of unknown quantities. It is therefore possible, in principle, to determine the distribution of stress in a body when the distribution of tractions on the surface is given. In practice, however, the solution of a boundary value problem in three dimensions, involving as many simultaneous equations as a problem in elasticity, is a nearly impossible task. There is no general method

TABLE III.1   Elastic Solutions for a Few Common Engineering Problems

Geometry	Governing equations

1.

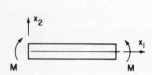

Pure bending:

$$\epsilon_{11} = -\frac{x_2}{\rho} \; ; \; \sigma_{11} = \frac{-Mx_2}{I_{22}}$$

$$\sigma_{12} = \sigma_{22} = \sigma_{13} = \sigma_{23} = \sigma_{33} = 0$$

$$I_{22} = \int_A x_2{}^2 \, dA$$

ρ is the radius of curvature.

2.

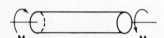

Torsion of a circular rod:

$$\epsilon_{z\theta} = \frac{r}{2} \frac{d\phi}{dz} \; ; \; \sigma_{z\theta} = \frac{Mr}{I_p}$$

$d\phi/dz$ is the angle of twist per unit length.

3.

Thin-walled tube with internal or external pressure:

$$\sigma_{\theta\theta} = (p_i - p_0) \frac{r}{t}$$

$$\sigma_{rr} \simeq \frac{1}{2} (p_i + p_0)$$

$$\sigma_{zz} = \frac{F}{2\pi rt}$$

for a tube with closed ends, $F = \pi r^2 (p_i - p_0)$ and $\sigma_{zz} = (p_i - p_0)r/2t$.

4.

Thin-walled sphere with internal and external pressure:

$$\sigma_{\theta\theta} = \sigma_{\phi\phi} = \frac{(p_i - p_0)r}{2t}$$

$$\sigma_{rr} \simeq \frac{p_i + p_0}{2}$$

TABLE III.1 Elastic Solutions for a Few Common Engineering Problems (*Continued*)

Geometry	Governing equations

5.

Approximate stress distributions for beams with transverse loads:

$$V = -\frac{\partial M}{\partial x_1}$$

$$\sigma_{11} = \frac{-Mx_2}{I_{22}} \qquad q = \frac{\partial^2 M}{\partial x_1^2}$$

$$M = EI \frac{\partial^2 v}{\partial x^2}$$

v is the displacement of the neutral axis of the beam in the x_2-direction.

for solving problems in elasticity, and the methods which have been most successful in generating either exact or approximate solutions to new problems involve mathematical techniques which are beyond the scope of this book. However, there is an extensive literature devoted to such exact and approximate solutions for a great variety of problems of engineering interest, including the torsion of circular shafts, the transverse bending of beams, internal and external pressures in tubes and disks, the stress distribution under point loads, and many others (see, for example, the references given at the end of this chapter). A few examples of solutions to elastic problems which are of particular importance are given in Table III.1.

III-4 ST. VENANT'S PRINCIPLE

A study of the various elastic solutions for structural elements such as beams and shafts reveals that the solutions are exact only if the external loads are applied in the form of very specific distributions of tractions on the external surfaces. For example, the solution for a circular rod in torsion requires that the load be applied to the rod in the form of shear tractions, as shown in Fig. III.1. In actual practice, shafts are usually loaded by means of keys or splines, which are used to attach gears and pulleys to the shafts. Similar examples can be found in the way that loads are actually applied to the ends of

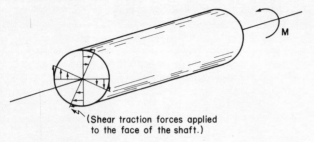

(Shear traction forces applied
to the face of the shaft.)

Fig. III.1 Required distribution of tractions on the face of a circular rod subjected to torsion, using the exact elastic solution.

beams. The question therefore arises as to how to adapt solutions of the idealized problems to the evaluation of real situations.

The general guideline in this respect is called St. Venant's principle which states:

> When two sets of tractions applied to a portion of the boundary of a body are statically equivalent (i.e., have the same resultant force and moment), the difference between the elastic solutions for the two cases decreases with distance from the loaded portion of the boundary.

St. Venant's principle is strictly qualitative and expresses only a trend. The details of the distribution of external loads become less important as the distance from the point of application increases. However, the principle does not allow us to determine at what distance the differences become negligible, or what the magnitude of these differences are. For slender members of reasonable cross section, such as those generally used for structural applications, the rule of thumb is that the stress distribution is within a few percent of the idealized problem at distances more than two or three diameters or thicknesses away from the point of loading. This is mainly a matter of experience and cannot always be assumed to be true. For certain types of cross-sectional shapes, significant differences in stress distribution can persist even at great distances from the point of loading [see Fig. III.2(a)]. This is also true of fiber composite structures consisting of high-modulus fibers bonded with a low-modulus matrix. This allows irregularities in stress distribution in directions perpendicular to the fiber axis to extend over large distances [see Fig. III.2(b)].

The characteristics of elastic behavior described by St. Venant's principle contain several important implications. For most bodies, the overall deflections of the body will be largely unaffected by local changes in loads which do not change the resultant forces and moments applied to the body. Furthermore, small localized disruptions of the stress which result from holes or cracks in the body do not affect the overall deflections and have only a localized effect on the stress distribution. Conversely, the presence of small defects in the body, or a nonuniform application of the load, may cause the stress to be locally increased to several times the nominal stress without changing the overall elastic response of the body. This may have important consequences for other types of behavior. As discussed in a later section, this is the principle of the stress concentration factor.

III-5 SUPERPOSITION

All of the governing equations for small elastic deformations are linear. Therefore, any linear combination of functions, each of which satisfy the equations of elasticity, also satisfies the equations. This is the principle of superposition. If σ_{ij}^{I} is a stress distribution which satisfies the equilibrium equations and corresponds to a prescribed set of tractions on the boundary of the body, and if σ_{ij}^{II} is another stress distribution which satisfies equilibrium and corresponds to a different set of tractions on the surface, then $\sigma_{ij}^{I} + \sigma_{ij}^{II}$ corresponds

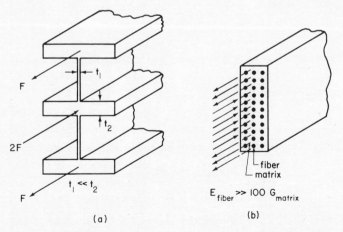

(a)

(b)

Fig. III.2 Examples of bodies in which irregularities in the applied loads affect the stresses at great distances from the point of application of the loads.

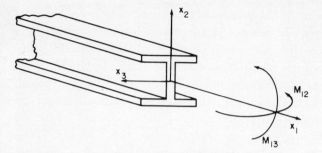

Fig. III.3 Bending of a beam about an arbitrary axis, considered as a superposition of bending moments about the two principal axes of the beam.

to the stress distribution when the body is loaded by both sets of tractions simultaneously.

EXAMPLE III.1—Superposition of Bending
Moments in an Elastic
Beam

As an example of superposition consider a beam subjected to a bending moment M_{12} about the x_2-axis and a moment M_{13} about the x_3-axis, as shown in Fig. III.3, where x_2 and x_3 are the principal axes of the beam cross section. Find the stress distribution in the beam.

Solution:

This problem can be split into its two parts as shown. Since the stress distribution for the beam subjected to bending about the x_2-axis is

$$\sigma_{11} = \frac{M_{12}x_3}{I_{33}} \tag{a}$$

and the stress distribution for the beam subjected to bending about the x_3-axis is given by

$$\sigma_{11} = -\frac{M_{13}x_2}{I_{22}} \tag{b}$$

the stress distribution for the case when both moments act simultaneously is

$$\sigma_{11} = \frac{M_{12}x_3}{I_{33}} - \frac{M_{13}x_2}{I_{22}} \tag{c}$$

Thus, the problem of finding the stresses and deflections of a beam subjected to loads which are not parallel to the principal axes of the beam can always be resolved into two simple problems involving loading along each of the principal axes of the beam.

EXAMPLE III.2—Bending of a Tube with
Internal Pressure

A pipe, shown in Fig. III.4, which is built-in at one end, contains a pressure p. The other end of the pipe is closed and unsupported. The pipe is loaded by its own weight, which is w lb per unit length. What is the stress in the pipe?

Solution:

Divide the problem into two parts: 1) A pipe with internal pressure and no vertical loads, and 2) a cantilever beam loaded by its own weight with no internal pressure.

For 1) the stress are

$$\sigma_{rr} \simeq \frac{p}{2}$$

$$\sigma_{\theta\theta} = \frac{pr}{t} \tag{a}$$

$$\sigma_{zz} = \sigma_{33} = \frac{pr}{2t}$$

For 2) the bending moment is given by

$$M = -\left(\frac{wL^2}{2} - \frac{wx_3^2}{2}\right) = -\frac{w}{2}(L^2 - x_3^2)$$

$$\sigma_{zz} = \sigma_{33} = -\frac{Mx_1}{I} = \frac{w}{2I}(L^2 - x_3^2)x_1 \tag{b}$$

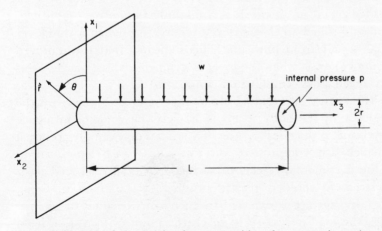

Fig. III.4 Method of determining by superposition the stresses in a pipe which contains internal pressure and acts as a beam carrying a uniform load per unit length.

Therefore, the total stress in the pipe is

$$\sigma_{rr} \simeq \frac{p}{2} \qquad \sigma_{\theta\theta} = \frac{pr}{t}$$

$$\sigma_{zz} = \sigma_{33} = \frac{pr}{2t} + \frac{w}{2I}(L^2 - x_3{}^2)x_1 \tag{c}$$

III-6 UNIQUENESS AND RESIDUAL STRESS

Using the principle of superposition, the uniqueness of elastic solutions can be investigated. To do this we shall adopt the method of proof by contradiction. Assume, first, that two solutions to the same problem exist. Let the stress distribution for one of these solutions be σ_{ij}^{I} and the stress distribution for the other solution be σ_{ij}^{II}. Since both of these stress distributions are assumed to be solutions to the same elastic problem, the surface tractions for σ_{ij}^{I} are the same as those for σ_{ij}^{II}. Using superposition, the stress distribution $\sigma_{ij}^{I} - \sigma_{ij}^{II}$ is the solution to an elastic problem where the surface tractions are zero. Two solutions to a given elastic problem can therefore differ only by a stress distribution which has no associated surface tractions.

Such a field of stress is called residual stress. Residual stress may result from changes in temperature or be due to prior inelastic deformation, or it may be built in by a fabrication process. A familiar manifestation of residual stress arises in cold-rolled or cold-extruded metal products. It is commonly found that an initially flat bar of cold-rolled steel will develop curvature when a thin layer of material is cut from one side. This indicates that the removed layer was under stress; when the layer is stripped off, the bar deforms in order to reach a new equilibrium condition.

It is in the nature of residual stress that it has no effect on the elastic response of the body. It might however have a large influence on the inelastic properties of the material. For example, residual stresses which are compressive on the surface are beneficial in reducing the detrimental effects of surface cracks in tempered glass. Residual stress in molded plastic products cause them to break suddenly if the surface is exposed to a solvent. Residual stress may also change the load necessary to reach the point of transition from elastic to plastic behavior in ductile materials.

EXAMPLE III.3—Residual Stress in Seamed
                    Tubing

As an example of the nature of residual stress produced by fabrication, consider the case of welded seam tubing. Such tubing is produced by bending a flat sheet into the shape of the pipe and welding the edges together with a seam parallel to the tube axis. The completed tube is free of external loads but is not free from stress. If the tube is slit along the seam, it is found to spring open by an angle α as shown in Fig. III.5. Find the change in the stress distribution which occurs during unloading.

Solution:

The opening can be prevented if equal and opposite moments M of sufficient magnitude are applied to the edges of the slit, as indicated in Fig. III.5. Note that the effect of slitting the tube was only to release stress on the newly formed free edges. A uniform pressure applied to the outside of the tube would also prevent opening, but this is not consistent with the loading which existed in the unslit tube. The residual stress in the unslit tube must correspond to the stress distribution in a tube with a uniform bending moment M acting in the tube wall.

The solution for pure bending of curved beams or plates is known and can be applied to this case. Housner and Vreeland (Ref. 8., p. 200) gives the stress distribution in a curved beam subjected to bending as

$$\sigma_{rr} = C_2 \left[\frac{\ln(b/a)}{a^2/b^2 - 1} \left(1 - \frac{a^2}{r^2} \right) + \ln \frac{r}{a} \right]$$

$$\sigma_{\theta\theta} = C_2 \left[\frac{\ln(b/a)}{a^2/b^2 - 1} \left(1 + \frac{a^2}{r^2} \right) + \ln \frac{r}{a} + 1 \right]$$

(a)

Since the tube is long, any strain in the z-direction must be independent of r, θ, and z. Assume plane strain, i.e., $\epsilon_{zz} = 0$. This allows calculation of σ_{zz}.

$$\sigma_{zz} = C_2 \nu \left[\frac{2 \ln(b/a)}{a^2/b^2 - 1} + 2 \ln \frac{r}{a} + 1 \right]$$

(b)

 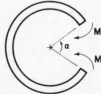

Fig. III.5 Opening of a seamed tube as a result of the residual stresses when it is slit along its axis.

Since the tube is not constrained axially, the resultant axial force must be equal to zero, if the assumption that $\epsilon_{zz} = 0$ is correct. Otherwise, a constant term must be added to σ_{zz} to satisfy the boundary conditions. Integration of the stress over the end of the tube yields

$$F_{zz} = \int_a^b \sigma_{zz} 2\pi r \, dr = \int_a^b C_2 \nu \left[\frac{2 \ln(b/a)}{a^2/b^2 - 1} + 2 \ln \frac{r}{a} + 1 \right] 2\pi r \, dr = 0 \quad \text{(c)}$$

This gives no net resultant force. Thus, the assumption that the problem is plane strain was, indeed, correct.

Hooke's law can be used to find ϵ_{rr} and $\epsilon_{\theta\theta}$.

$$\epsilon_{rr} = \frac{C_2(1 + \nu)}{E} \left[(1 - 2\nu) \left(\frac{\ln(b/a)}{a^2/b^2 - 1} + \ln \frac{r}{a} \right) - \frac{\ln(b/a)}{a^2/b^2 - 1} \frac{a^2}{r^2} - \nu \right]$$

$$\epsilon_{\theta\theta} = \frac{C_2(1 + \nu)}{E} \left[(1 - 2\nu) \left(\frac{\ln(b/a)}{a^2/b^2 - 1} + \ln \frac{r}{a} \right) + \frac{\ln(b/a)}{a^2/b^2 - 1} \frac{a^2}{r^2} - \nu + 1 \right]$$

(d)

Integration of the strains gives

$$u_r = \frac{C_2(1 + \nu)}{E} \left[(1 - 2\nu) \left(\frac{r \ln(b/a)}{a^2/b^2 - 1} + r \ln \frac{r}{a} \right) + \frac{\ln(b/a)}{a^2/b^2 - 1} \frac{a^2}{r} - (1 - \nu)r \right]$$

$$u_\theta = \frac{C_2(1 + \nu)2(1 - \nu)r\theta}{E}$$

(e)

The constant C_2 can be evaluated such that $u_\theta = \alpha r$ for $\theta = 2\pi$.

$$C_2 = \frac{\alpha E}{4\pi(1 - \nu^2)}$$

(f)

Substitution of this expression for C_2 into the expressions for the stresses completes the description of the residual stress which resulted from processing. Careful note should be taken of the fact that evidence of the residual stress, in this case the opening of the tube, could only be detected by cutting the tube and allowing the stresses to relax. Note that the displacement component u_θ depends linearly on θ. Therefore, in the complete tube, this solution is not allowed because the displacement at a point determined by integrating the strains depends on the path of integration (i.e., if two paths circle the origin a different number of times, the θ-values are different). The strains in this solution can only occur in the slit tube. This points up the fact that residual stress exists because the strain which must occur to relieve the stress does not satisfy compatibility.

EXAMPLE III.4—Residual Stress in
  Tempered Glass

Tempered glass is made by giving it either a suitable heat treatment or a chemical treatment to produce a residual stress distribution consisting of a biaxial compression at the surfaces and a tension in the interior. Assume that a sheet of glass 1/2 in. thick has a residual stress distribution away from the edges given by

$$\sigma_{11} = \sigma_{22} = \left[10^4 - 1.2 \times 10^5 \left(\frac{x_3}{t} \right)^2 \right] \text{psi}$$

$$\sigma_{33} = \sigma_{12} = \sigma_{23} = \sigma_{31} = 0$$

(a)

Here, t is the thickness of the glass, the x_1- and x_2-axes are in the plane of the glass, and the origin is on the centerline of the plate.

a. Show that this stress distribution is in equilibrium and that no resultant forces or moments are required on the glass to maintain this stress distribution. (Neglect edge effects.)
b. If the top 1/16 in. of the plate is softened by heating so that the stress in this region is relieved, what is the new stress distribution in the plate and what deformation occurs?

Solution:
a) Since σ_{11} and σ_{22} are the only nonzero stresses, and they depend only on x_3, all of the terms in the three equilibrium equations are zero, and the equations are identically satisfied. The resultant force on a surface in the glass perpendicular to x_1 is

$$F_{11} = w_2 \int_{-t/2}^{t/2} \left[10^4 - 1.2 \times 10^5 \left(\frac{x_3}{t} \right)^2 \right] dx_3$$

$$= w_2 \left(10^4 t - 1.2 \times 10^5 \frac{1}{3t^2} \frac{t^3}{4} \right) = w_2 (10^4 t - 10^4 t) = 0$$

(b)

where w_2 is the width of the plate in the x_2-direction. The calculation of the resultant force F_{22} on a surface perpendicular to the x_2-axis is identical. The resultant bending moment on a surface perpendicular to the x_1-axis is

$$M_{12} = -w_2 \int_{-t/2}^{t/2} \left[10^4 - 1.2 \times 10^5 \left(\frac{x_3}{t} \right)^2 \right] x_3 \, dx_3 = 0$$

(c)

b) If the surface of the glass is softened (1/16 in. deep), the original stress distribution in the remainder of the plate now has resultant forces and moments, F_{11}, F_{22}, M_{12}, and M_{21}, which are no longer zero. Therefore, in the absence of external constraints, the glass will deform elastically to find a new equilibrium stress distribution with zero resultants. The resulting stress distribution can be found by *superimposing* other elastic stress fields which compensate for the net resultant force and moment of the original stress distribution in the unsoftened portion of the glass.

The net force and moment in the unsoftened glass with the original stress field are calculated from the stress distribution

$$\sigma_{11} = \sigma_{22} = \begin{cases} 10^4 - 1.2 \times 10^5 (x_3/t)^2 \text{ psi for } -t/2 < x_3 < 3t/8 \\ 0 \text{ for } x_3 > 3t/8 \text{ because the top } 1/16 \text{ in. is softened} \end{cases} \tag{d}$$

The net resultant force per unit width is

$$\frac{F_{22}}{w_1} = \frac{F_{11}}{w_2} = \int_{-t/2}^{3t/8} \left[10^4 - 1.2 \times 10^5 \left(\frac{x_3}{t} \right)^2 \right] dx_3 \tag{e}$$

$$= (0.875 - 0.711) \, 10^4 t = 820 \text{ lb/in.}$$

The resultant moment per unit width is

$$\frac{M_{21}}{w_1} = \frac{M_{12}}{w_2} = -\int_{-t/2}^{3t/8} \left[10^4 - 1.2 \times 10^5 \left(\frac{x_3}{t} \right)^2 \right] x_3 \, dx_3$$

$$= -10^4 t^2 \left[\frac{1}{2} \left(\frac{9}{64} - \frac{1}{4} \right) - 3 \left(\frac{3^4}{8^4} - \frac{1}{2^4} \right) \right] \tag{f}$$

$$= -10^4 \, \frac{1}{4} \, (0.0735) \text{ lb} = -184 \text{ lb}$$

By adding to this a uniform stress distribution in the unsoftened part of the glass which has resultants $-F_{11}$ and $-F_{22}$ one achieves a stress distribution with zero resultant. The necessary stresses are

$$\sigma_{11}^{I} = \sigma_{22}^{I} = -\frac{F_{11}}{w_2} \left(\frac{1}{7t/8} \right) = -820 \text{ lb/in.} \left(\frac{16}{7} \text{ in.}^{-1} \right) \tag{g}$$

$$= -1,880 \text{ psi}$$

By also adding a linear stress distribution with resultant moments $-M_{21}$ and $-M_{12}$, the net moment can be eliminated. The required stress distribution is

$$\sigma_{11}^{II} = \sigma_{22}^{II} = -\frac{M_{21}}{w_1} \frac{x_3 + t/16}{(7t/8)^3/12} = \frac{(184 \text{ lb})(12)(x_3 + 1/32 \text{ in.})}{(7/16)^3} \tag{h}$$

$$\sigma_{11}^{II} = \sigma_{22}^{II} = 26{,}300 \ (\text{lb/in.}^3)(x_3 + 1/32 \text{ in.})$$

The total stress distribution $\sigma_{11}{}^t = \sigma_{22}{}^t = \sigma_{11} + \sigma_{11}^{I} + \sigma_{11}^{II}$ has no resultant forces or moments. The deformation of the plate is associated with the stress distributions σ_{11}^{I}, σ_{11}^{II} and σ_{22}^{I}, σ_{22}^{II}. The additional strains in the plate are

$$\epsilon_{11}^{I} = \epsilon_{22}^{I} = \frac{\sigma_{11}^{I}(1 - \nu)}{E}$$

$$\epsilon_{11}^{II} = \epsilon_{22}^{II} = \frac{\sigma_{11}^{II}(1 - \nu)}{E} \tag{i}$$

For $E = 10^7$ psi and $\nu = 0.3$, the bending strains give the plate a radius of curvature in both directions of

$$\rho = 543 \text{ in.} \tag{j}$$

III.7 THERMAL STRAINS

The expression for Hooke's law given in Eq. (III.8) neglects the effect of temperature. It is well known that most materials expand when heated. Therefore, if strains are to be defined relative to a reference state at a fixed temperature, a term must be included in the expressions for Hooke's law which describe the change in dimensions of a body due to heating alone. When this is done, Hooke's law for an isotropic material becomes

$$\epsilon_{11} = \frac{1}{E}\left[\sigma_{11} - \nu(\sigma_{22} + \sigma_{33})\right] + \alpha\Delta T$$

$$\epsilon_{22} = \cdots\cdots\cdots\cdots\cdots + \alpha\Delta T$$

$$\epsilon_{33} = \cdots\cdots\cdots\cdots\cdots + \alpha\Delta T \tag{III.10}$$

$$\epsilon_{12} = \frac{\sigma_{12}}{2G}, \ \ldots \text{ etc.}$$

where α is called the coefficient of thermal expansion or coefficient of linear expansion.*

*For anisotropic materials, the coefficients of expansion α will depend on direction. The temperature change can also produce shear strain in an anisotropic body.

If an unstressed body with no surface tractions and no displacement constraints is heated uniformly from its reference temperature by an amount ΔT, the state of strain can be taken as uniform throughout the body and equal to

$$\epsilon_{11} = \epsilon_{22} = \epsilon_{33} = \alpha \Delta T \qquad * \qquad \text{(III.11)}$$

which from the above form of Hooke's law implies that the stresses everywhere remain zero. By superposition, any stress-strain distribution caused by surface tractions can be added to the above solution to give the combined solution for stress and strain in a body subjected to both loads and change of temperature. It is clear that uniform heating of an isotropic body has no effect on the stress distribution in the absence of external constraints on the size of the body.

When the temperature change in a body is not uniform, a stress may be produced in an initially stress free body even in the absence of surface tractions. Once again this comes about because the thermal strains alone do not satisfy compatibility. The following examples illustrate some of the properties of these thermal expansion induced stresses, often called thermal stresses.

EXAMPLE III.5–Stresses in a Constrained Heated Pipe

It is well established plumbing practice that a pipe is never run in a straight line between two fixed points. The reason for this is to avoid temperature-induced stresses in the pipe or excessive loads applied to the anchor points. Estimate these stresses for a copper hot-water pipe run in a straight line between two rigid walls. Assume that the pipe was stress free when installed at room temperature ($70°$F) and that the hot water temperature is $150°$F. Neglect the pressure in the pipe.

Solution:
Because of the constraints

$$\epsilon_{zz} = 0 \qquad \text{(a)}$$

From the boundary conditions

$$\sigma_{rr} = \sigma_{\theta\theta} = 0 \qquad \text{(b)}$$

---

*This is a valid state of strain because it satisfies equilibrium, Hooke's law, the boundary conditions (i.e., stress free) and can be shown to correspond to displacements $u_1 = \alpha \Delta T \, x_1$, $u_2 = \alpha \Delta T \, x_2$, $u_3 = \alpha \Delta T \, x_3$, which are single-valued and continuous.

Eq. (III.10) yields

$$\epsilon_{zz} = 0 = \frac{\sigma_{zz}}{E} - \alpha\Delta T \tag{c}$$

$$\sigma_{zz} = E\alpha(80°F) = 17 \times 10^6 (9.4 \times 10^{-6})80 = 12{,}800 \text{ psi}$$

This stress can easily be eliminated by putting a jog in the pipe. This allows the thermal expansion to be accommodated by bending deflections. This greatly reduces the resulting stresses.

In this example, the temperature of the body was assumed to be uniform so that stresses arose only because of external constraints on the length of the pipe. The following example demonstrates how thermal stresses arise in an unconstrained body when temperature gradients are present.

EXAMPLE III.6—Stresses in a Reactor
Fuel Element

The fuel element for a fast breeder reactor* is a slender rod of uranium or uranium oxide 1/4″ in diameter by 2–3 feet long encased in a thin stainless steel tube. Fission of the uranium occurs uniformly throughout the core so that heat is generated at all points in the core. Since the core is long compared to its diameter, all heat flow is radial. The outside temperature is kept constant at T_0 by the coolant flowing around the outside of the tube. Since the temperature distribution must be a function of r only and since the rate of heat flow through the wall of any cylinder must be equal to the heat generated inside the cylinder, the temperature distribution is easily calculated.

$$-2\pi r k \frac{\partial T}{\partial r} = \pi r^2 g = \frac{qr^2}{R^2} \tag{a}$$

where g is the rate of heat generation per unit volume, q is the rate of heat generation per unit length of fuel element, R is the radius of the element and k is the thermal conductivity. Thus,

$$T = -\frac{q}{4\pi k}\left(\frac{r}{R}\right)^2 + C \tag{b}$$

But $T = T_0$ at $r = R$, so that $C = T_0 + q/4\pi k$ and, therefore,

*A breeder reactor is a second-generation device for the production of nuclear energy, which differs from the first-generation thermal reactors in that the chain reaction is sustained by fast neutrons rather than by thermal neutrons which have been slowed down by collisions with light atoms in moderator rods. As a result, the fast breeder reactor yields a much greater output of power per unit volume and has the very attractive feature that, besides the power produced by fission of the sparse quantity of isotope U_{235} in natural uranium, yields the reaction with the massively preponderant U_{238} Pu_{239}, which can be used to fuel conventional reactors and provide additional nuclear power.

$$T = \frac{q}{4\pi k}\left(1 - \frac{r^2}{R^2}\right) + T_0 \tag{c}$$

Assume this to be a given condition of the problem.

Assume that the fuel element is elastic and ignore the contact stress between the core and the cladding. Calculate the stresses introduced by the temperature distribution.

Solution:

Because of the symmetry, $u_\theta = 0$, all shear strains are zero, and all strains are independent of θ and z. The axial strain must be constant.

$$\epsilon_{rr} = \frac{\partial u_r}{\partial r}, \quad \epsilon_{\theta\theta} = \frac{u_r}{r}, \quad \epsilon_{zz} = C_1 \tag{d}$$

For convenience assume that the length of the element remains unchanged so that $C_1 = 0$. The equilibrium equations reduce to

$$\frac{\partial \sigma_{rr}}{\partial r} + \frac{\sigma_{rr} - \sigma_{\theta\theta}}{r} = 0 \tag{e}$$

Using the form of Hooke's law which expresses the stress in terms of strain

$$\sigma_{rr} = (\lambda + 2G)\epsilon_{rr} + \lambda\epsilon_{\theta\theta} - (3\lambda + 2G)\alpha\Delta T \qquad *$$
$$\sigma_{\theta\theta} = (\lambda + 2G)\epsilon_{\theta\theta} + \lambda\epsilon_{rr} - (3\lambda + 2G)\alpha\Delta T \tag{f}$$

Substituting Eq. (d) into Eq. (f), and the result into Eq. (e) gives

$$\frac{\partial^2 u_r}{\partial r^2} + \frac{1}{r}\frac{\partial u_r}{\partial r} - \frac{u_r}{r^2} = \left(\frac{3\lambda + 2G}{\lambda + 2G}\right)\alpha\frac{\partial \Delta T}{\partial r} = \left(\frac{3\lambda + 2G}{\lambda + 2G}\right)\alpha\left(-\frac{qr}{2\pi k R^2}\right) \tag{g}$$

which has the general solution

$$u_r = C_2 r + \frac{C_3}{r} + C_4 r^3 \tag{h}$$

$C_3 = 0$ since u_r is finite at $r = 0$. The constant C_4 is found by substituting Eq. (h) into Eq. (g) and solving for C_4 to obtain

$$C_4 = -\frac{1}{16}\left(\frac{3\lambda + 2G}{\lambda + 2G}\right)\frac{\alpha q}{\pi k R^2} \tag{i}$$

---

*Here, $\lambda = E\nu/[(1 + \nu)(1 - 2\nu)]$; λ is called Lame's constant. See Appendix III-A for alternative forms of Hooke's law.

$\sigma_{rr} = 0$ at $r = R$ implies that

$$(\lambda + 2G)(C_2 + 3C_4 R^2) + \lambda(C_2 + C_4 R^2) = 0 \tag{j}$$

since $\Delta T = 0$ at $r = R$. Then

$$2C_2(\lambda + G) = -C_4 R^2(4\lambda + 6G)$$

$$C_2 = -\frac{C_4 R^2(2\lambda + 3G)}{\lambda + G} = \frac{(2\lambda + 3G)(3\lambda + 2G)\alpha q}{16(\lambda + G)(\lambda + 2G)\pi k} \tag{k}$$

Substituting into Eq. (h) and then into Eq. (d) gives the strains

$$\epsilon_{rr} = \frac{(3\lambda + 2G)\alpha q}{16(\lambda + 2G)\pi k}\left(\frac{2\lambda + 3G}{\lambda + G} - \frac{3r^2}{R^2}\right)$$

$$\epsilon_{\theta\theta} = \frac{(3\lambda + 2G)\alpha q}{16(\lambda + 2G)\pi k}\left(\frac{2\lambda + 3G}{\lambda + G} - \frac{r^2}{R^2}\right) \tag{l}$$

Substitution of Eq. (l) into Eq. (f) gives

$$\sigma_{rr} = \frac{(3\lambda + 2G)\alpha q}{4\pi k}\left[\left(\frac{2\lambda + 3G}{2(\lambda + 2G)} - 1\right)\left(1 - \frac{r^2}{R^2}\right)\right]$$

$$\sigma_{\theta\theta} = \frac{(3\lambda + 2G)\alpha q}{4\pi k}\left[\frac{2\lambda + 3G}{\lambda + 2G} - 1 + \frac{r^2}{R^2}\left(1 - \frac{2\lambda + G}{\lambda + 2G}\right)\right] \tag{m}$$

In this solution there is a net axial force in the fuel element because of the assumption $\epsilon_{zz} = 0$. If a solution is desired for no axial force in the element it can easily be obtained by superposition.

III-8 STRAIN ENERGY

When a body is deformed by external forces, these forces do work on the body. Since the elastic deformation is reversible, this work is recovered when the loads are removed. The work must therefore be stored in the deformed body as internal energy, as required by the conservation of energy (the first law of thermodynamics). This type of internal energy is called strain energy. The strain energy stored in a small element of volume which has been deformed elastically can be calculated from the work done by the stresses acting on the surfaces of the elements during loading. Consider a volume element such as the one shown in Fig. II.6 acted on by stresses σ_{ij}. Let this element undergo an additional increment of strain $\delta\epsilon_{ij}$. Then the work done by the stresses is simply the stress times the area of the surface on which it acts times the relative displacement of opposite sides of the element in the direction in which the stress acts. Thus, the increment of work, $\delta[dW(\sigma_{11})]$, done by the stress component σ_{11} is

$$
\begin{aligned}
\delta[dW(\sigma_{11})] &= (\sigma_{11}\,dx_2 dx_3)\,\delta\epsilon_{11}\,dx_1 \\
&= \sigma_{11}\delta\epsilon_{11}\,dx_1 dx_2 dx_3 = \sigma_{11}\delta\epsilon_{11}\,dV
\end{aligned}
\qquad \text{(III.12)}
$$

For a stress component σ_{ij} the increment of work is given by

$$
\delta[dW(\sigma_{11})] = \sigma_{ij}\delta\epsilon_{ij}\,dV \qquad \text{(III.13)}
$$

and the total work increment is

$$
\delta(dW) = \sum_{i=1}^{3}\sum_{j=1}^{3}\sigma_{ij}\delta\epsilon_{ij}\,dV \qquad \text{(III.14)}
$$

To get the total work done in loading the element to a stress level σ_{ij} one must integrate the work increments $\delta(dW)$ over the loading path. For some choices of loading path, such as increasing each stress component sequentially, the calculation is algebraically difficult because changing one stress σ_{ij} can effect other strain components ϵ_{kl}. However, for elastic behavior, the state of the material is a function of σ_{ij} only and independent of the path of loading. Therefore, a particular loading path can be chosen to simplify the calculation of the strain energy. Consider for example the case of proportional loading such that the stresses on the element of the body are increased from zero to a final value, $\sigma_{ij}{}^{f}$, over a period t_0 such that the stress $\sigma_{ij}(t)$ at any time $t \leq t_0$:

$$
\sigma_{ij}(t) = \frac{t}{t_0}\,\sigma_{ij}{}^{f} \qquad \text{(III.15)}
$$

Since the strains are linearly related to the stresses by Eq. III.2 in the most general linearly elastic case, the strains at time t can be related to the final strain values $\epsilon_{ij}{}^{f}$ by

$$\epsilon_{ij}(t) = \frac{t}{t_0} \epsilon_{ij}{}^f \tag{III.16}$$

The work done during loading can then be calculated from Eq. III.14, since

$$\delta\epsilon_{ij} = \frac{d\epsilon_{ij}(t)}{dt} dt = \frac{\epsilon_{ij}{}^f}{t_0} dt \tag{III.17}$$

The work done in the volume element is then

$$
\begin{aligned}
dW &= \int_0^{t_0} \sum_{i=1}^{3} \sum_{j=1}^{3} \left(\frac{t}{t_0}\right) \sigma_{ij}{}^f \left(\frac{1}{t_0}\right) \epsilon_{ij}{}^f \, dt \, dV \\
&= \sum_{i=1}^{3} \sum_{j=1}^{3} \sigma_{ij}{}^f \epsilon_{ij}{}^f \left(\frac{1}{t_0{}^2}\right) \int_0^{t_0} t \, dt \, dV
\end{aligned}
\tag{III.18}
$$

Since the work dW done on the element equals the energy dU stored in the element, the strain energy per unit volume is given by

$$\frac{dU}{dV} = \sum_{i=1}^{3} \sum_{j=1}^{3} \frac{1}{2} \sigma_{ij} \epsilon_{ij} \tag{III.19}$$

and the total strain energy in the body is

$$U = \int_{\text{Vol}} \sum_{i=1}^{3} \sum_{j=1}^{3} \frac{1}{2} \sigma_{ij} \epsilon_{ij} \, dV \tag{III.20}$$

where the integral is taken over the volume of the body.

When the strain energy of any elastic bodies in a system is included in the potential energy of the system, the energy principles such as virtual work and minimization of potential energy which are used in other areas of physics can be applied to elastic problems.

One example of such an energy principle, and a powerful tool for many calculations, is Castigliano's theorem. If an elastic body is acted upon by N applied loads F_i, the body contains an elastic strain energy U which is a function of the forces F_i. This elastic energy is also equal to the work done by the applied forces during loading. If v_i is the displacement of the body in a direction parallel to F_i at the point where F_i is applied, then the work done by the applied forces is

$$W = U = \sum_{i=1}^{N} \int_{0}^{v_i} F_i \cdot dv_i \qquad (III.21)$$

Consequently, the magnitude of the force F_i required to produce a given displacement v_i can be calculated by

$$\frac{\partial U}{\partial v_i} = |F_i| = F_i \qquad (III.22)$$

This form of Castigliano's theorem is valid for any elastic body, linear or nonlinear. In the case of a linear system, another form of Castigliano's theorem is possible. In such a system, where the displacements are linearly related to the loads, the work done can be written as

$$W = U = \sum_{i=1}^{N} \int_{0}^{F_i} v_i \cdot dF_i \qquad (III.23)$$

Therefore in this linear case, the displacement at any point of application of a load can be calculated using

$$\frac{\partial U}{\partial F_i} = v_i \ * \qquad (III.24)$$

The following example illustrates the use of Castigliano's theorem.

EXAMPLE III.7—Strain Energy in a Coiled Spring

Consider a coil spring made from wire of circular cross section. When a force is applied to the spring along the axis of the coil, the resultant forces and

*If some of the loads are applied moments M_i, the theorem can be used to find the rotation θ_i which occurs at the point of application of the moment, i.e.,

$$\frac{\partial U}{\partial M_i} = \theta_i$$

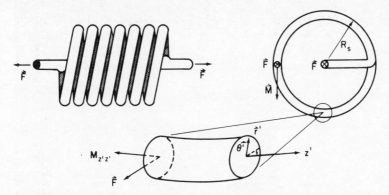

Fig. III.6 Decomposition of an axial force through a spring into a torsional moment and a shear force.

moments in the wire of the spring are a twisting moment FR_s about the axis of the wire and a shear force F perpendicular to the wire, as shown in Fig. III.6. The moment and the shear force are the same everywhere in the spring. Determine the spring constant of the spring from its dimensions. The pitch angle α of the spring is small.

Solution:

It is customary in the deformation of slender members to ignore the deflections resulting from shear stresses and consider only bending or in this case torsional deformation. This procedure will be followed here. The stress and strain in the wire are therefore given by the formula for simple torsion

$$\sigma_{\theta'z'} = \frac{Mr'}{I_p} = \frac{2Mr'}{\pi R^4}; \quad \epsilon_{\theta'z'} = \frac{Mr'}{G\pi R^4} \tag{a}$$

where R is the radius of the wire and the coordinates r', θ', and z' are oriented along the wire and not along the spring.

Even when the strain is known, the task of integrating to find the displacements is formidable, since it would require referring the strains to a fixed coordinate system. However, calculation of the strain energy is quite easy.

$$\frac{dU}{dV} = \frac{1}{2} \sum_{i=1}^{3} \sum_{j=1}^{3} \sigma_{ij} \epsilon_{ij} = \frac{1}{2} (\sigma_{\theta'z'} \epsilon_{\theta'z'} + \sigma_{z'\theta'} \epsilon_{z'\theta'})$$

$$\frac{dU}{dV} = \frac{1}{2G} \left(\frac{2Mr'}{\pi R^4} \right)^2 \tag{b}$$

$$U = \int_0^R \frac{2}{G}\left(\frac{M}{\pi R^4}\right)^2 r'^2 (2\pi L r') dr' = \frac{4\pi L}{G}\left(\frac{M}{\pi R^2}\right)^2 \frac{R^4}{4} \qquad \begin{matrix}(b)\\(\text{Cont.})\end{matrix}$$

where L is the total length of wire in the spring or

$$L = 2\pi R_s N \qquad (c)$$

where N is the number of coils and R_s is the radius of the spring. Equation (b) then becomes

$$U = \frac{\pi L}{G\pi^2 R^4} M^2 = \frac{\pi L (F R_s)^2}{G\pi^2 R^4} \qquad (d)$$

Using Castigliano's theorem (Eq. III.24), the displacement of the spring, δ, is

$$\delta = \frac{\partial U}{\partial F} = \frac{2 L R_s^2 F}{G\pi R^4} \qquad (e)$$

and the spring constant, $k = F/\delta$, is

$$k = \frac{G\pi R^4}{2 L R_s^2} = \frac{G\pi R^4}{4\pi N R_s^3} = \frac{G R^4}{4 N R_s^3} \qquad (f)$$

by substituting Eq. (c).

Note that, in the above example, the alternate form of Castigliano's theorem, Eq. III.22, could be used to calculate k. By substituting $k\delta = F$ in Eq. (d) and taking the derivative with respect to δ, an expression for F can be obtained in terms of δ. The problem can also be solved by applying conservation of energy directly, i.e., by equating the strain energy U to the work done externally, $W = F^2/2k$, and solving for k.

III-9 STRESS CONCENTRATIONS

When the problem is to calculate overall elastic deflections, St. Venant's principle justifies neglecting the effect of local irregularities in the stress distribution when they do not produce any resultant force. However, if the problem concerns the stress level in the body, it is just these local irregularities which are of greatest interest. This is because irregularities in the body, such as holes, cracks, or notches, always induce an elastic stress near the irregularity which is greater than the nominal stress calculated neglecting the irregularity.

One example of this stress-concentrating effect of an irregularity is the effect of a small hole in a very large plate which is subjected to uniaxial tension (see Fig. III.7). Since the surface of the hole must be stress free, the stress in the vicinity of the hole must be modified from the uniaxial tension. Finding the solution to this problem involves techniques beyond the scope of this book. The solution is however one of the many elastic solutions which are well known and may be found in many books on elasticity (see, for example, Ref. 5). In cylindrical coordinates, the stresses, for plane stress, are

$$\sigma_{rr} = \frac{\sigma_{nom}}{2}\left[1 - \frac{a^2}{r^2} + \left(1 - 4\frac{a^2}{r^2} + 3\frac{a^4}{r^4}\right)\cos 2\theta\right]$$

$$\sigma_{\theta\theta} = \frac{\sigma_{nom}}{2}\left[1 + \frac{a^2}{r^2} - \left(1 + 3\frac{a^4}{r^4}\right)\cos 2\theta\right] \qquad \text{(III.25)}$$

$$\sigma_{r\theta} = -\frac{\sigma_{nom}}{2}\left(1 + 2\frac{a^2}{r^2} - 3\frac{a^4}{r^4}\right)\sin 2\theta$$

where σ_{nom} is the nominal stress when the hole is absent. Note that this solution demonstrates a case where St. Venant's principle is valid. The introduction of the stress-free hole does not change the resultant forces on the plate, so that the effect of the hole should decrease with increasing distance from the hole. You will note that the constant terms in the expression represent the unperturbed uniform stress distribution (i.e., $\sigma_{11} = \sigma_{nom}$, $\sigma_{22} = \sigma_{12} = 0$) expressed in cylindrical coordinates, while the terms which are negative powers of r represent the effect of the hole.

As required by the boundary conditions,

$$\sigma_{rr} = \sigma_{r\theta} = 0 \qquad \text{(III.26)}$$

at the edge of the hole. The other stress components at the edge of the hole are given by

$$\sigma_{\theta\theta} = \sigma_{nom}(1 - 2\cos 2\theta)$$

$$\sigma_{\theta\theta} = \begin{cases} -\sigma_{nom} & \text{for } \theta = 0, \pi \\ 3\sigma_{nom} & \text{for } \theta = \pm\pi/2 \end{cases} \qquad \text{(III.27)}$$

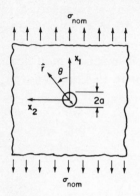

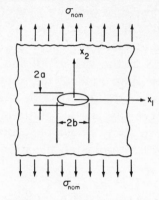

Fig. III.7 Stress concentration at a hole with a stress-free boundary in an infinite plate subjected to a uniaxial stress.

Fig. III.8 Stress concentration at an elliptic hole with a stress-free boundary in an infinite plate subjected to a uniaxial stress.

The maximum stress with the hole present is therefore three times the nominal stress or, stated another way, the hole has a stress concentration factor of 3 in a uniaxial state of stress. This solution for the effect of a hole in uniaxial tension in plane stress is very useful, since by adding two of these solutions directed along the two principal stress directions, one can determine the stress concentration factor for a hole in any plane-stress condition, provided that other boundaries of the body are sufficiently far from the hole.

The above solution indicates that local perturbations such as holes tend to raise the stress locally. However, the specialized case of a circular hole cannot be directly extended to other types of holes and notches. The case of the elliptical hole adds further insight into the nature of stress concentrations.

Consider a sheet in plane stress with an elliptical hole oriented so that the axes of the ellipse are parallel and perpendicular to the tension direction (see Fig. III.8). The stress concentration factor for this case is

$$\frac{\sigma_{max}}{\sigma_{nom}} = 1 + 2\frac{b}{a} \qquad ((\text{III}.28)$$

and the maximum stress is σ_{22} at the ends of the ellipse, $x_1 = \pm b$. Note that if $b > a$ the elliptic hole has a greater stress concentration factor than a circular hole, but if $b < a$ the stress concentration is less than for a circular hole.

The solution for an elliptic hole can be used to obtain approximate stress concentration factors for other shapes of holes. For this purpose, expression (III.28) can be changed so that the stress concentration factor is expressed as a function of the radius of curvature ρ at the end of the ellipse and the width of the hole perpendicular to the stress axis, $2b$. This is easily done, since the equation for the boundary of the elliptical hole is

$$\frac{x_1^2}{b^2} + \frac{x_2^2}{a^2} = 1 \qquad\qquad \text{(III.29)}$$

The radius of curvature at the end $x_1 = b$ is given by

$$\frac{1}{\rho} = -\left.\frac{\partial^2 x_1}{\partial x_2^2}\right|_{x_2 = 0} \qquad\qquad \text{(III.30)}$$

Solving Eq. (III.29) for x_1 yields

$$x_1 = b\left(1 - \frac{x_2^2}{a^2}\right)^{1/2} \qquad\qquad \text{(III.31)}$$

Differentiating the above equation twice and evaluating at $x_2 = 0$, we obtain

$$\frac{\partial^2 x_1}{\partial x_2^2} = \frac{b}{2}\left[-\frac{1}{2}\left(1 - \frac{x_2^2}{a^2}\right)^{-3/2}\left(\frac{2x_2}{a^2}\right)^2 + \left(1 - \frac{x_2^2}{a^2}\right)^{-1/2}\left(-\frac{2}{a^2}\right)\right]$$

$$\left.\frac{\partial^2 x_1}{\partial x_2^2}\right|_{x_2 = 0} = -\frac{b}{a^2} \qquad\qquad \text{(III.32)}$$

Therefore, $\rho = a^2/b$ or $a = \sqrt{b\rho}$. Substituting this expression for a in Eq. (III.28) gives the desired formula for the stress concentration factor:

$$\frac{\sigma_{max}}{\sigma_{nom}} = 1 + 2\sqrt{\frac{b}{\rho}} \qquad\qquad \text{(III.33)}$$

St. Venant's principle indicates that the stress concentration factor should be a strong function of the local radius of curvature at the point of maximum stress, but it should be much less sensitive to the shape of the boundary at large distances from this point. For this reason, the elliptic hole formula can be used to get approximate values of the stress concentration factor for other shapes of holes by approximating the latter with an elliptic hole having the same radius of curvature at the ends and the same overall length. Examples of such an approximation are shown in Fig. III.9.

Neuber (Ref. 6) has used essentially this procedure to find analytic solutions for stress concentrations in a number of geometries and states of loading. In his work, Neuber uses elliptic shapes for internal holes and hyperbolic shapes for external notches, since solutions are most readily obtainable for these geometries. Table III.2 summarizes some of the most useful results from Neuber. (Neuber's book, Ref. 6, contains a number of nomographs which can be used to find stress concentration factors for a larger number of cases than given here.)

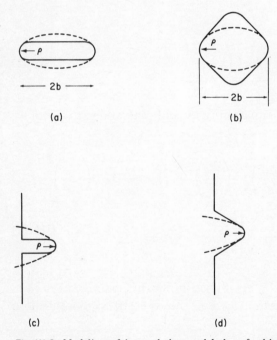

Fig. III.9 Modeling of internal slots and holes of arbitrary shape using ellipses with the same minimum radius of curvature and overall length. (External notches can be modeled by hyperbolas with the same minimum radius of curvature.)

TABLE III.2 Stress Concentration Factors

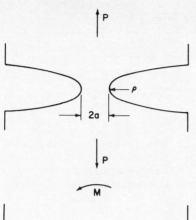

1. Deep edge notches in tension.

$$\sigma_{nom} = \frac{P}{2ad}$$

$$\frac{\sigma_{max}}{\sigma_{nom}} = \frac{2(a/\rho + 1)\sqrt{a/\rho}}{\sqrt{a/\rho} + (a/\rho + 1)\tan^{-1}\sqrt{a/\rho}}$$

(d is the depth of the part).

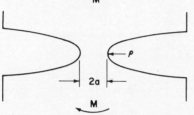

2. Deep edge notches in bending.

$$\sigma_{nom} = \frac{3M}{2a^2d}$$

$$\frac{\sigma_{max}}{\sigma_{nom}} = \frac{4(a/\rho)^{3/2}}{3(\sqrt{a/\rho} + (a/\rho - 1)\tan^{-1}\sqrt{a/\rho})}$$

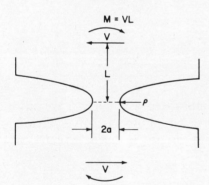

3. Deep edge notches in shear.

$$\tau_{nom} = \frac{V}{2ad}$$

$$\frac{\sigma_{max}}{\tau_{nom}} = \frac{a/\rho\sqrt{a/\rho + 1}}{-\sqrt{a/\rho} + (a/\rho + 1)\tan^{-1}\sqrt{a/\rho}}$$

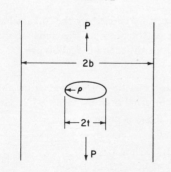

4. Center hole for tension or bending.

$$\sigma_{nom} = \frac{P}{2bd} \text{ for tension}$$

$$\sigma_{nom} = \frac{3Mt}{2b^3d} \text{ for bending}$$

$$\frac{\sigma_{max}}{\sigma_{nom}} = 1 + C\sqrt{t/\rho}$$

where $C = 2$ for tension
$C = 1$ for bending

TABLE III.2 Stress Concentration Factors (*Continued*)

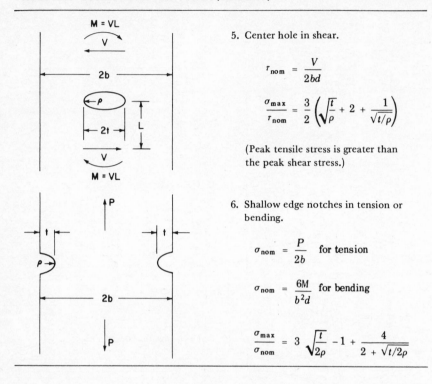

5. Center hole in shear.

$$\tau_{nom} = \frac{V}{2bd}$$

$$\frac{\sigma_{max}}{\tau_{nom}} = \frac{3}{2}\left(\sqrt{\frac{t}{\rho}} + 2 + \frac{1}{\sqrt{t/\rho}}\right)$$

(Peak tensile stress is greater than the peak shear stress.)

6. Shallow edge notches in tension or bending.

$$\sigma_{nom} = \frac{P}{2b} \quad \text{for tension}$$

$$\sigma_{nom} = \frac{6M}{b^2 d} \quad \text{for bending}$$

$$\frac{\sigma_{max}}{\sigma_{nom}} = 3\sqrt{\frac{t}{2\rho}} - 1 + \frac{4}{2 + \sqrt{t/2\rho}}$$

A number of cases of more complicated geometry have also been analyzed experimentally by use of photoelasticity. A summary of many of these experimental results is given in Peterson (Ref. 11).

EXAMPLE III.8—Stress Concentration of a Slot

Estimate the maximum shear force F which can be carried by the linkage bar shown in Fig. III.10 without causing local plastic flow at the ends of the elongated slot in the center. The yield strength of the metal is 75,000 psi.

Solution:

The maximum stress at the end of the slot can be determined from the formula for bending of a wide bar with an internal hole in Table III.2. The bending moment $M = FL_1$. The nominal stress σ_{nom} is given by

$$\sigma_{nom} = \frac{3}{2}\frac{Mt}{b^3 d} = \frac{3}{2}\frac{FL_1 t}{b^3 d} \tag{a}$$

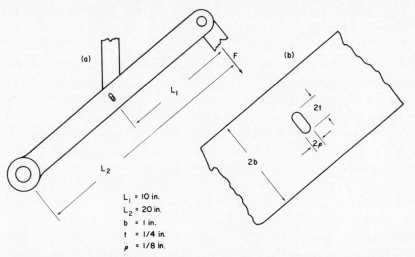

Fig. III.10 (a) Machine linkage containing an elongated hole. (b) Detail of the hole in the linkage.

and the maximum stress, or yield strength, is

$$\sigma_{11_{max}} = \sigma_{nom}\left(1 + \sqrt{\frac{t}{\rho}}\right) = \frac{3FL_1 t}{2b^3 d}\left(1 + \sqrt{\frac{t}{\rho}}\right) \qquad (b)$$

Substituting $L_1 = 10$ in., $t = 1/4$ in., $\rho = 1/8$ in., $b = 1$ in., and $d = 1/4$ in., we obtain

$$\sigma_{11_{max}} = 75{,}000 \text{ psi} = \frac{7.5F}{0.5}(1 + \sqrt{2}) \text{ in.}^{-2} \qquad (c)$$

$$F = \frac{(0.5)\,10^4}{1 + \sqrt{2}} \text{ lb} = 2{,}070 \text{ lb} \qquad (d)$$

III-10 BUCKLING OF COLUMNS—ELASTIC INSTABILITY UNDER COMPRESSION

All of the problems considered in the previous sections involved small deformations, a single equilibrium state, and the validity of the uniqueness theorem. An interesting example of a class of problems which does not satisfy these conditions is the buckling of a column. A small elastic deformation induces an instability, which, when propagated, leads to a large-scale change in geometry (buckling) and the violation of uniqueness. Elastic buckling of a column is just one of several types of instability with important engineering applications. Some other types will be considered in Chapter IV.

The buckling phenomenon is associated with the transition of the column configuration from a stable equilibrium condition to an unstable equilibrium condition as the load exceeds a critical load. Buckling is typified by sudden collapse or a large deflection of a structure when a critical load is reached. The concept may be illustrated using the model shown in Fig. III.11.* Two rigid rods are pinned near their ends, and a spring is attached laterally to the pinned joint. The spring constant is k. When the rod assumes a straight configuration, the spring does not exert any lateral force. When it is bent, the spring exerts a restoring force to the rod while the applied load tends to bend the rod. The bent configuration is an equilibrium state whenever the applied load satisfies the equilibrium condition, i.e., (for small deflections),

$$ku_2 = 2P \tan\alpha = 2P\left(\frac{u_2}{L/2}\right) \qquad \text{(III.34)}$$

*The discussion of this model follows that of Shanley (Ref. 9).

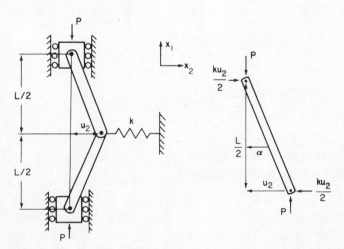

Fig. III.11 A hinged rigid bar model for buckling. (For small deflection angles, the vertical distance $L/2$ is nearly equal to the length of one of the rods.)

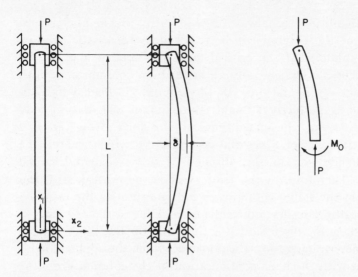

Fig. III.12 Buckling of an elastic column which is pin connected at its ends.

or
$$P_{crit} = \frac{kL}{4}$$
(III.34)
(Cont.)

If the applied load is less than the critical value given above, the restoring force exerted by the spring is larger than the upsetting (applied) force component, and the rod assumes a straight configuration. On the other hand, if the applied load exceeds the critical value, and the rods are accidentally displaced laterally or have an initial eccentricity, the rods will continue to bend farther and farther. At the critical load, the straight configuration and the bent configuration are in neutral equilibrium. The critical load at which bifurcation of the equilibrium positions takes place is called the buckling load. It is also interesting to note that, in Eq. (III.34), P_{crit} is independent of the deflection of the spring (for small deflections), indicating that the lateral motion of the pinned joint is not inhibited by the spring, when the applied load is equal to the critical load. For large deflections, the load actually rises very slowly with increasing deflection.

Now let us consider a simple, axially loaded elastic column, as shown in Fig. III.12. In this case, the restoring force is exerted by the elasticity of the column material, rather than the elasticity of the spring in the preceding example. When the applied load is small, the straight configuration is always stable. Any accidental lateral displacement of the column will be restored by the elastic bending stress in the column. At the critical load, the applied load can be carried by either the straight column or the bent column, since both are equilibrium configurations. If the column is loaded axisymmetrically and is perfectly straight, the column will deform only axially, regardless of the magnitude of the load. However, real structures always have a certain eccentricity, which causes the column to bend even in the absence of any accidental lateral displacement. Therefore, as the load exceeds the critical load, the transition from the stable equilibrium configuration to the unstable equilibrium position occurs and results in buckling of the column.

From the foregoing discussion it is clear that the buckling load can be found by determining the critical load at which the applied load is in equilibrium with the bent configuration of the column. From the free-body diagram shown in Fig. III.12, the governing equations may be written as

$$EI_{22} \frac{d^2u_2}{dx_1^2} = M = -M_0 = -Pu_2 \quad *$$

(III.35)

The boundary conditions are

$$\begin{aligned} u_2 &= 0 \\ \frac{d^2u_2}{dx_1^2} &= 0 \end{aligned} \quad \text{at } x_1 = 0 \quad \text{and} \quad x_1 = L$$

(III.36)

Solving Eq. (III.20) and applying the boundary conditions, we obtain

$$u_2 = C \sin \frac{n\pi}{L} x_1$$

(III.37)

*Note that the positive bending moment is defined as

where C is a constant if the applied load is equal to the characteristic values given by

$$P = (n\pi)^2 \frac{EI_{22}}{L^2} \qquad (III.38)$$

The smallest elastic buckling load corresponds to the case $n = 1$:

$$P^e_{crit} = \pi^2 \frac{EI_{22}}{L^2}$$

or

$$\sigma_{crit} = \frac{\pi^2 E}{(L/k)^2} \qquad (III.39)$$

where k is defined as the radius of gyration given by $I_{22} = Ak^2$. This notation will be used for convenience. The trivial solution $n = 0$ corresponds to the straight configuration of the column. It should be noted that the buckling load is very sensitive to the column length. Other constants in place of π^2 give the buckling loads for columns with different end conditions.

EXAMPLE III.9—Buckling of a Column Structure

A structure consists of a heavy rigid slab supported by four steel columns, as shown in Fig. III.13. The columns are built into the ground but are pin-connected to the slab. The columns are 10-ft long and have the dimensions 2 in. OD and 1.92 in. ID. What is the maximum load, including the weight of the slab, which the columns can support?

Solution:

Consider first the possibility of buckling. The boundary conditions are not the same as in the discussion above, since the columns must remain vertical at the bottom but are free to move horizontally at the top, if they buckle cooperatively as shown in Fig. III.13b. Assuming from symmetry that each column carries a vertical load P, the equilibrium equation for the column is given by

$$EI_{22} \frac{d^2 u_2}{dx_1^2} = M = P(\delta - u_2) \qquad (a)$$

where δ is the deflection at $x_1 = L$. The boundary conditions are

$$\frac{du_2}{dx_1} = 0 \Bigg\} \quad \text{for} \ x_1 = 0 \tag{b}$$

$$u_2 = 0 \tag{c}$$

$$M = EI_{22}\frac{d^2u_2}{dx_1^{\,2}} = 0 \Bigg\} \quad \text{for} \ x_1 = L \tag{d}$$

$$u_2 = \delta \tag{e}$$

The general solution to the differential equation (a) is

$$u_2 = A \sin \alpha x_1 + B \cos \alpha x_1 + \delta \tag{f}$$

where $\alpha = \sqrt{P/EI_{22}}$. Also

$$\frac{du_2}{dx_1} = A\alpha \cos \alpha x_1 - B\alpha \sin \alpha x_1 \tag{g}$$

The boundary condition in Eq. (b) implies $A = 0$, so that

$$\frac{d^2u_2}{dx_1^{\,2}} = -B\alpha^2 \cos \alpha x_1 \tag{h}$$

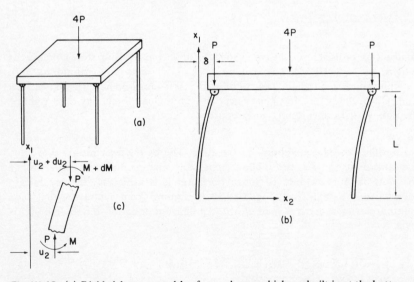

Fig. III.13 (a) Rigid slab supported by four columns which are built in at the bottom and pin connected to the slab at the top. (b) Cooperative buckling of the four columns into a shape which is one-quarter of a sine wave. (c) Action of the resultant forces and moments on an element of one of the columns.

Using Eq. (d) one finds

$$\cos \alpha L = 0 \tag{i}$$

which implies that

$$\alpha L = \sqrt{\frac{P}{EI_{22}}} L = \frac{\pi}{2} + n\pi \tag{j}$$

The lowest buckling load occurs for $n = 0$, giving a value for the critical load of

$$P_{\text{crit}} = \frac{\pi^2 EI_{22}}{4L^2} \tag{k}$$

To complete the solution, we evaluate the boundary condition in Eqs. (c) and obtain

$$B = -\delta \tag{l}$$

Equation (e) is identically satisfied when $\cos \alpha L = 0$. Thus, the complete solution is

$$u_2 = \delta \left(1 - \cos \frac{\pi x_1}{2L} \right) \tag{m}$$

when $P = P_{\text{crit}}$.

Using the dimensions given in the problem

$$P_{\text{crit}} = \frac{\pi^2 E \pi (R_0^4 - R_i^4)/4}{4L^2} = \frac{\pi^3 (3 \times 10^7)(1 - (0.96)^4)}{16(120)^2} \text{ lb} \tag{n}$$

$$P_{\text{crit}} = 625 \text{ lb}$$

Since the compressive stress in the tube at this load is only

$$\sigma_{11} = -\frac{P}{\pi (R_0^2 - R_i^2)} = -\frac{625}{\pi (1 - 0.922)} = -2,500 \text{ psi} \tag{o}$$

failure will certainly be by buckling.

III-11 DEVIATIONS OF REAL METALS FROM PERFECT ELASTICITY

In the preceding sections of this chapter it was assumed that elastic deformation is completely reversible, i.e., that strain is a single-valued function of the stress and that the loading and unloading paths are exactly the same. The corollary of these assumptions is that there is no energy loss during a loading-unloading cycle of elastic deformation. These assumptions are reasonable in structural applications of metals, where the rate of loading is nearly quasi-static. However in dynamic applications where high-frequency vibration occurs, as, for example, in displacement amplifiers used for ultrasonic grinding, the deviations from perfect elasticity become important. The departures cause only a small amount of energy to be lost per cycle, but, since the cycle rate is high, the energy loss per unit time may be appreciable, and the resulting internal damping may account for a significant portion of the power requirement of the device. Thus, if the displacement amplifier of an ultrasonic grinder is made of steel, it becomes red hot and fails.

When a metal specimen is loaded from a stress-free state to an elastic state (indicated by A in Fig. III.14) and then unloaded again to a stress-free state B, close examination will usually show that the loading path OA does not exactly coincide with the unloading path AB and that B does not coincide with O.* Sometimes the strain may asymptotically approach O with time after unloading. This is called the *elastic aftereffect*. In other cases the strain is never recovered, and a permanent deformation is incurred. During cyclic loading, the specimen may follow a path, such as ABCD of Fig. III.14, which forms a hysteresis loop enclosing a nonzero area. The area enclosed in the loop is then proportional to the net energy lost per cycle. Although this energy loss is quite small for each cycle, the power dissipation at high frequencies can be very large.

Deviations from perfect elasticity are caused by a number of different physical phenomena: dependence of the strain on temperature; coupling between the strain and electromagnetic fields, as in magnetostriction; microplasticity; grain boundary sliding; and other similar effects which cause the strain to be a multivalued function of the stress.

Any permanent component of anelastic strain (OB in Fig. III.14) is caused by microplasticity, resulting from small motions of dislocations and the shearing of grain boundaries, especially at elevated temperatures. As discussed further in Chapter V, a limited amount of

---

*Zener (Ref. 10) defined as "anelasticity" the property of a solid by virtue of which stress and strain are not uniquely related in the elastic regime.

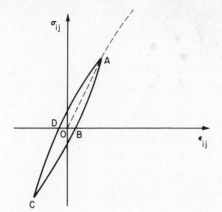

Fig. III.14 Loading and unloading path of a metal, showing a hysterisis loss.

dislocation motion can take place at stresses well below the yield stress of the metal. At elevated temperatures, the grain boundaries can also slip by viscous sliding, causing additional permanent deformation. At stresses below the yield stress, continuous dislocation motion and generation cannot occur, and the permanent strain is limited. Incompatibility at grain boundary corners and irregularities halts grain boundary sliding at low stress, after a small offset.

The time-dependent but reversible deviation from perfect elasticity is a consequence of the fact that the elastic strain at a given stress can depend slightly on the temperature, impurity distribution, and magnetic field, all of which can be changed by the state of strain. (It can be generally stated that a change in any thermodynamic potential of the material may be coupled to the elastic deformation such as to make strain a multivalued function of stress.) This can be illustrated using the energy loss due to thermoelastic effects as an example, as follows.

When a polycrystalline piece of metal is strained, the stress and strain vary from grain to grain because of the anisotropy of the individual grains. In addition to this, there may be stress and strain gradients in the part as a result of the geometry, as, for example, in a beam in bending. The elastic deformation causes a slight temperature change in the material such that the temperature decreases under tensile strain and increases under compressive strain. Since the strain is not uniform, the temperature changes in different parts of the sample are not identical, and, immediately after straining, there is a nonequilibrium temperature distribution. Under suitable circumstances this effect can cause internal damping. When the frequency of a strain cycle is very high, the deformation of each part of the body takes place adiabatically, since there is insufficient time for the heat to be conducted from one part of the body to another. Unloading causes the strain to be released, and each part of the body returns to its original temperature. In this case, unloading follows the same path as loading, and no energy is lost. Conversely, at low frequencies where the loading is quasi-static, heat can be conducted from one part of the body to another with sufficient speed that the temperature of the body remains uniform throughout the deformation. In this case, the loading and unloading paths are again the same, and little damping occurs. At intermediate frequencies, the loading is rapid enough for significant temperature differences to develop in the material and slow enough for heat transfer by conduction to take place before unloading occurs. In this case, a circulating heat flow occurs along with the straining, and the strain cycle is no longer in phase with the stress cycle. The resulting hysteresis cycle causes dissipation of energy and damping. The peak damping will occur when the period of the vibration is about equal to the characteristic time for the thermal diffusion process. This characteristic time depends on the average grain diameter d and the thermal diffusivity D. Since the thermal diffusivity is equal to the thermal conductivity divided by the heat capacity and has dimensions of $(length)^2/time$, the frequency of the peak loss must be given by

$$f_{max} = A \frac{D}{d^2} \tag{III.40}$$

where the proportionality constant A is about 2.16.

A number of other possible mechanisms for internal damping by time-dependent processes are discussed by Zener (Ref. 10).

REFERENCES

1. Crandall, S. H., N. C. Dahl, and T. J. Lardner (eds.): "An Introduction to the Mechanics of Solids," 2d ed., McGraw-Hill, New York, 1972.
2. Flügge, W., (ed.): "Handbook of Engineering Mechanics," McGraw-Hill, New York, 1962.
3. McClintock, F. A., and A. S. Argon: "Mechanical Behavior of Materials," Addison-Wesley, Reading, Mass., 1966.
4. Sokolnikoff, I. S.: "Mathematical Theory of Elasticity," McGraw-Hill, New York, 1956.

5. Timoshenko, S., and J. N. Goodier: "Theory of Elasticity," 2d ed., McGraw-Hill, New York, 1951.
6. Neuber, H.: "Theory of Notch Stresses," Edwards, Ann Arbor, Mich., 1946.
7. Wang, C.: "Applied Elasticity," McGraw-Hill, New York, 1953.
8. Housner, G. W., and T. Vreeland, Jr.: "Analysis of Stress and Deformation," Macmillan, New York, 1966.
9. Shanley, F. R.: "Strength of Materials," McGraw-Hill, New York, 1957.
10. Zener, C.: "Elasticity and Anelasticity of Metals," University of Chicago Press, Chicago, 1948.
11. Peterson, R. E.: "Stress Concentration Design Factors," John Wiley, New York, 1953.

PROBLEMS

III.1   A coil spring is wound from wire of diameter $2r$. The diameter of the coil is $2R$ and the number of turns is N. A bending moment M is applied to the ends of the spring as shown. What is the strain energy in the spring? What is the spring constant k which relates the moment M to the relative rotation θ between the ends?

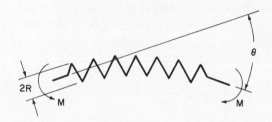

III.2   An article in the Boston Globe of Sunday, September 5, 1971, states that the front bumpers of 1972 model General Motors cars are designed to withstand a car-to-car collision of 5 mph relative velocity without sustaining any damage. (You may assume that this is equivalent to a 2 1/2-mph collision into a solid object.) The article further states that Oldsmobiles will do this by absorbing the kinetic energy of the car in two springs which can deflect 1 1/2 in. We shall assume that the springs are coil springs, since these are usually most efficient when the spring must fit into a small space.

1) Consider a coil spring similar to the one discussed in Example III.7. If the maximum allowable shear stress is σ_{max}, show that the maximum energy which can be stored in the spring depends only on the volume of the spring,

but not on the details of its shape. How much material is required to absorb the kinetic energy of a 2-ton car moving at 2 1/2 mph? Assume that the springs are made of steel with a maximum shear stress of 150,000 psi.

2) Starting with the condition above, design a spring which might reasonably be expected to fit behind an automobile bumper which can absorb the necessary energy in a 1 1/2-in. deflection.

3) Describe *briefly* any changes in your design which would be required to allow for impact on a corner of the bumper, or on the bumper guard above or below the bumper.

III.3   Simplified representations of two proposed shapes for the cutting end of a pair of wire cutters are shown below. Estimate the ratio of failure loads F for the two cutters, assuming that they fail by brittle fracture at the location indicated. (Note: You are *not* asked to determine the failure loads themselves.) Both cutters are 1/4-in. thick. Indicate with a sketch the important dimensions used in your analysis. What can you say about the ratio of the limiting loads, if the part fails by plastic deformation? (Give a quick answer; no analysis is required.)

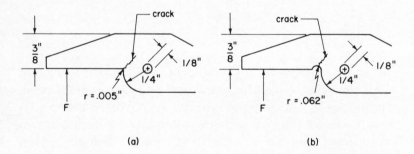

(a)   (b)

III.4   Sandwich beams are usually made of a soft core with outer skin layers of a more rigid material. Its advantages include increased sectional modulus, structural damping, and insulation effects.

Consider the sandwich beam shown below. The core is a foamed polymer with elastic modulus E_f, and the skin is aluminum with modulus E_A. A load per unit width of F is applied to the end of the beam. What is the shear stress distribution in the beam? First, find the normal stress in the two sections of the beam, and then apply the equilibrium equations in two dimensions to find the shear stresses.

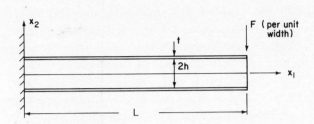

III.5   For various types of loading of a cantilever beam of constant height h, the normal distribution of material is not efficient in terms of carrying the load at each cross section. For example, under a point load applied to the end of the beam, the "weakest" cross section is that near the wall (since the moment is highest there). At the same time, the material at the extreme end of the beam contributes nothing in terms of flexural strength to the beam. If the flexure of each cross section can be described in terms of the radius of curvature, then one definition of an efficient distribution of material could be achieved (for a particular loading situation) by maintaining a constant radius of curvature along the beam length using an appropriate choice of beam thickness $h(x_1)$. Determine the most "efficient" beam thickness to carry:

a) A pure bending moment M applied at $x_1 = L$.
b) A point load F at $x_1 = L$.
c) A distributed load w (lb/unit length).

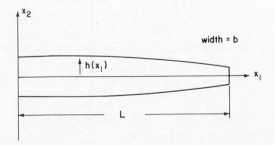

III.6   Two O-ring seals at the ends of a steel pressure vessel are compressed by .014-in. when the ends seat onto a cylinder. The ends of the vessel are held onto the cylinder by eight 3/4-in.-diameter steel rods. Assuming that the pretension on the rods is 2,000 psi (applied by tightening the nuts to a specified torque), at what internal pressure will the O-rings no longer be compressed? Obviously, the vessel will begin to leak at a pressure slightly less than this value. Consider only elastic deformation. The modulus of steel is 30×10^6 psi, and Poisson's ratio is 0.3. Use a thin-walled tube approximation for the cylinder and neglect deformation of the end plates.

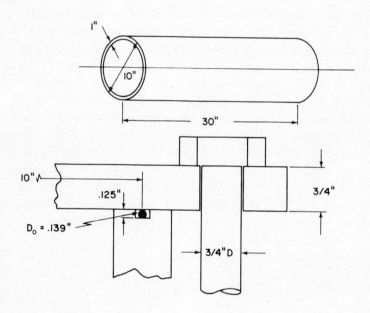

III.7   A ring with a circular cross section is made of steel with Young's modulus $E = 30 \times 10^6$ psi and Poisson's ratio $\nu = 0.29$. If a force is applied in diametrically opposite directions, as shown, determine the moment at $\theta = 0$ and the change in the distance between A and A'.

(Hint: Express the strain energy stored in the ring in terms of the applied force and the moment at $\theta = \pi$ or 0, and then use the energy theorem, noting that $\partial U / \partial M = 0$ at $\theta = 0$ or π, by symmetry.)

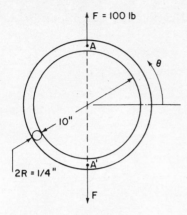

III.8   A square tensile specimen with two notches on opposing sides is subjected to an axial load of 500 lb. Determine the maximum tensile stress before plastic deformation occurs. The notch configuration is shown.

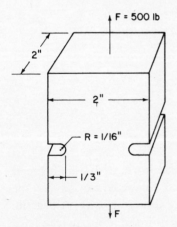

III.9   An aluminum I-beam which supports the back of the seat in an airplane has the dimensions shown below. Although this beam is more than sufficient to carry the loads it is designed for, it is too flexible to provide a comfortable ride. It is proposed that it be stiffened by putting a layer of graphite-epoxy composite material on the outside of either flange. The Young's modulus of the composite is 30×10^7 psi. The beam is loaded in service as a cantilever beam, with the loads applied near the end of the beam. How thick must the

composite be to decrease the deflection at a given load by a factor of 2?

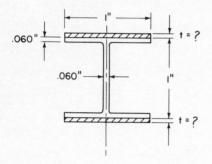

III.10 The illustrated aluminum and steel bars are of equal lengths and have equal cross-sectional areas. At zero temperature they just fit, stress free, between the two rigid boundaries. The temperature of the bars is then raised $100°F$.
a) What is the axial stress in the bars?
b) What is the deflection of the interface (in terms of L)?

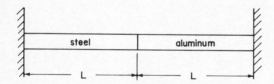

III.11 A proposed design for a paper punch for use with a high speed card punch will eliminate the need for separate drivers for each of the multiple punches in the following way. All of the punching fingers are mounted in a common block driven by a single drive mechanism. The punching fingers are long slender rods 20 cm long with cross-sectional dimensions 1.5×3.0 mm. When a hole is not desired at a given spot in a column, that punch is blocked by a metal gate. When the punch hits the gate, it buckles, so that the punching head can drive the other punches through the card. The buckling mode is as if the punch were built in at both ends. What is the maximum force the punch can exert on a card during the punching operation? The punches are made of steel.

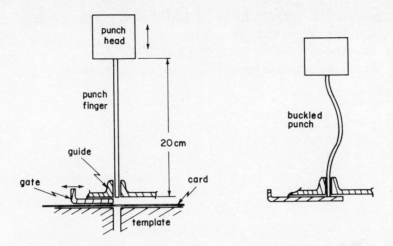

III.12 If a column of stiffness EI is subjected to its critical buckling load P and buckles elastically, it assumes a sinusoidal shape given by

$$u_1 = C \sin \frac{n\pi x_2}{L}$$

where n is a constant which depends on the end conditions at the end of the column. If the end undergoes a displacement $u_2 = \delta$ parallel to the axis of the column, what is the amplitude constant C?

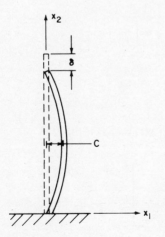

III.13 A semicircular length of pipe is loaded with a pure moment
M, as shown. Using Castigliano's theorem:
a) Find the rotation of the loaded end of the pipe.
b) Find the displacement of the loaded end of the pipe.

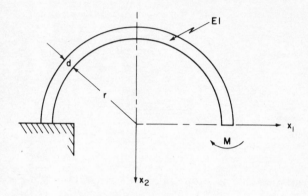

III.14 A simple prestressed beam is made by inserting a cable in the
center of a rectangular concrete beam and tightening it to a
tension T. Assuming that concrete cannot resist tension,
calculate the maximum load P that a beam of length l can
support. Since the distance a is not known, the most
unfavorable location must be assumed. The maximum allowable compressive stress is σ_0.

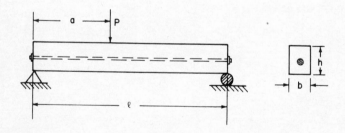

III.15 A steel block is subjected to a state of stress given by

$$\sigma_{ij} = \begin{bmatrix} 20{,}000 & 2{,}000 & 0 \\ 2{,}000 & 25{,}000 & 0 \\ 0 & 0 & 15{,}000 \end{bmatrix} \text{psi}$$

Determine the volume change (dilatation) of the block due to the applied load. The uniaxial yield stress of the steel is 125,000 psi. What would the volume change be, if the yield stress were 20,000 psi? Think about this question and answer it after you have read Chapter IV.

III.16 A flywheel with moment of inertia I is welded onto a thin rod of radius r, shear modulus G, and length l. If the flywheel is turned θ degrees from the equilibrium position and then released, determine the maximum rotational velocity of the flywheel. Assume the inertia of the rod is negligible in comparison to that of the flywheel.

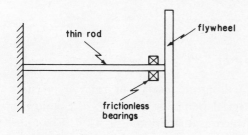

APPENDIX III-A

Alternative Formulations of Elastic Constitutive* Relations

There are several different ways to formulate the elastic constitutive relations which lead naturally to the definition of different elastic constants. Since only two elastic constants can be unique for an isotropic material, these other constants can always be defined in terms of E and G.

It is often convenient to use a formulation which places the stresses on the left-hand side of the equation and the strains on the right-hand side. When thermal strains are included, the expressions are

$$\sigma_{11} = (\lambda + 2G)\epsilon_{11} + \lambda\epsilon_{22} + \lambda\epsilon_{33} - (3\lambda + 2G)\alpha\Delta T$$

$$\sigma_{22} = (\lambda + 2G)\epsilon_{22} + \lambda\epsilon_{33} + \lambda\epsilon_{11} - (3\lambda + 2G)\alpha\Delta T \qquad \text{(A1)}$$

$$\sigma_{33} = (\lambda + 2G)\epsilon_{33} + \lambda\epsilon_{11} + \lambda\epsilon_{22} - (3\lambda + 2G)\alpha\Delta T$$

---

*Any expression which relates stress to strain is a constitutive relation.

$$\sigma_{12} = 2G\,\epsilon_{12}$$
$$\sigma_{23} = 2G\,\epsilon_{23}$$
$$\sigma_{31} = 2G\,\epsilon_{31}$$

$$\text{(A1)}$$
$$\text{(Cont.)}$$

The constant λ can be evaluated in terms of E and ν by solving the other Hooke's law equations for the streses and is found to be

$$\lambda = \frac{E\nu}{(1 + \nu)(1 - 2\nu)} \tag{A2}$$

Here, G is the shear modulus previously defined as

$$G = \frac{E}{2(1 + \nu)} \tag{A3}$$

Another formulation which will prove useful in relating elastic behavior to other types of material behavior separates the volume change portion of the strain from the distortional portion. For any state of strain, the change in volume per unit volume $(\Delta V/V)$ is the sum of the three normal strain components. This quantity is an invariant of strain, as discussed in Appendix II-B.

$$\frac{\Delta V}{V} = I_1 = \epsilon = \epsilon_{11} + \epsilon_{22} + \epsilon_{33} \tag{A4}$$

The symbol ϵ is also used to denote the first strain invariant.

A general state of strain may be decomposed into its volume-changing portion and its distortional portion by subtracting $1/3\ \epsilon$ from each of the normal states of strain, so that

$$
\begin{bmatrix} \epsilon_{11} & \epsilon_{12} & \epsilon_{13} \\ \epsilon_{21} & \epsilon_{22} & \epsilon_{23} \\ \epsilon_{31} & \epsilon_{32} & \epsilon_{33} \end{bmatrix}
=
\begin{bmatrix} \epsilon_{11} - \dfrac{\epsilon}{3} & \epsilon_{12} & \epsilon_{13} \\ \epsilon_{21} & \epsilon_{22} - \dfrac{\epsilon}{3} & \epsilon_{23} \\ \epsilon_{31} & \epsilon_{32} & \epsilon_{33} - \dfrac{\epsilon}{3} \end{bmatrix}
+
\begin{bmatrix} \dfrac{\epsilon}{3} & 0 & 0 \\ 0 & \dfrac{\epsilon}{3} & 0 \\ 0 & 0 & \dfrac{\epsilon}{3} \end{bmatrix}
$$

$$\text{(A5)}$$

The first matrix on the right-hand side represents a state of strain for which the volume change is zero, while the second matrix represents a pure change in volume with no distortion. The state of strain represented by the second matrix will be called the *dilatation strain*. The terms of the first matrix on the right-hand side are called *deviator strain* components and they are denoted by the symbol ϵ'_{ij} defined by

$$\epsilon'_{11} = \epsilon_{11} - \frac{\epsilon}{3} \qquad \epsilon'_{12} = \epsilon_{12}$$

or
$$\epsilon'_{ij} = \epsilon_{ij} - \delta_{ij}\frac{\epsilon}{3}^* \qquad\qquad\qquad (A6)$$

A state of stress may be similarly decomposed into deviator and hydrostatic components. The *hydrostatic or mean normal stress* is defined as

$$\sigma = \tfrac{1}{3}J_1 = \tfrac{1}{3}(\sigma_{11} + \sigma_{22} + \sigma_{33}) \qquad\qquad (A7)$$

where J_1 is the first invariant of the stress tensor. A state of stress can then be decomposed into the deviator stress and the hydrostatic stress.

$$\begin{bmatrix} \sigma_{11} & \sigma_{12} & \sigma_{13} \\ \sigma_{21} & \sigma_{22} & \sigma_{23} \\ \sigma_{31} & \sigma_{32} & \sigma_{33} \end{bmatrix} = \begin{bmatrix} \sigma_{11} - \sigma & \sigma_{12} & \sigma_{13} \\ \sigma_{21} & \sigma_{22} - \sigma & \sigma_{23} \\ \sigma_{31} & \sigma_{32} & \sigma_{33} - \sigma \end{bmatrix} + \begin{bmatrix} \sigma & 0 & 0 \\ 0 & \sigma & 0 \\ 0 & 0 & \sigma \end{bmatrix}$$

$$(A8)$$

The second matrix on the right-hand side is called the hydrostatic stress. It is the same in all directions, which is characteristic of the pressure in a static fluid.†

The first matrix on the right-hand side is the deviator stress, which is a state of stress with zero hydrostatic component. The deviator stress components (σ'_{ij}) are therefore defined as

---

*The symbol δ_{ij} is called the Kronecker delta. It has values $\delta_{ij} = 1$ for $i = j$ and $\delta_{ij} = 0$ for $i \neq j$.

†Note that a positive pressure in a fluid is a negative stress. That is, in a fluid under pressure p, the state of stress is $\sigma_{11} = \sigma_{22} = \sigma_{33} = -p$.

$$\sigma'_{11} = \sigma_{11} - \sigma \qquad \sigma'_{12} = \sigma_{12}$$

or $$\sigma'_{ij} = \sigma_{ij} - \delta_{ij}\sigma \tag{A9}$$

In an isotropic elastic body, a state of stress with a zero hydrostatic component produces only distortion and no volume change in the body. A purely hydrostatic state of stress produces only a volume change with no distortion. Hooke's law, expressed in terms of the decomposed stresses and strains, is

$$\epsilon'_{ij} = \frac{\sigma'_{ij}}{2G} \qquad \epsilon = \frac{\sigma}{B} \tag{A10}$$

where B is the bulk modulus,

$$B = \frac{E}{3(1 - 2\nu)} \tag{A11}$$

Notice that in this formulation the form of the deviator stress-deviator strain relationship is the same for all components [Eq. (A10)], and that the dilatation strain is a function of only the hydrostatic stress.

It is interesting to relate the properties of a fluid to the above formulation of the stress-strain relations. One way of defining the fluid state is to say that the static shear modulus of a fluid is zero. The equation above clearly implies that the deviator stresses in a static fluid must also be zero. Therefore, $\sigma = \sigma_{11} = \sigma_{22} = \sigma_{33} = -p$, where p is the hydrostatic pressure in the fluid, which is the same in all directions. The fluid is elastic only for changes in volume. The static behavior of a fluid can be described by a single constant, B, the bulk modulus, or by its reciprocal, $X = 1/B$, the compressibility.

In later chapters it will be shown that, in general, the change in volume of a material is elastic, while the distortions of the body may be inelastic. The form of the relationship between σ and ϵ is general for all linear materials. The relationship between the deviator stress and strain components depends on the type of material behavior in question.

It should be noted that the deviator stress and the deviator strain and the hydrostatic stress and volume change strain are second-order tensor quantities. They therefore have all of the usual tensor properties summarized below:

$$\epsilon'_{i'j'} = \sum_{m=1}^{3} \sum_{n=1}^{3} l_{i'm} l_{j'n} \epsilon'_{mn}$$

(A12)

$$\sigma'_{i'j'} = \sum_{m=1}^{3} \sum_{n=1}^{3} l_{i'm} l_{j'n} \sigma'_{mn}$$

The hydrostatic stress and the dilatation strain are invariant for coordinate rotation but do follow the above transformation. The deviator stress and strain also have principal axes and principal values. For the hydrostatic stress and dilatation strain, any set of axes are principal axes, and the principal values are $\epsilon/3$ and σ.

Deviator stresses and deviator strains also have invariants, but I'_1 and J'_1 are zero.

The other invariants of the deviator stresses and deviator strains are

$$J'_2 = \sigma'_{11}\sigma'_{22} + \sigma'_{22}\sigma'_{33} + \sigma'_{33}\sigma'_{11} - \sigma'_{12}{}^2 - \sigma'_{23}{}^2 - \sigma'_{31}{}^2$$

$$= \sigma_{11}\sigma_{22} + \sigma_{22}\sigma_{33} + \sigma_{33}\sigma_{11} - \sigma_{12}{}^2 - \sigma_{23}{}^2 - \sigma_{31}{}^2 - 3\sigma^2$$

$$= J_2 - \frac{1}{3}J_1{}^2$$

$$J'_3 = \sigma'_{11}\sigma'_{22}\sigma'_{33} + 2\sigma'_{12}\sigma'_{23}\sigma'_{31} - \sigma'_{11}\sigma'_{23}{}^2 - \sigma'_{22}\sigma'_{31}{}^2 - \sigma'_{33}\sigma'_{12}{}^2$$

$$= \sigma_{11}\sigma_{22}\sigma_{33} + 2\sigma_{12}\sigma_{23}\sigma_{31} - \sigma_{11}\sigma_{23}{}^2 - \sigma_{22}\sigma_{31}{}^2 - \sigma_{33}\sigma_{12}{}^2$$

$$\qquad - \sigma(\sigma_{11}\sigma_{22} + \sigma_{22}\sigma_{33} + \sigma_{33}\sigma_{11} - \sigma_{12}{}^2 - \sigma_{23}{}^2 - \sigma_{31}{}^2) + 2\sigma^3$$

$$= J_3 - \frac{1}{3}J_1 J_2 + \frac{2}{27}J_1{}^3$$

$$I'_2 = \epsilon'_{11}\epsilon'_{22} + \epsilon'_{22}\epsilon'_{33} + \epsilon'_{33}\epsilon'_{11} - \epsilon'_{12}{}^2 - \epsilon'_{23}{}^2 - \epsilon'_{31}{}^2 \qquad \text{(A13)}$$

$$= \epsilon_{11}\epsilon_{22} + \epsilon_{22}\epsilon_{33} + \epsilon_{33}\epsilon_{11} - \epsilon_{12}{}^2 - \epsilon_{23}{}^2 - \epsilon_{31}{}^2 - \frac{1}{3}\epsilon^2$$

$$= I_2 - \frac{1}{3}I_1{}^2$$

$$I'_3 = \epsilon'_{11}\epsilon'_{22}\epsilon'_{33} + 2\epsilon'_{12}\epsilon'_{23}\epsilon'_{31} - \epsilon'_{11}\epsilon'_{23}{}^2 - \epsilon'_{22}\epsilon'_{31}{}^2 - \epsilon'_{33}\epsilon'_{12}{}^2$$

$$= \epsilon_{11}\epsilon_{22}\epsilon_{33} + 2\epsilon_{12}\epsilon_{23}\epsilon_{31} - \epsilon_{11}\epsilon_{23}{}^2 - \epsilon_{22}\epsilon_{31}{}^2 - \epsilon_{33}\epsilon_{12}{}^2$$

$$\qquad - \frac{\epsilon}{3}(\epsilon_{11}\epsilon_{22} + \epsilon_{22}\epsilon_{33} + \epsilon_{33}\epsilon_{11} - \epsilon_{12}{}^2 - \epsilon_{23}{}^2 - \epsilon_{31}{}^2) + \frac{2\epsilon^3}{27}$$

$$= I_3 - \frac{1}{3}I_1 I_2 + \frac{2}{27}I_1{}^3$$

For the hydrostatic stress and dilatation strain the invariants are

$$
\begin{array}{ll}
I_1^D = \epsilon & J_1^H = 3\sigma \\[2mm]
I_2^D = \dfrac{\epsilon^2}{3} & J_2^H = 3\sigma^2 \\[2mm]
I_3^D = \dfrac{\epsilon^3}{27} & J_3^H = \sigma^3
\end{array}
\qquad \text{(A14)}
$$

CHAPTER FOUR

Plastic Response— Continuum Treatment

IV-1 INTRODUCTION

The fact that metals can deform permanently upon loading after a certain initial elastic deformation is well known. In Chapter I, time-independent permanent deformation was defined as a plastic deformation. *The purpose of this chapter is to develop physical insights into the mechanics of plastic deformation.* Problems involving plastic deformation commonly occur in the materials processing field, i.e., in wire drawing, rolling, extrusion, coining, and forging. In fact, most of the useful items made of metals have undergone some form of permanent deformation during the manufacturing process. Unwanted plastic deformation in structures can be a mechanism of failure, but it can also be beneficial in relieving stress concentrations. Plastic deformation also plays a role in the fracture of metals.

Several interesting differences between elastic and plastic deformation should be noted. As discussed extensively in the previous chapters, elastic deformation, which occurs as soon as any load is applied, involves only the stretching of atomic bonds and is a reversible process, i.e., the work done in deforming the body can be completely recovered by unloading the body. Upon unloading, the state of the metal, which may be defined in terms of the stress, the strain, and the temperature, returns to its original configuration. Furthermore, the state of an elastic solid can always be uniquely defined when two of the variables, e.g., strain and temperature, are known, regardless of the particular path of loading.* The work done in deforming an elastic body is stored in the form of the strain energy and also in the form of a thermal energy change. The temperature change is positive when the solid is compressed and negative when it is stretched, provided the application of the load is carried out so rapidly that adiabaticity prevails during the loading process.† However, the total energy involved in an elastic deformation is so small that the maximum temperature rise is only on the order of 0.1°C. Most elastic properties, such as the modulus, are affected only very slightly by such small temperature changes. Therefore, it can be said that elastic deformation is *strain rate independent*.

Plastic deformation is a permanent change in atomic positions. This occurs when one layer of atoms slips over the adjacent layer in such a way that an atom moves from its initial position to an equivalent adjacent position. Such a shift can occur whenever the *shear stress* on a plane in the body reaches a critical value. Unlike elastic deformation, plastic deformation is very sensitive to discrete defects in the material. If the slipping process can be stopped completely at one point on a plane in the body, a large portion of the plane can be kept from slipping. The stress necessary to cause plastic deformation is therefore quite sensitive to localized irregularities in the material, such as crystal lattice defects, impurity atoms, second-phase particles, and grain boundaries. The plastic properties of the material

---

*In this case, stress and strain may be considered to be thermodynamic properties, just as temperature is, and either of the variables may be used to define the state of a simple thermodynamic system at a given temperature.

†All metals which elongate on heating behave in this manner. The sign of the temperature change is reversed in the case of the few materials that have a negative coefficient of thermal expansion.

are, for this reason, much more strongly affected by composition and prior history than is the elastic behavior. One important example of this is the phenomenon of *work hardening*. During deformation, defects, primarily dislocations, accumulate in the lattice structure of the material and inhibit the slipping process. This causes an increase in the stress required for continuous deformation.

Plastic deformation is *not reversible*. During plastic deformation the equilibrium position of the atoms is permanently changed. Also plastic deformation *depends on loading history*, because the structure of the material changes continuously during the deformation. Consequently, the plastic state of a metal *cannot be defined uniquely* in terms of the stress, the current strain, and the temperature. The nonuniqueness of the plastic state and the dependence of plastic deformation on the loading path imply that it is no longer possible to determine the total plastic strain from the state of stress and the temperature, whereas in the elastic case, the total strain is directly related to these two variables. Therefore, in plasticity it is necessary to deal with *incremental* deformation at a given state of loading and obtain the total plastic deformation by summing up all of the incremental loading. At this point it should again be emphasized that the strain increments are tensors, as defined in Chapter II, but the total strains are not tensors, when the total deformation is large.

Because of the nonreversible nature of plastic deformation, 90 to 95% of the work done by an applied load is transformed into thermal energy. When metals are rapidly deformed through the plastic regime, the adiabatic temperature rise can be significant. For example, in metal cutting, where the strain rate is as high as 10^4 to 10^5 per second, the average temperature rise at the chip—tool contact area due to the plastic deformation of the work material and the interfacial friction can be as high as $1300°C$.

The basic assumptions of the continuum treatment of plasticity to be presented in this chapter are: that the volume of the body does not change during plastic deformation; and that plastic deformation is solely caused by shear stresses. These assumptions imply that hydrostatic stresses alone cannot cause plastic deformation. Furthermore, since elastic volume changes in metals are small as a result of their high bulk moduli, *the critical shear stresses for plastic deformation are assumed to be independent of the magnitude of the hydrostatic pressure.** Throughout this chapter all materials will be assumed to be isotropic. Furthermore, a number of second-order effects, to be discussed later, are disregarded in the treatment of plasticity here.

When the deformation of a metal is such that the plastic and the elastic deformations are nearly the same order of magnitude, both have to be considered simultaneously. In this case, it will be assumed that there is a distinct line of demarcation between the elastic and plastic deformation of a metal, defined as the *yield point*, and that an incremental total strain is the sum of incremental elastic and plastic strains. On the other hand, when the magnitude of plastic deformation is much larger than that of elastic deformation, as is the case in wire drawing and extrusion, the elastic deformation may be neglected.

In order to obtain a solution to a problem involving plastic deformation, the entire deformation history must be traced. The solution is often complicated because accurate constitutive relations for real materials are highly intractable mathematically, and even using the idealized constitutive relations, exact solutions to many real problems are difficult, if not impossible, to obtain. Because of these difficulties, many of the available solutions are approximate. One of the most useful approximate methods, which is based on the theorems of limit analysis, can be used to obtain an approximation to the limiting load which a part can carry (the limit load). The limiting solutions are obtained in the form of the upper and lower bounds on the limit load, which, respectively, provide values larger and smaller than the exact value. The smaller the difference between the upper- and the lower-bound solutions, the closer is the approximation to the exact value. When the bounding values are exactly identical, the limit load is uniquely determined, although the accompanying deformation is not. The method of limit analysis is a very powerful and useful tool for mechanical engineers.

---

*Actually, many metals exhibit a difference between their yield strengths in tension and compression. Since the maximum shear stress is the same in tension as in compression, this effect indicates that plastic deformation is influenced by the hydrostatic pressure. However, the difference between tensile and compressive yield strength rarely exceeds 5%, which is not considered to be a significant effect for the purposes of this treatment and will thus be ignored.

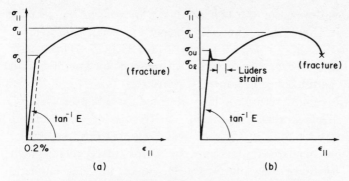

Fig. IV.1 Typical engineering stress-engineering strain curves for metals. (a) For metals without a distinct yield point, the yield stress is often taken to be the stress at 0.2% plastic strain. (b) For metals with a sharp discontinuous yield, σ_{ou} and σ_{ol} are called the upper and lower yield stresses, respectively; σ_u is the ultimate tensile strength. The tensile sample begins to form a neck at σ_u.

IV-2 PHENOMENOLOGICAL REPRESENTATION OF CONSTITUTIVE RELATIONS UNDER UNIAXIAL LOADING

The constitutive relations are obtained experimentally by deforming a standard tensile specimen and are commonly plotted in terms of the engineering normal stress and the engineering normal strain. The engineering normal stress is defined as the current force acting on a unit original area perpendicular to the direction of the force. The engineering normal strain is defined as the change in the length divided by the original length. The uniaxial stress-strain curves of typical metals are shown in Fig. IV.1. The curve shown in Fig. IV.1(a) is typical of metals which exhibit gradual yield, whereas the curve in (b) is typical of metals with unstable yield points. The first group includes such metals as aluminum, copper, and cold-worked steel; the best known member of the second group is annealed low-carbon steel. At the onset of loading, metals deform elastically until a critical strain or stress is reached, after which they deform plastically. The magnitude of the yield stress is nearly the same for both tensile and compressive loading. Certain metals have unstable yield points at which the stress drops sharply when the deformation changes from elastic to plastic. The maximum stress at the end of the elastic deformation region is called the upper yield point, while the stress minimum that follows is called the lower yield point. For other metals the transition from the elastic to the plastic deformation is more gradual. In these cases, the yield strength is usually defined for convenience to be the stress at a plastic strain of 0.2%. The plastic portions of these curves are greatly influenced by metallurgical variables such as grain size and impurity content.

In the elastic region, Hooke's law (defined in Chapter III) can describe the stress-strain relationship in terms of two elastic constants, such as Young's modulus E and Poisson's ratio ν. Similar relationships are needed to describe the constitutive relations of metals which are deforming plastically. However, the mathematical description of the stress-strain relations in the plastic region is complicated by the fact that the stress is not linearly dependent on the strain and that the constitutive relationship is a function of the prior deformation history. In the plastic strain range, the flow stress depends on the entire plastic strain history and the plastic strain increment $d\epsilon_{11}{}^{p}*$ is related to an increment of stress through the local slope of the work-hardening curve $d\sigma_{11}/d\epsilon_{11}{}^{p}$. Plastic stress-strain curves must be determined experimentally just as elastic constants must be.

When plastic deformation is very large, the change in the cross-sectional area and the change in the length of the element may be so large that the engineering normal stress and normal strain may not truly represent the state of loading and deformation. Therefore, it is sometimes more convenient to discuss the plastic state of deformation in terms of the true normal stresses and true normal strains. The true normal stress is defined as the normal force divided by the true cross-sectional area. The true normal strain is defined as the change in the length of the specimen over the *current* length. For uniform deformation, the true normal strain becomes

$$\epsilon_{11} = \frac{l_1 - l_0}{l_0} + \frac{l_2 - l_1}{l_1} + \frac{l_3 - l_2}{l_2} + \cdots + \frac{l_f - l_{f-1}}{l_{f-1}}$$

$$\epsilon_{11} = \sum_{i=0}^{i=f} \frac{\Delta l_i}{l_i} = \int_{l_0}^{l_f} \frac{dl}{l} = \ln \frac{l_f}{l_0}$$

(IV.1)

where l_1, l_2, etc. are the lengths of the element at each incremental step of loading. Since the final expression is given in terms of a

---

*The superscript p denotes plastic strain.

logarithmic function, the true strain is sometimes referred to as the logarithmic strain,* or, occasionally, as the natural strain. Example IV.1 shows that the true strains are much more convenient to use than the engineering strains when deformation is large.

True strains for the uniform deformation of the circular solid rod shown in Fig. IV.2(a) are

$$\epsilon_{rr} = \ln \frac{r_f}{r_0} = \epsilon_{\theta\theta}$$

$$\epsilon_{zz} = \ln \frac{l_f}{l_0}$$

(IV.2)

where the subscripts 0 and f refer to the initial and final dimensions of the solid. Similarly, the true strains for the deformation of the thin-walled tube in Fig. IV.2(b) are

$$\epsilon_{rr} = \ln \frac{t_f}{t_0}$$

$$\epsilon_{\theta\theta} = \ln \frac{r_f}{r_0}$$

(IV.3)

$$\epsilon_{zz} = \ln \frac{l_f}{l_0}$$

For the parallelepiped in Fig. IV-2(c), the true strains are

$$\epsilon_{11} = \ln \frac{l_{1f}}{l_{10}}$$

$$\epsilon_{22} = \ln \frac{l_{2f}}{l_{20}}$$

(IV.4)

$$\epsilon_{33} = \ln \frac{l_{3f}}{l_{30}}$$

The derivations of these true strain expressions are left to the reader as an exercise (see Problem IV.2).

---

*It should be noted that, for multiaxial deformation, the true strain components are not always logarithmic.

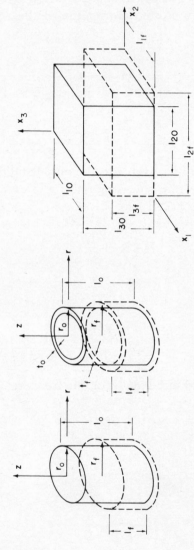

Fig. IV.2 True strains for a right circular cylinder (solid rod), thin-walled tube, and parallelepiped block, expressed in terms of the initial dimensions, subscript 0, and the final dimensions, subscript f.

*EXAMPLE IV.1—True Strain vs. Engineering
          Strain*

Consider a uniform, large plastic deformation of the parallelepiped body with
sides l_{10}, l_{20}, l_{30} into l_1, l_2, l_3 as shown in Fig. IV.3. Assuming that there is no
volume change during the plastic deformation, derive expressions for engineer-
ing strains and true strains. Derive the expressions for the volume change in
terms of these strains, if it is possible.

Solution:

   The engineering strains are given by

$$\epsilon_{11} = \frac{l_1 - l_{10}}{l_{10}}, \quad \epsilon_{22} = \frac{l_2 - l_{20}}{l_{20}}, \quad \epsilon_{33} = \frac{l_3 - l_{30}}{l_{30}} \tag{a}$$

The true strains are

$$\epsilon_{11} = \ln \frac{l_1}{l_{10}}, \quad \epsilon_{22} = \ln \frac{l_2}{l_{20}}, \quad \epsilon_{33} = \ln \frac{l_3}{l_{30}} \tag{b}$$

The volume change is given by

$$dV = l_1 l_2 l_3 - l_{10} l_{20} l_{30} = 0 \tag{c}$$

Dividing Eq. (c) by the total volume, we obtain

$$\frac{dV}{V} = \frac{l_1 l_2 l_3}{l_{10} l_{20} l_{30}} - 1 = 0 \tag{d}$$

The zero volume change during plastic deformation is one of the basic
assumptions of this treatment of plasticity. Equation (d) is difficult to express
in terms of engineering strains. In terms of the true strains, the constancy of
volume condition becomes

$$\ln \left(\frac{l_1 l_2 l_3}{l_{10} l_{20} l_{30}} \right) = \ln \left(\frac{l_1}{l_{10}} \right) + \ln \left(\frac{l_2}{l_{20}} \right) + \ln \left(\frac{l_3}{l_{30}} \right) = 0$$

$$\epsilon_{11} + \epsilon_{22} + \epsilon_{33} = \sum_{i=1}^{3} \epsilon_{ii} = 0 \tag{e}$$

   The stress-strain curves of Fig. IV.1, replotted in terms of true
stress and true strain, are shown in Fig. IV.4. Note that in the true
stress-true strain curves the stress continues to increase until fracture,
instead of dropping off after necking as it does in the engineering
stress-strain curve. For operational simplicity, actual stress-strain
curves are often idealized to fit simple mathematical relations. One

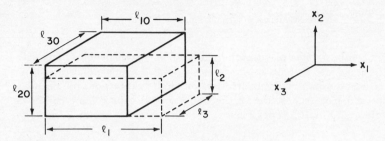

Fig. IV.3 Deformation of a rectangular parallelepiped block.

of the most commonly used idealizations is the rigid-perfectly plastic curve shown by the dotted lines A in Fig. IV.4. In this case, the effect of work hardening is disregarded by choosing a flow stress level such that the plastic work done (i.e., the area under the stress-strain curve) is equal to the true plastic work done according to the real stress-strain curve. Another commonly employed idealization is the constant work hardening rate, which is shown by the bilinear stress-strain curve (curves B) in Fig. IV.4. A more realistic stress-strain relationship is one of the power-law type (shown by the curves C), i.e.,

$$\sigma_{11} = \sigma_0(\epsilon_{11})^n \tag{IV.5}$$

The exponent n indicates the work hardening rate, and σ_0 corresponds to the flow stress at unit strain.

Many other idealizations have been tried in the past. Any of these models which has been based upon reasonable assumptions should be suitable for practical work.

A few words of caution should be stated here. The strain tensor defined in Chapter II is only valid for small deformations. Therefore, large true strains are not tensor quantities. However, the incremental true strains are always tensor quantities, regardless of the absolute magnitude of the plastic strains, i.e.,

$$d\epsilon_{11} = \frac{dl_1}{l_1} = \frac{\partial du_1}{\partial x_1} \tag{IV.6}$$

It is also important to emphasize that both the engineering and true strains are only valid at a dimensionless point, not over a finite element. Wherever we have discussed the strains in terms of the latter, it has been assumed that the deformation was uniform over the entire element. The strains are everywhere identical only when a body deforms homogeneously.

The stress-strain relationships are drastically altered by cold working, as the following consideration of the deformation cycle of a metal indicates (Fig. IV.5). When a metal is deformed to plastic state A and unloaded, the unloading follows the elastic curve AB. When this cold-worked metal is reloaded, it deforms elastically almost to point A, and then deforms plastically. Once the plastic deformation is initiated at A, the stress-strain relationship nearly coincides with an extrapolation of the curve of the original annealed metal. Therefore, every point on the stress-strain curve of the plastically deformed region may be considered to be a yield point for continued plastic deformation. It is sometimes useful to treat the stress-strain curve in the plastic region as the locus of yield points. Since the cold working alters the constitutive relations significantly, it is not possible to develop a universal stress-strain relationship even for chemically identical metals.

When a metal is deformed first in tension and then in compression, the magnitude of the flow stress in compression is found to be less than it was in tension. This is known as the Bauschinger effect. However, for ease of mathematical manipulation, it will be assumed that the deformation of the metal is always invariant, i.e., independent of the elastic unloading history and the sense of the normal

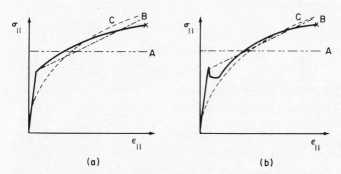

(a)   (b)

Fig. IV.4 True stress-true strain curves for the materials for which engineering stress-engineering strain curves were shown in Fig. IV.1. The dashed lines represent various idealizations of the real true stress-true strain curves. Curve A is elastic-perfectly plastic. Curve B is elastic linear work hardening. Curve C is a power law relationship.

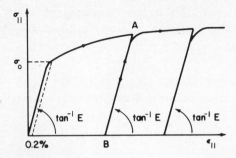

Fig. IV.5 Effect of cold working on the stress-strain relationship. Note that when the metal is unloaded from A, it unloads elastically. When it is reloaded from B, it behaves elastically until it almost reaches the stress it had (i.e., the stress at A) prior to the onset of plastic deformation.

stress. This assumption is reasonable, since the Bauschinger strain is of the order of the elastic strains, which are usually much smaller than the plastic strains. Thus, when the metal shown in Fig. IV.5 is compressed after it is unloaded from A to B, it begins to yield when the magnitude of the compressive stress equals the tensile stress at A. Since plastic deformation is irreversible, it is further assumed that the effect of cold working accumulates, being dependent only on the total plastic work done, and not on the sequences or the history of loading.

When the plastic deformation is nearly equal in magnitude to the elastic deformation, both the elastic and the plastic deformations have to be included in the constitutive relationship. This can be done by writing the total strain as the sum of the elastic and plastic strains, i.e.,

$$\epsilon_{ij} = \epsilon_{ij}{}^e + \epsilon_{ij}{}^p \tag{IV.7}$$

For uniaxial loading, $\epsilon_{11}{}^e = \sigma_{11}/E$ and $\epsilon_{11}{}^p$ can be replaced by any suitable relationship discussed in this section.

IV-3 CORRELATIONS BETWEEN UNIAXIAL LOADING AND MULTIAXIAL LOADING

The stress-strain relationships presented in the preceding section were given in terms of the normal stress and normal strain. Such a representation was possible since only uniaxial loading was considered. However, in many applications of engineering materials the loading is multiaxial. The immediate question is then: "How can the constitutive relations for a material under multiaxial loading be determined and represented without plotting all of the stresses and all of the corresponding strains?" An answer will be provided by an extension of the considerations for the uniaxial case discussed in the preceding section.

The uniaxial behavior of a material represented by the tensile stress-strain curve can be generalized to cases of multiaxial loading by making use of the following assumptions: *plastic deformation occurs in response to shear stress*; and *plastic deformation does not change the volume of a body*. The tensile stress-strain curves of Figs. IV.1 and IV.4 could also have been presented as plots of the maximum shear stress versus the maximum shear strain, since, in uniaxial loading, the latter are proportional to the tensile stress and strain. The maximum shear stress for other states of loading could then be compared to the maximum shear stress for tension. The concept that the behaviors in different cases are identical when the maximum shear stresses are identical could be used to develop constitutive relations for the more general state of loading. Unfortunately, as we shall demonstrate later, this procedure does not lend itself to simple mathematical description. Therefore, it is more common to compare different states of stress and strain by using as parameters the equivalent (or effective) stress $\bar{\sigma}$ and the equivalent plastic strain increment $d\bar{\epsilon}^p$. Before proceeding with the development of the multiaxial plastic constitutive relations, it will be shown that these two parameters are closely related to the maximum shear stress and the maximum shear strain increment.

The *equivalent (or effective) stress* $\bar{\sigma}$ may be written as the square root of minus three times the second invariant of the deviator stresses, i.e.,

$$\bar{\sigma} = \left(\sum_{i=1}^{3} \sum_{j=1}^{3} \frac{3}{2} \sigma'_{ij} \sigma'_{ij} \right)^{1/2} = (-3J'_2)^{1/2}$$

$$\bar{\sigma} = \left\{ \frac{1}{2} \left[(\sigma_{11} - \sigma_{22})^2 + (\sigma_{22} - \sigma_{33})^2 + (\sigma_{33} - \sigma_{11})^2 \right] \right.$$
$$\left. + 3(\sigma_{12}^2 + \sigma_{13}^2 + \sigma_{23}^2) \right\}^{1/2} \quad \text{(IV.8)}$$

Since the effective stress $\bar{\sigma}$ is nearly proportional to the maximum resultant shear stress, $\bar{\sigma}$ can be used in place of the uniaxial normal stress σ_{11} (or the maximum shear stress) to characterize the nature of loading on a multiaxially loaded member.

The *equivalent strain increment* $d\bar{\epsilon}$ may be defined as

$$(d\bar{\epsilon})^2 = \frac{4}{9}\left\{\frac{1}{2}\left[(d\epsilon_{11} - d\epsilon_{22})^2 + (d\epsilon_{11} - d\epsilon_{33})^2 + (d\epsilon_{22} - d\epsilon_{33})^2\right]\right.$$
$$\left. + 3(d\epsilon_{12}^2 + d\epsilon_{13}^2 + d\epsilon_{23}^2)\right\} \quad \text{(IV.9)}$$

The constant 4/9 is inserted in the definition of the equivalent strain so that the incremental effective strain $d\bar{\epsilon}$ becomes the uniaxial strain $d\epsilon_{11}$ under uniaxial loading conditions for incompressible materials. The incremental equivalent plastic strain $d\bar{\epsilon}^p$ is similarly given by

$$(d\bar{\epsilon}^p)^2 = \frac{4}{9}\left\{\frac{1}{2}\left[(d\epsilon_{11}^p - d\epsilon_{22}^p)^2 + (d\epsilon_{22}^p - d\epsilon_{33}^p)^2 + (d\epsilon_{11}^p - d\epsilon_{33}^p)^2\right]\right.$$
$$\left. + 3(d\epsilon_{12}^{p2} + d\epsilon_{13}^{p2} + d\epsilon_{23}^{p2})\right\} \quad \text{(IV.10)}$$

Using an argument similar to that used for the resultant shear stress, it can be shown that the effective strain provides a reasonable approximation to the maximum resultant shear strain. The equivalent or effective strain increment has additional significance in that it is always a positive quantity, regardless of the actual sense of the strain components. Therefore, the total effective plastic strain, defined as $\bar{\epsilon}^p = \int d\bar{\epsilon}^p$, is a measure of the complete plastic loading history.

EXAMPLE IV.2—Equivalent Stress and the
Maximum Resultant
Shear Stress

Show that the equivalent stress is always nearly twice the maximum resultant shear stress a) under uniaxial loading, b) under pure shear loading, and c) under the state of stress given by

$$\sigma_{ij} = \begin{bmatrix} \sigma_{11} & 0 & 0 \\ 0 & \sigma_{22} & 0 \\ 0 & 0 & \sigma_{33} \end{bmatrix}$$

Solution:
a) Under uniaxial loading, all the stress components are equal to zero except σ_{11} which is equal to twice the maximum shear stress. Therefore,

$$\bar{\sigma} = \sigma_{11} = 2\tau_{max} \tag{a}$$

b) Under pure shear loading, again all of the stress components are equal to zero, except one of the shear stress components, say, $\sigma_{12} = \tau_{max}$. The equivalent stress is

$$\bar{\sigma} = \sqrt{3}\tau_{max} = 1.733\tau_{max} \tag{b}$$

c) For the triaxial loading case, assume for simplicity that $\sigma_{11} > \sigma_{22} > \sigma_{33}$. Then, the maximum shear stress is

$$\tau_{max} = \frac{\sigma_{11} - \sigma_{33}}{2} \tag{c}$$

The equivalent stress is

$$\bar{\sigma} = \left\{ \frac{1}{2}\left[(\sigma_{11} - \sigma_{22})^2 + (\sigma_{22} - \sigma_{33})^2 + (\sigma_{33} - \sigma_{11})^2 \right] \right\}^{1/2} \tag{d}$$

Consider how the ratio $\tau_{max}/\bar{\sigma}$ depends on the value of σ_{22}. At the extreme values $\sigma_{22} = \sigma_{11}$ or $\sigma_{22} = \sigma_{33}$

$$\bar{\sigma} = 2\tau_{max} \tag{e}$$

For intermediate values of σ_{22} the ratio of $\bar{\sigma}/\tau_{max}$ is less than 2. The minimum value is found by solving the expression

$$\frac{d(\bar{\sigma}/\tau_{max})}{d\sigma_{22}} = \frac{-(\sigma_{11} - \sigma_{22}) + (\sigma_{22} - \sigma_{33})}{2\bar{\sigma}\tau_{max}} = 0 \tag{f}$$

for σ_{22}. The solution is

$$\sigma_{22} = \frac{\sigma_{11} + \sigma_{33}}{2} \tag{g}$$

By substituting this value of σ_{22} into Eq. (d), we obtain the minimum value of the ratio which can occur.

$$\bar{\sigma} = \sqrt{3}\tau_{max} \quad \text{when} \quad \sigma_{22} = \frac{\sigma_{11} + \sigma_{33}}{2} \tag{h}$$

Since the state of stress in part (c) is a general state of stress, $\bar{\sigma}$ is always between the bounds obtained from parts (a) and (b).

$$\sqrt{3}\tau_{max} \leq \bar{\sigma} \leq 2\tau_{max} \tag{i}$$

One sees that the upper and lower extremes correspond to uniaxial stress and pure shear, respectively.

Since the proportionality between $\bar{\sigma}$ and the maximum shear stress and that between $\bar{\epsilon}$ and the maximum shear strain are similar, a representation of the stress-strain relationship in terms of $\bar{\sigma}$ and $\bar{\epsilon}$ is nearly equivalent to one expressed in terms of the maximum shear stress and maximum shear strain. Therefore, the uniaxial data shown in Fig. IV.4 can be used to represent the stress-strain relationship under multiaxial loading by simply substituting $\bar{\sigma}$ for σ_{11} and $\bar{\epsilon}$ for ϵ_{11}. In fact, when the $\bar{\sigma}-\bar{\epsilon}$ relationship obtained from various tests (both uniaxial and multiaxial) are plotted together, all the experimental points lie within the bounds of the normal experimental scatter in the plastic regime.* The usefulness of the equivalent stress and strain expressions is that flow stresses obtained under many different loading conditions can be correlated reasonably well in terms of $\bar{\sigma}$ and $\bar{\epsilon}$.

In the plastic range the ratio $(d\bar{\epsilon}^p/\bar{\sigma})$ is used to characterize the plastic behavior in relating the individual incremental plastic strains to stresses (see Sec. IV-6 for details), just as E is used to characterize the elastic behavior in the elastic range. It should be noted that the value of $(d\bar{\epsilon}^p/\bar{\sigma})$ varies continuously as a function of the deformation and loading history, whereas E remains constant.

The physical significance of the equivalent stress and equivalent strain is further illustrated by the fact that the elastic distortion energy† per unit volume is

$$U_s = \sum_{i=1}^{3} \sum_{j=1}^{3} \tfrac{1}{2} \sigma'_{ij} \epsilon'^{e}_{ij} = \tfrac{1}{2} \bar{\sigma} \bar{\epsilon}^{e} = \frac{\bar{\sigma}^2}{6G} \qquad (IV.11)$$

From Eq. (IV.11) it is seen that the equivalent stress is also proportional to the square root of the distortional energy. Similarly, it can be shown that in plastic deformation the incremental plastic work done per unit volume is given by

---

*However, it should be cautioned that the slope of the $\bar{\sigma}-\bar{\epsilon}$-curve in the elastic regime depends on the testing technique and therefore, should not be used to compute the modulus without knowing how the experimental data have been obtained.

†The deformation of a solid body may be divided into two components: one which involves only volume change called dilatation (or dilation), and another which involves only shear deformation, called distortion. Dilatation is given by the first strain invariant I. The elastic distortion energy refers to the energy stored in an elastic body due to shear deformation only.

$$dU_p = \sum_{i=1}^{3} \sum_{j=1}^{3} \sigma'_{ij} \, d\epsilon'_{ij}{}^p = \bar{\sigma} d\bar{\epsilon}^p \qquad\qquad \text{(IV.12)}$$

IV-4 CONSTITUTIVE RELATIONS UNDER MULTIAXIAL LOADING CONDITIONS— YIELD CRITERIA

In Sec. IV-3 above, it was proposed that the transition from elastic to plastic deformation occurs when the maximum resultant shear stress reaches a critical value. This critical value is the shear strength of the material, k. In Sec. IV-2, it was noted that every point in the work-hardening region of the stress-strain curve is a yield point for further deformation, since a metal unloaded from a plastic state behaves elastically until the shear stress again exceeds the value it had prior to unloading. In this section, these ideas will be used to develop mathematical expressions or yield criteria for the conditions of plastic yield. There are two common yield criteria for metals, one based on the maximum shear stress, and the other on the equivalent stress $\bar{\sigma}$.

We will begin the discussion by considering the uniaxial loading case shown in Fig. IV.6. When the axial stress σ_{11} is increased, the maximum shear stress which acts on the plane 45° away from the loading direction also increases. When this maximum resultant shear stress reaches a critical value, plastic deformation occurs. The state of

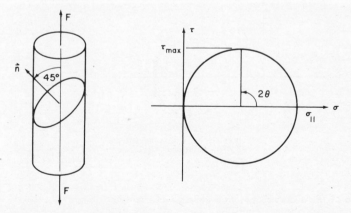

Fig. IV.6 Mohr's circle representation of the state of stress in uniaxial tension.

stress can be represented by a Mohr's circle, as shown in Fig. IV.6. When the radius of the Mohr's circle reaches the critical value, the metal deforms plastically. Since there are an infinite number of such 45° planes, plastic deformation is clearly not confined to any one particular plane.

On the specimen shown in Fig. IV.6, let a tensile stress be superimposed radially, as shown in Fig. IV.7, where the Mohr's circle for this case is also presented. Note that the additional radial stress reduces the shear stress, unloading the metal from the plastic state. Therefore, σ_{zz} must be further increased to reach the critical shear stress for plastic deformation. When the principal stresses have three different magnitudes, the plastic state is reached when the radius of the largest Mohr's circle (Fig. IV.8) is again equal to the critical shear stress k. In terms of the principal stresses, a yield criterion based on the critical shear stress concept may be written as

$$\frac{\sigma_I - \sigma_{III}}{2} = \tau_{max} = k \ \ \text{(at yield)} \tag{IV.13}$$

where σ_I and σ_{III} are the maximum and the minimum principal stresses, respectively. The yield criterion given by Eq. (IV.13) is known as the *maximum shear criterion* or the *Tresca yield criterion*.

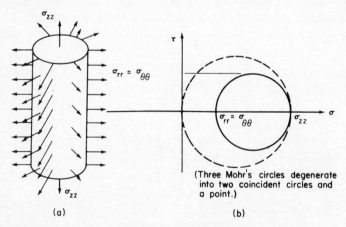

Fig. IV.7 Decrease of the diameter of Mohr's circle by superposition of radial stress on a tensile specimen initially loaded uniaxially. This means that the maximum resultant shear stress is reduced by the application of the radial stress.

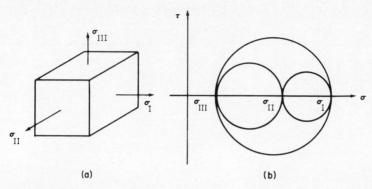

Fig. IV.8 Mohr's circle representation of the state of stress in a triaxially loaded element.

In Example IV. 2 it was shown that the equivalent stress $\bar{\sigma}$ is nearly equal to twice the maximum resultant shear stress under various loading conditions. Since metals deform plastically when the resultant shear stress reaches a critical value, another yield criterion may be established in terms of $\bar{\sigma}$ as*

$$\bar{\sigma} = \text{constant} = 2k \quad \text{(at yield)} \tag{IV.14}$$

As shown in Eq. (IV.11), $1/2 \, \bar{\sigma} \, \bar{\epsilon}^{e}$ is equal to the shear strain energy stored during elastic deformation. Therefore, the yield criterion given by Eq. (IV.14) is known as the *Mises yield criterion or the maximum distortion energy criterion*.

Because $\bar{\sigma}$ is not exactly equal to twice the resultant shear stress, the predictions of the Tresca and the Mises yield criteria do not agree everywhere, the maximum difference being about 15.5% when a simple shear stress is applied. The experimental results lie somewhere between these two criteria. Although the experimental results tend to follow the Mises yield criterion, the choice between the two criteria should be dictated by mathematical convenience. In developing the Tresca and the Mises yield criteria, it was assumed that the material is fully isotropic, independent of the sense of normal stresses, and independent of the hydrostatic stress.

*In some formulations, depending upon how k is defined, Eq. (IV.14) will be $\bar{\sigma} = \sqrt{3} \, k$. This discrepancy occurs since $\tau_{\max}$ and $\bar{\sigma}$ are not completely equivalent in all reference loading cases, as shown in Example IV.2.

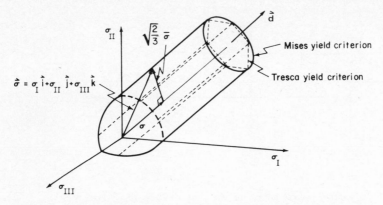

Fig. IV.9 Graphical representation of the Mises and Tresca (maximum shear) yield criteria in principal stress space. A vector $\sigma_I i + \sigma_{II} j + \sigma_{III} k$ can be separated into its hydrostatic component along the axis $d = i + j + k$ and its distortional component perpendicular to this direction. The yield locus described by the Mises criterion is a circular cylinder of radius $\sqrt{2/3}\ \bar{\sigma}$ with its axis parallel to the hydrostatic axis. The Tresca criterion is the inscribed hexagonal cylinder.

IV-5 FURTHER DISCUSSION OF THE YIELD CRITERIA

The yield criteria given by Eqs. (IV.13) and (IV.14) may be represented graphically in principal-stress space (see Fig. IV.9). The vector from the origin of the stress space to the point σ_I, σ_{II}, σ_{III} can be separated into two components. One component along the axis $\sigma_1 = \sigma_2 = \sigma_3$ has magnitude equal to the hydrostatic or mean normal stress $\sigma = (\sigma_1 + \sigma_2 + \sigma_3)/3$. The perpendicular component from the hydrostatic axis to the point $\sigma_I i + \sigma_{II} j + \sigma_{III} k$ has magnitude equal to $\sqrt{2/3}$ times the equivalent stress $\bar{\sigma}$. Therefore the cylindrical surface generated by a finite plane rotating about the hydrostatic axis is a surface of constant equivalent stress. Since the Mises yield criterion states that when $\bar{\sigma}$ is equal to the yield strength in tension, Y (or twice the shear yield strength, $2k$), the Mises criterion in principal stress space is represented by a circular cylinder with its axis parallel to the hydrostatic stress axis. The Tresca yield criterion can be shown to be a right hexagonal cylinder in the principal stress space with its edges coinciding with the Mises criterion as shown in Fig. IV.9.*

*The argument given here does not follow the traditional development of the subject in which the Mises yield criterion reflects the distortional energy of the material. Thus, in the

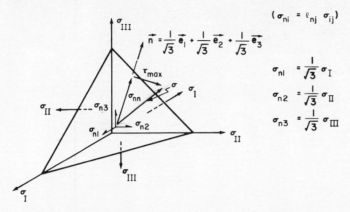

Fig. IV.10 Stress components acting on the octahedral plane.

In Example IV.3 it will be shown that the normal stress acting on a plane perpendicular to the direction vector $(1/\sqrt{3},\ 1/\sqrt{3},\ 1/\sqrt{3})$ in principal coordinates is the hydrostatic stress σ, and that the shear stress on this plane is proportional to the equivalent stress $\bar{\sigma}$. This plane is often called the octahedral plane and the shear stress on it is called the octahedral shear stress.

EXAMPLE IV.3—Stresses on Octahedral Planes

Show that in the principal-stress space $(\sigma_I,\ \sigma_{II},\ \sigma_{III})$ the normal stress on a plane perpendicular to the direction $(1/\sqrt{3},\ 1/\sqrt{3},\ 1/\sqrt{3})$ is equal to the hydrostatic stress and that the resultant shear stress on this plane is nearly equal to one-half of the equivalent stress.

Solution:

Under triaxial loading, the stress components acting on the plane directed along the three principal axes are given by the expressions in Fig. IV.10. The resultant stress is

$$\sigma_{nr}^2 = \sigma_{n1}^2 + \sigma_{n2}^2 + \sigma_{n3}^2$$
$$= \tfrac{1}{3}(\sigma_I^2 + \sigma_{II}^2 + \sigma_{III}^2) \tag{a}$$

The normal stress is

---

usual exposition, the Tresca yield criterion has been related to the yielding of nonuniformly deforming metals, such as annealed low-carbon steel, which show deformation bands indicating the occurrence of plastic shear when the critical shear stress on a given plane reaches a critical value, while the Mises yield criterion has been associated with the yielding of uniformly deforming metals, such as cold-worked low-carbon steel and polycrystalline face-centered cubic metals. Though quite different, the approach we have chosen is nevertheless thoroughly consistent with the physics of deformation.

$$\sigma_{nn} = \sum_{i=1}^{3} \sum_{j=1}^{3} l_{ni} l_{nj} \sigma_{ij} = \frac{1}{3}(\sigma_I + \sigma_{II} + \sigma_{III}) \tag{b}$$

The maximum shear stress is

$$\tau_{max}^2 = \sigma_{nr}^2 - \sigma_{nn}^2$$
$$= \frac{1}{9}\left[(\sigma_I - \sigma_{II})^2 + (\sigma_{II} - \sigma_{III})^2 + (\sigma_I - \sigma_{III})^2\right] \tag{c}$$

Therefore,

$$\bar{\sigma} = \frac{3}{\sqrt{2}}\tau_{max} = 2.12\tau_{max} \tag{d}$$

EXAMPLE IV.4—Strength of a Solder Joint

Discuss why a soldered joint can withstand a tensile load much greater than the uniaxial tensile strength of the solder.

Solution:

The soldered joint is under a triaxial state of stress with a large hydrostatic component of stress due to the constraints imposed by the parts joined. Since plastic deformation is caused only by shear stresses, the axial stress must be increased to compensate for this hydrostatic component. The proof is found by considering a tensile specimen (Fig. IV.11) which consists of a solder rod bonded to steel. The yield strength of the solder is much less than that of the steel. When the solder section is long, the state of loading is approximated to be that of uniaxial tension at the center of the long test section, and the stress state is given by

$$\sigma_{zz} = \frac{F_0}{A_0}, \quad \sigma_{rr} = 0, \quad \sigma_{\theta\theta} = 0 \tag{a}$$

This approximation is valid for regions away from the constrained ends of the solder section, where end effects can be neglected. In this case, yielding will occur when

$$\bar{\sigma} = \sigma_{zz} = 2k \tag{b}$$

since the center of the section is free to expand or contract.

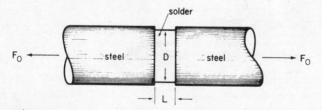

Fig. IV.11 Speciman for testing constrained solder disks.

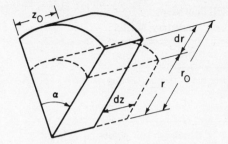

Fig. IV.12 Deformation geometry of unconstrained disk deforming in response to a uniaxial tension, showing the lateral contraction.

When the length of the solder section is made very short, the ends of the specimen become constrained laterally. Consider a wedge-shaped section from the specimen, as shown in Fig. IV.12. As tension is applied, the section tries to expand to $z + dz$ and contract radially to radius r. The angle α will remain the same by symmetry. In the absence of constraints, the strains are given by

$$\epsilon_{zz} = \frac{dz}{z}, \quad \epsilon_{rr} = -\frac{dr}{r_0}$$

$$\epsilon_{\theta\theta} = -\frac{d(\alpha r)}{\alpha r_0} = -\frac{dr}{r_0} = \epsilon_{rr}$$

(c)

The induced radial and circumferential stresses in the solder due to the constraints will be of the same sign as the axial stress. The constrained stress state is given by

$$\sigma_{zz} = \frac{F}{A}, \quad \sigma_{rr} = \sigma_{\theta\theta} > 0$$

(d)

The equivalent flow stress in this case shows the influence of the induced stress

$$\bar{\sigma} = \left\{ \frac{1}{2} \left[(\sigma_{zz} - \sigma_{rr})^2 + (\sigma_{rr} - \sigma_{\theta\theta})^2 + (\sigma_{\theta\theta} - \sigma_{zz})^2 \right] \right\}^{1/2}$$

$$= \sigma_{zz} - \sigma_{rr}$$

(e)

and yielding will occur when

$$\sigma_{zz} - \sigma_{rr} = 2k$$

(f)

Since $\sigma_{rr} > 0$, we find that the magnitude of the applied force necessary to cause yielding is greater than was required in the original (long) specimen.

$$\frac{F}{A} = \frac{F_0}{A} + \sigma_{rr}$$

(g)

TABLE IV.1—Results of Yielding Under Physical Constraints
(Results from tests run in the Materials Processing Laboratory, M.I.T.)*

Sample*	F(lbf)†	$\sigma_{zz}(\times 10^3$ psi)	L/D
S-1-1	−372	−3.37	2.27
S-2-4	−370	−3.36	1.00
S-3-2	−610	−5.52	.50

*Specimens designated by S consisted of a "sandwich" of solder between two steel rods placed end to end and were tested in compression. The aspect ratio L/D relates to the relative length of the test section to diameter and indicates the degree of physical constraint imposed by the ends.

†Forces were measured at the yield point.

Table IV.1 gives experimental results of the compressive and tensile tests with specimens designed to show the effect of hydrostatic stress on plastic deformation. The force required to initiate plastic deformation in the triaxial loading condition was as much as twice that of the uniaxial case.

The observed results can be explained by using a two-dimensional Mohr's circle representation of stress [since in both cases (a) and (b) $\sigma_{\theta\theta} = \sigma_{rr}$], as shown in Fig. IV.13. If radial (and circumferential) stress is negligible, as in the first case, the applied force will increase until the diameter of the Mohr's circle is equal to the yield stress. This will happen also in case (b), but since the lateral stresses also increase with the axial stress, the entire loading state will be shifted far to the right before yield is reached.

This shift is caused by the ability of the short specimen to impose a hydrostatic state of stress upon itself when a load is applied. The induced hydrostatic component of stress enables the specimen to support a greater axial load than otherwise possible. This is the reason soldered joints can carry a load larger than the uniaxial yield strength of the solder material.

If a state of stress is such that it lies inside the Mises cylinder or the Tresca hexagon, the deformation is purely elastic. If it lies on the

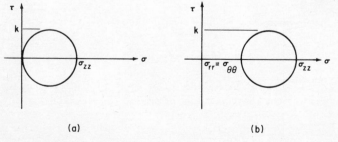

Fig. IV.13 Mohr's circle representation for the state of stress at the center of a long (a) and a short (b) solder joint.

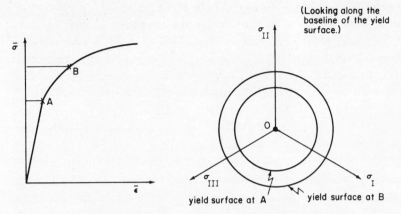

Fig. IV.14 Expansion of the yield surface of a work-hardening material without change of shape during plastic deformation. The rate of expansion is assumed to be a function only solely of the integrated equivalent plastic strain $\bar{\varepsilon}^p$.

surface, plastic deformation is initiated. If the material is rigid-perfectly plastic, continuous plastic deformation can occur as long as the state of stress remains on any part of the yield surface. If the state of stress varies such that it moves toward the inside of the surface, elastic unloading occurs. In Fig. IV.14, which represents a work-hardening material, the deformation from A to B can occur only if the loading changes in such a way that as the state of stress moves it will remain on the current yield surface produced by the process of work hardening. Since every point on the stress-strain curve in the work-hardening region is a new yield point for continued plastic flow, there are an infinite number of such yield surfaces.

The changes which occur in the yield surface due to work hardening depend on the amount of plastic flow which has occurred and on the assumptions made about the nature of the work-hardening process. Two basic questions arise with respect to the process of work hardening: "What is the effect of work hardening on the shape of the yield surface?" and "How rapidly does the size of the yield surface increase with plastic strain?" These questions can only be answered by experimental observation.

Experiments show that when yield is defined as 0.2% deviation from elastic behavior, changes in the shape of the yield surface caused by prior work hardening are quite small. Therefore, it will be assumed that the effect of work hardening is simply to enlarge the yield locus without changing its shape. This assumption is often called the isotropic work-hardening hypothesis.

The rate of change of the size of the yield locus with plastic strain is well defined for uniaxial tension by the tensile stress-strain curve. When the state of strain differs from this case, it is necessary to determine how much work-hardening effect the strain has. Two possible procedures have been suggested in previous sections. One method is to express the work-hardening effect of the plastic strain in terms of the total maximum shear strain accumulated over the deformation history. Since this procedure is cumbersome mathematically, a second method will be adopted here.

This alternative procedure makes use of the concept of the equivalent plastic strain increment, defined in Eq. (IV.10), and the equivalent plastic strain, defined by $\bar{\epsilon}^p = \int d\bar{\epsilon}^p$. Since $d\bar{\epsilon}^p$ is always positive, $\bar{\epsilon}^p$ increases for each increment of plastic deformation. It can therefore be used as a measure of the cumulative effect of the plastic strain, independent of the sign of the plastic strain components. It can then be assumed that the size of the yield locus defined by the current value of the flow stress in tension is solely a function of $\bar{\epsilon}^p$.

If it is assumed that 1) the work hardening is isotropic and 2) the size of the yield surface is a function of only the equivalent plastic strain $\bar{\epsilon}^p$, then the conditions for plastic yield under any state of loading can be determined from the value of $\bar{\epsilon}^p$ integrated over the deformation history of the body and the tensile stress-strain curve, as illustrated in Fig. IV.4. These assumptions predict yield stresses which agree with those found by experiment to within 10% for most metals. This precision is considered to be quite adequate for calculations of the type discussed in the remainder of this chapter.

It should be noted that engineering materials may not be fully isotropic, independent of hydrostatic pressure, and independent of the sense of normal stresses, which are the underlying assumptions for the Tresca and Mises yield criterion. For example, wood has different properties along different directions, and the flow stress of plastics is a sensitive function of hydrostatic pressure. However, even for this type of engineering material an approximate yield surface can be constructed by performing a limited number of experiments.

The procedure will be illustrated by reference to a two-dimensional representation of the yield locus. Figure IV.15 shows this projection, which is formed by the intersection of the yield locus with the plane

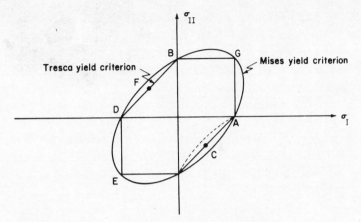

Fig. IV.15 Intersection of the yield surfaces for the Mises and Tresca yield criteria with the plane $\sigma_{III} = 0$.

$\sigma_{III} = 0$. Several points can be located on the yield locus by a few experiments. Point A is determined by the yield stress in a tensile test with the tensile axis along the σ_I-direction. If the σ_{II}-axis is equivalent to the σ_I-axis, then point B is determined by the same experiment as it is an isotropic material. If the yield in the material is the same for tension and compression, point D is given by the experiment used to determine point A. If the σ_{III}-axis is equivalent to the σ_I-axis and the yield in the material is independent of hydrostatic stress, point E is determined by superimposing a hydrostatic stress on a uniaxial stress along the σ_{III}-direction, as shown in Fig. IV.16. The yield stress at C can be determined by a pure shear experiment. For an isotropic material which is pressure independent, the location of all these points requires only two experiments, i.e., a tensile test and a shear test. For anisotropic materials a larger but still limited number of tests is required.

After these points have been located, an approximate yield surface may be drawn by connecting the points with straight lines. As shown in Appendix IV-A, the yield surface must be convex; the straight lines are therefore the limiting case. The actual yield locus lies outside the polygon and is convex everywhere. The dotted line in Fig. IV.15 is not an acceptable yield criterion, since it provides a concave yield surface.

The concept of a yield criterion thus far discussed can be generalized by asking some basic questions. The Tresca and the Mises yield criteria are functions of invariants only. This should not be surprising since the yielding of fully isotropic materials must be independent of any mathematical coordinate system that happens to be chosen. Therefore, it is possible that a suitable yield criterion might be developed by considering other functions of invariants.

The stress invariants considered in Chapters II and III were: the invariants of the stress tensor, J_1, J_2, and J_3; the invariants of the deviator stress tensor, J'_2 and J'_3; and the principal stresses σ_I, σ_{II}, and σ_{III}. If we wish to develop a yield function f based on stress invariants, the generalized yield criteria may be written as

$$f(J_1, J_2, J_3) = 0 \qquad\qquad \text{(IV.15a)}$$

$$f(J'_2, J'_3) = 0 \qquad\qquad \text{(IV.15b)}$$

$$f(\sigma_I, \sigma_{II}, \sigma_{III}) = 0 \qquad\qquad \text{(IV.15c)}$$

It is easy to show that the Mises yield criterion can be deduced from Eq. (IV.15b) by assuming that the material is fully isotropic and independent of the hydrostatic stress and the sign of the normal stresses. The Tresca yield criterion is consistent with Eq. (IV.15c). The plastic deformation of plastics and granular materials such as soil and metal powders is a sensitive function of hydrostatic pressure. The yield criterion for such a material may be written as

$$f(J_1, J_2) = \alpha J_1 + (-J'_2)^{1/2} = k \qquad\qquad \text{(IV.16)}$$

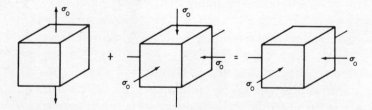

Fig. IV.16 Superposition of a hydrostatic compression on a state of uniaxial tension to give a state of biaxial compression. Since the hydrostatic stress does not affect the plastic deformation, the state of biaxial compression is equivalent to uniaxial tension, as far as plastic behavior is concerned.

since J_1 is equal to the hydrostatic stress and $(-J_2')^{1/2}$ is proportional to the resultant shear stress. Equation (IV.16) is known as the Mohr-Coulomb yield criterion and is used in soil mechanics.

IV-6 FLOW RULE–STRESS-STRAIN RELATIONSHIP IN THE PLASTIC REGIME

The previous sections were confined to a discussion of the problem of determining which states of stress can cause plastic deformation and how they depend on the previous plastic strain history. In these sections yield criteria were established and a rule was presented for handling the effects of work hardening. However, nothing has yet been said about what plastic strain increments occur when the yield condition is satisfied. The purpose of this section is to develop relationships between various stress components and various plastic strain components, similar to the generalized Hooke's laws for elastic deformation, assuming that the material behavior as characterized by the equivalent stress and the equivalent strain is known.

Since the plastic deformation of metals is due *only* to the shear stress components, and since the deviator stress represents the shear component (i.e., the distortional component of stress) as discussed in Chapter III, it is reasonable to expect that the incremental deviatoric plastic strain is proportional to the deviator stress, i.e.,

$$d\epsilon_{ij}^{p} = d\epsilon_{ij}^{p'} = dc \, \sigma_{ij}' \qquad (IV.17)$$

The proportionality constant dc can be expressed in terms of the equivalent stress and strain by considering the special case of a tensile test where $d\epsilon_{11}^{p} = d\bar{\epsilon}^{p}$ and $\sigma_{11} = \bar{\sigma}$. Therefore, for $i = j = 1$, Eq. IV.17 can be written as

$$d\epsilon_{11}^{p} = d\bar{\epsilon}^{p} = dc \, \sigma_{11}' \qquad (IV.18)$$

Since $\sigma_{11}' = 2/3 \, \sigma_{11} = 2/3 \, \bar{\sigma}$ under uniaxial loading, Eq. (IX.18) can be solved for dc in terms of $\bar{\sigma}$ and $d\bar{\epsilon}^{p}$. Substituting this value for dc in Eq. (IV.17), the resulting stress-strain relation for plastic deformation may be written as

$$d\epsilon_{ij}^{p} = \frac{3}{2} \frac{d\bar{\epsilon}^{p}}{\bar{\sigma}} \sigma'_{ij} \tag{IV.19}$$

The constant of proportionality for the plastic regime, $dc = 3/2\ d\bar{\epsilon}^{p}/\bar{\sigma}$, is equivalent to the proportionality constant for elastic distortion, $1/2G$, except that dc is a function of strain.* Equation (IV.19) may be written along the x_1-, x_2-, and x_3-axes as

$$d\epsilon_{11}^{p} = \frac{d\bar{\epsilon}^{p}}{\bar{\sigma}}\left[\sigma_{11} - \frac{1}{2}(\sigma_{22} + \sigma_{33})\right]$$

$$d\epsilon_{22}^{p} = \frac{d\bar{\epsilon}^{p}}{\bar{\sigma}}\left[\sigma_{22} - \frac{1}{2}(\sigma_{11} + \sigma_{33})\right]$$

$$d\epsilon_{33}^{p} = \frac{d\bar{\epsilon}^{p}}{\bar{\sigma}}\left[\sigma_{33} - \frac{1}{2}(\sigma_{11} + \sigma_{22})\right]$$

$$d\epsilon_{12}^{p} = \frac{3}{2}\frac{d\bar{\epsilon}^{p}}{\bar{\sigma}}\sigma_{12} \tag{IV.20}$$

$$d\epsilon_{13}^{p} = \frac{3}{2}\frac{d\bar{\epsilon}^{p}}{\bar{\sigma}}\sigma_{13}$$

$$d\epsilon_{23}^{p} = \frac{3}{2}\frac{d\bar{\epsilon}^{p}}{\bar{\sigma}}\sigma_{23}$$

Equations (IV.20) are known as the *Prandtl-Reuss equations* or the *proportionality flow rule*. Note the similarity with the generalized Hooke's laws. Poisson's ratio is replaced by $1/2$ to reflect the incompressibility during plastic deformation. Again note the fact that the nonlinear nature of plastic deformation and the dependence of plastic deformation on the loading path preclude the writing of stress-strain relationships in terms of total plastic strains. Also note that the plastic strain increments and the deviator plastic strain increments are equal, since the hydrostatic component of plastic strain increment is equal to zero, i.e., $d\epsilon_{ij}^{p} = d\epsilon_{ij}^{p'}$. The constant $d\bar{\epsilon}^{p}/\bar{\sigma}$ can be evaluated from any of the various constitutive relations discussed in Secs. IV-3 and IV-4, e.g., $\bar{\sigma} = C(\bar{\epsilon}^{p})^{n}$. The total strain increments are the sum of elastic and plastic increments, which can be expressed as

*We repeat that $(d\bar{\epsilon}^{p}/\bar{\sigma})$ is simply an experimentally determined constant that relates incremental plastic strain to deviator stress, just as G or E does in the elastic case.

$$d\epsilon_{ij} = d\epsilon_{ij}{}^e + d\epsilon_{ij}{}^p \qquad\qquad (IV.21)$$

A more rigorous derivation of Eq. (IV.19), based on the principle of maximum plastic work, is given in Appendix IV-A at the end of this chapter. The constitutive relations discussed for the plastic case in the preceding sections may be summarized by comparing them with those of the elastic case. In the elastic case, the constitutive relation is completely described by Hooke's law, with two elastic constants, e.g., E and ν, to characterize the material properties. In the plastic case, the flow rule, such as the Prandtl-Reuss equations, substitutes for Hooke's law in the elastic case, and a work-hardening relationship, such as $\bar{\sigma} = C(\bar{\epsilon}^p)^n$, is used to characterize the material properties. In the plastic regime however, the complete description of the constitutive relations requires an additional rule which describes the elastic-plastic transition, i.e., the yield criterion. Again, note that the constitutive relations for all these cases are determined experimentally.

EXAMPLE IV.5—Plastic Deformation of a Slab

A commercially pure aluminum (1100) slab of $1 \times 10 \times 50$ in. is to be stretched uniformly to a length of 57 in., maintaining its width at 10 in. Determine the final thickness of the slab and the maximum force necessary for the stretching operation. The stress-strain relationship for the aluminum is given by

$$\bar{\sigma} = 26 \times 10^3 (\bar{\epsilon})^{0.20} \text{ psi} \qquad\qquad (a)$$

Solution:

Since the deformation is large, the elastic deformation may be neglected. A rectangular coordinate system with its x_1- and x_3-axes parallel to the thickness and lengthwise directions, respectively, will be used. The final thickness of the slab may be determined from volume constancy as

$$1 \text{ in.} \times 10 \text{ in.} \times 50 \text{ in.} = t \times 10 \text{ in.} \times 57 \text{ in.} \quad \text{or} \quad t = 0.878 \text{ in.} \qquad (b)$$

Equation (b) could also be obtained by considering the strain as

$$\epsilon_{11} + \epsilon_{22} + \epsilon_{33} = 0$$

or

$$\ln \frac{t}{t_0} + \ln \frac{w}{w_0} + \ln \frac{l}{l_0} = 0 \qquad\qquad (c)$$

where t, w and l are thickness, width, and length of the slab, respectively. The subscript 0 designates the original state. Since the width does not change, Eq. (c) yields

$$t = \frac{l_0}{l} \, t_0 = 0.878 \text{ in.} \tag{d}$$

In order to determine the force the stress must be determined using the stress-strain relations given by Eq. (IV.20), i.e.,

$$d\epsilon_{11} = -\frac{d\bar{\epsilon}}{2\bar{\sigma}}(\sigma_{22} + \sigma_{33})$$

$$d\epsilon_{22} = \frac{d\bar{\epsilon}}{\bar{\sigma}}\left(\sigma_{22} - \frac{1}{2}\sigma_{33}\right) \tag{e}$$

$$d\epsilon_{33} = \frac{d\bar{\epsilon}}{\bar{\sigma}}\left(\sigma_{33} - \frac{1}{2}\sigma_{22}\right)$$

Since the slab deforms uniformly and is thin, σ_{11} is assumed to be equal to zero. Since $\epsilon_{22} = 0$, we can use the second equation of Eq. (e) to obtain

$$\sigma_{22} = \frac{1}{2}\sigma_{33} \tag{f}$$

Then, Eq. (e) may be written as

$$d\epsilon_{11} = -\frac{3}{4}\frac{d\bar{\epsilon}}{\bar{\sigma}}\sigma_{33} \; ; \quad d\epsilon_{33} = \frac{3}{4}\frac{d\bar{\epsilon}}{\bar{\sigma}}\sigma_{33} \tag{g}$$

The equivalent stress is obtained from Eq. (IV.8) as

$$\bar{\sigma} = \left\{\frac{1}{2}\left[(\sigma_{22} - \sigma_{33})^2 + \sigma_{33}^2 + \sigma_{22}^2\right]\right\}^{1/2}$$

$$= \frac{\sqrt{3}}{2}\sigma_{33} \tag{h}$$

The substitution of Eq. (h) into Eq. (g) yields

$$d\epsilon_{11} = -\frac{\sqrt{3}}{2}d\bar{\epsilon} \; ; \quad d\epsilon_{33} = \frac{\sqrt{3}}{2}d\bar{\epsilon} \tag{i}$$

Integrating Eq. (i), we get

$$\epsilon_{11} = -\frac{\sqrt{3}}{2}\bar{\epsilon} \; ; \quad \epsilon_{33} = \frac{\sqrt{3}}{2}\bar{\epsilon}$$

$$\bar{\epsilon} = \frac{2}{\sqrt{3}}\ln\frac{l}{l_0} \tag{j}$$

Substituting Eq. (j) into Eq. (a), we find

$$\bar{\sigma} = 26 \times 10^3 \left(\frac{2}{\sqrt{3}} \ln \frac{l}{l_0} \right)^{0.2} = 27.6 \times 10^3 \text{ psi} \tag{k}$$

The maximum stress occurs at the end of the stretching operation and is given, from Eq. (h), as

$$\sigma_{33} = \frac{2}{\sqrt{3}} \bar{\sigma} = 31,800 \text{ psi} \tag{l}$$

The maximum force has to be applied at the end of the stretching operation for a work-hardening material, unless necking occurs before the completion of the operation (see Sec. IV-7-b). Therefore,

$$F_{max} = (\sigma_{33})_{1=1_f} A_f = 31,800 \times 0.878 \text{ in.} \times 10 \text{ in.}$$
$$F_{max} = 280,000 \text{ lb} \tag{m}$$

EXAMPLE IV.6—Elastoplastic Deformation of an Oceanographic Research Vessel

A thin-walled sphere made of steel is gradually immersed in the Atlantic Ocean near Puerto Rico, where the maximum hydrostatic pressure is 2,000 psig. After the vessel is immersed, the internal pressure of the vessel is gradually raised to 2,000 psig in order to equalize the internal pressure with the outside pressure. The outside diameter of the sphere is 30 in. and the wall thickness is 1/4 in. Assume that the sphere does not buckle or neck. The stress-strain relationship for the plastic regime is given by

$$\bar{\sigma} = 10^5 \, (\bar{\epsilon}^p)^{0.2} \text{ psi} \quad {}^* \tag{a}$$

and the yield stress is 30,000 psi. Poisson's ratio and Young's modulus for steel are 0.29 and 30×10^6 psi, respectively. Find the dimensions of the sphere after it is pressurized in the ocean.

Solution:
A) *Governing Relations*
A spherical coordinate system (r, θ, ϕ) with its origin at the center of the sphere will be used. An element from the shell of the sphere may be isolated as shown in Fig. IV.17(b). In a thin shell one may assume that the stresses are constant throughout the thickness. The equilibrium equation for the element along the radial direction may be written as

$$2 (rt \, d\phi) \sin \frac{d\theta}{2} \sigma_{\theta\theta} + 2 (rt \, d\theta) \sin \frac{d\phi}{2} \sigma_{\phi\phi} = (p_i - p_0) rd\theta \, rd\phi \tag{b}$$

---

*This type of relation is only accurate when the plastic strains are much greater than the elastic strains. This assumption must be checked after the problem has been solved.

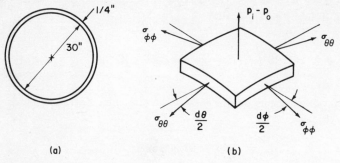

(a)   (b)

Fig. IV.17   Geometry of a spherical diving vessel.

From the symmetry condition of the sphere, $\sigma_{\theta\theta} = \sigma_{\phi\phi}$. Equation (b) may be simplified using the small-angle approximation, i.e., $\sin d\theta = d\theta$, as

$$\sigma_{\theta\theta} = \sigma_{\phi\phi} = (p_i - p_0)\frac{r}{2t} \tag{c}$$

while σ_{rr} is bounded by $-p_i$ and $-p_0$. However, at pressures where $p_{max} \gg p_{min}$ the hoop stress $\sigma_{\theta\theta}$ is so much greater than σ_{rr} that any value of pressure in this range may be used. The strain-displacement relationships for a hollow sphere are

$$\epsilon_{rr} = \ln\frac{t}{t_0} ; \quad \epsilon_{\theta\theta} = \epsilon_{\phi\phi} = \ln\frac{d}{d_0} \tag{d}$$

where t_0 and d_0 are the original thickness and diameter of the sphere. The stress-strain relationships for elastic deformation are

$$\epsilon_{rr}{}^e = \frac{1}{E}[\sigma_{rr} - \nu(\sigma_{\theta\theta} + \sigma_{\phi\phi})] = \frac{1}{E}(\sigma_{rr} - 2\nu\sigma_{\theta\theta})$$

$$\epsilon_{\theta\theta}{}^e = \frac{1}{E}[\sigma_{\theta\theta} - \nu(\sigma_{\phi\phi} + \sigma_{rr})] = \frac{1}{E}[(1-\nu)\sigma_{\theta\theta} - \nu\sigma_{rr}] = \epsilon_{\phi\phi}{}^e \tag{e}$$

For plastic deformation,

$$d\epsilon_{rr}{}^p = \frac{d\bar{\epsilon}^p}{\bar{\sigma}}\left[\sigma_{rr} - \frac{1}{2}(\sigma_{\theta\theta} + \sigma_{\phi\phi})\right] = \frac{d\bar{\epsilon}^p}{\bar{\sigma}}(\sigma_{rr} - \sigma_{\theta\theta})$$

$$d\epsilon_{\theta\theta}{}^p = \frac{d\bar{\epsilon}^p}{\bar{\sigma}}\left[\sigma_{\theta\theta} - \frac{1}{2}(\sigma_{rr} + \sigma_{\phi\phi})\right] \tag{f}$$

$$= \frac{1}{2}\frac{d\bar{\epsilon}^p}{\bar{\sigma}}(\sigma_{\theta\theta} - \sigma_{rr}) = d\epsilon_{\phi\phi}{}^p$$

The Mises yield criterion states that

$$\bar{\sigma} = \left\{ \frac{1}{2} \left[(\sigma_{rr} - \sigma_{\theta\theta})^2 + (\sigma_{rr} - \sigma_{\phi\phi})^2 + (\sigma_{\theta\theta} - \sigma_{\phi\phi})^2 \right] \right\}^{1/2} \tag{g}$$

$$\bar{\sigma} = |\sigma_{rr} - \sigma_{\theta\theta}| = 2k = 30{,}000 \text{ psi}$$

This is identical to the Tresca yield criterion, since $\sigma_{\theta\theta} = \sigma_{\phi\phi}$.

B) *Elastic Deformation Prior to Yielding*

The pressure at yielding can be determined by substituting Eq. (c) into Eq. (g) to give

$$\left| \sigma_{rr} - (p_i - p_0) \frac{r}{2t} \right| = 2k \tag{h}$$

On the descent, p_0 increases from 0 to 2,000 psi, while p_i stays at atmospheric pressure (approximately). This loading history is shown by the path $\overline{OA}'$ in Fig. IV.18.

$$p_{\min} = p_i = 0$$

$$\frac{p_0 r}{2t} = 2k \; ; \quad p_0 = \frac{4kt}{r} = 1{,}000 \text{ psig} \tag{i}$$

The sphere deforms plastically at 1,000 psig. The dimensions of the sphere at this point (A') can be found from Eqs. (d) and (e) as

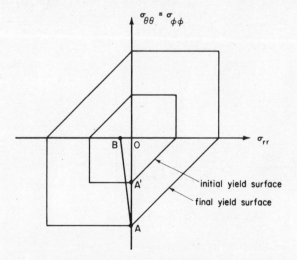

Fig. IV.18 Tresca yield criterion and the loading path for an oceanographic vessel. (The distance OB is exaggerated for clarity.)

$$\epsilon_{rr}{}^e = \frac{\Delta t}{t} = \frac{2\nu}{E} \cdot \frac{p_0 r}{2t}$$

(j)

$$\epsilon_{\theta\theta}{}^e = \frac{\Delta r}{r} = -\frac{1-\nu}{E}\frac{p_0 r}{2t}$$

The values for r' and t' at point A' are found to be

$$r' = 14.989 \text{ in.}; \quad t' = 0.2501 \text{ in.} \tag{k}$$

C) *Plastic Deformation During Submersion*

As the pressure p_0 increases from 1,000 to 2,000 psig, the sphere deforms plastically along the loading path $A'A$. The plastic strain increments can be computed from Eqs. (f) as

$$d\epsilon_{rr}{}^p = d\bar{\epsilon}^p \; ; \quad d\epsilon_{\theta\theta}{}^p = -\frac{1}{2}d\bar{\epsilon}^p \tag{l}$$

since $\bar{\sigma} = |\sigma_{rr} - \sigma_{\theta\theta}|$ and $\sigma_{\theta\theta} > \sigma_{rr}$. The integration of Eq. (1) yields

$$\epsilon_{rr}{}^p = \bar{\epsilon}^p \; ; \quad \epsilon_{\theta\theta}{}^p = -\frac{1}{2}\bar{\epsilon}^p \tag{m}$$

According to the stress-strain relationship of Eq. (a)

$$\bar{\epsilon}^p = \left(\frac{\bar{\sigma}}{10^5}\right)^5 = \left(\frac{p_0 r}{2t \times 10^5}\right)^5 \tag{n}$$

Assuming that the plastic strain will be much larger than the elastic strain, let $\bar{\epsilon}^p$ be the total strain at point A. The constancy of volume requires that

$$V_0 = 4\pi r_0{}^2 t_0 = 4\pi r^2 t$$

$$r = \frac{1}{2}\left(\frac{V_0}{\pi t}\right)^{1/2} \tag{o}$$

Combining Eqs. (n) and (o)

$$\ln \frac{t}{t_0} = \left(\frac{p_0 r}{2t \times 10^5}\right)^5 = \left[\frac{p_0}{2t \times 10^5}\left(\frac{V_0}{4\pi t}\right)^{1/2}\right]^5 \tag{p}$$

By a trial-and-error process, r and t are found at point A to be

$$r = 14.621 \text{ in.}; \quad t = 0.265 \text{ in.} \tag{q}$$

Note that $\epsilon_{ii}{}^p \gg \epsilon_{ii}{}^e$. Thus the initial assumption is valid.

D) *Deformation due to Internal Pressure Change*

As p_i is increased to 2,000 psi, σ_{rr} decreases from zero to $-2,000$ psi and $\sigma_{\theta\theta}$ unloads from its compressive state to zero. The effective stress becomes

$$
\begin{aligned}
\bar{\sigma} = |\sigma_{rr} - \sigma_{\theta\theta}| &= \left| -p_i + \frac{(p_0 - p_i)r}{2t} \right| \\
&= \left| -p_i\left(1 + \frac{r}{2t}\right) + p_0\frac{r}{2t} \right|
\end{aligned}
\tag{r}
$$

As p_i rises, $\bar{\sigma}$ falls to zero and then rises back to 2,000 psi. The shell behaves elastically throughout the unloading path to $p_i = p_0$. Since the elastic strains in the two states A and B are uniquely determined from the state of stress and Hooke's law, the difference in the elastic strains between the two states can be calculated by superposition, yielding

$$
\begin{aligned}
\epsilon_{rr}{}^e &= \frac{1}{E}\left[(\sigma_{rr}{}^B - \sigma_{rr}{}^A) - 2\nu(\sigma_{\theta\theta}{}^B - \sigma_{\theta\theta}{}^A)\right] \\
\epsilon_{\theta\theta}{}^e &= \frac{1}{E}\left[(1 - \nu)(\sigma_{\theta\theta}{}^B - \sigma_{\theta\theta}{}^A) - \nu(\sigma_{rr}{}^B - \sigma_{rr}{}^A)\right]
\end{aligned}
\tag{s}
$$

At point B, $p_i = p_0 = 2,000$ psi, which gives*

$$
\begin{array}{ll}
\sigma_{rr}{}^A = 0 & \sigma_{rr}{}^B = -p_i \\[2mm]
\sigma_{\theta\theta}{}^A = -p_0\dfrac{r_A}{2t_A} & \sigma_{\theta\theta}{}^B = 0
\end{array}
\tag{t}
$$

Equation (s) now becomes

$$
\begin{aligned}
\epsilon_{rr}{}^e &= \frac{1}{E}\left[-p_i - 2\nu\left(\frac{p_0 r_A}{2t_A}\right)\right] = \frac{\Delta t}{t} \\
\epsilon_{\theta\theta}{}^e &= \frac{1}{E}\left[(1 - \nu)\left(\frac{p_0 r_A}{2t_A}\right) - \nu(-p_i)\right] = \frac{\Delta r}{r}
\end{aligned}
\tag{u}
$$

Thus, after elastic unloading from the plastic region (from point A to point B), t and r can be found from Eq. (u) and

$$
t_B = t_A + \Delta t \;; \qquad r_B = r_A + \Delta r
\tag{v}
$$

*Note that unloading from the stress state A to B is the same as going from A to O and then from O to B since elastic deformation is path independent.

Thus,

$$r_B = 14.639 \text{ in.} \qquad t_B = 0.265 \text{ in.} \qquad \text{(w)}$$

It should be noted that the elastic strains are much smaller than the plastic strains, indicating that the elastic strains may be neglected.

IV-7 APPLICATIONS—LARGE PLASTIC DEFORMATION PROBLEMS

IV-7-a Methods of Solution

When the magnitude of plastic deformation is larger than the elastic strain by an order of magnitude or more, and if the particular phenomenon of interest is largely controlled by the plastic deformation, the elastic deformation may be neglected. The solution to many physical problems becomes much easier when this can be done. The purpose of this section is to review a few typical problems of large plastic deformation. Solutions can be obtained by using one of the following methods:

1) *Energy Method*—The work done to deform a body is equated to the amount of energy consumed in plastic deformation. It is difficult to simultaneously satisfy the equilibrium and compatibility conditions using this technique.

2) *Slab Method or the Uniform Stress Method*—The deformation of a body is assumed to occur in such a fashion that either plane sections remain plane or spherical surfaces remain spherical and that the stress is uniform on that plane. The solutions obtained by this technique do not usually satisfy the compatibility conditions. As will be shown in Sec. IV.9, this method normally provides lower-bound solutions.

3) *Slip Line Solutions*—For ideally plastic materials "exact" solutions can be obtained by considering the deformation along the maximum shear stress directions using the method of characteristics. These exact solutions can be obtained for only relatively simple plane strain and axisymmetric problems. Since the solution is obtained for ideally plastic materials, the deformation field predicted by this type of solution may not agree with the physically observed deformation field.

4) *Limit Analysis*—Approximate limiting solutions, which establish upper and lower bounds on the force necessary to plastically deform the body can be obtained for many practical problems using

the limit theorems. Limit analyses are relatively easy to carry out and thus are very useful.

The subject to be considered in this section is the use of the energy and slab methods. Limit theorems will be taken up in Sec. IV.9. Slip line methods will not be discussed, since they are outside the scope of this book.

IV-7-b   Wire Drawing

Wire drawing, which is one of the oldest metal forming processes, reduces and sizes the wire diameter. It is done by drawing wire through a die whose cross-sectional area changes gradually as shown in Fig. IV.19. The working section is a cone. The wire drawing is carried out at velocities ranging from a few feet per minute (fpm) under experimental conditions to several hundred fpm for small rods to several thousand fpm for fine wires. Most wire drawing is done at room temperature. Tungsten carbide dies are used in most cases, except when the wire diameter is less than 55×10^{-3} in., in which case diamond dies are used. Lubricants are applied to the entering wire to reduce the frictional force.

The purpose of this subsection is to analyze the mechanics of wire drawing and then discuss the analytical results in light of the industrial practices and experimental results.

Consider a wire drawing die whose inner geometry is a cone with an included angle of 2α, as shown in Fig. IV.19. (The total included

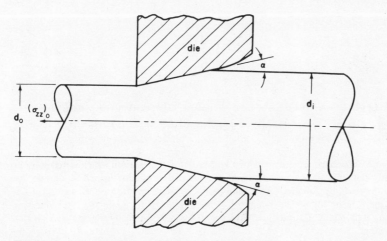

Fig. IV.19 Schematic diagram of a wire drawing operation.

angle 2α ranges from $5°$ to $25°$). The coefficient of friction at the die-work interface is μ. The axial stress required for wire drawing is to be determined. The energy method and the slab method will be used in the analyses.

1) *Energy Method*

The work done by the external force is dissipated by deforming the rod plastically and by overcoming the frictional force at the die-work interface. The work done by the external drawing force per unit time is

$$\left(\frac{\pi}{4}\right)d_0^2(\sigma_{zz})_0 V_0 \tag{IV.22}$$

where V_0 is the drawing velocity at the exit of the die. The work done in deforming the rod per unit time is

$$\frac{d}{dt} \iint \bar{\sigma} d\bar{\epsilon}\, dv \tag{IV.23}$$

where v is the volume of the material deforming in the die. If a uniform deformation is assumed (i.e., every particle undergoes the same amount of deformation), which is equivalent to neglecting the redundant work,* the above equation may be rewritten as

$$\left(\frac{\pi}{4} d_0^2 V_0\right) \int \bar{\sigma} d\bar{\epsilon} \tag{IV.24}$$

If the stress-strain relations are known, the integral $\int \bar{\sigma} d\bar{\epsilon}$ can be evaluated either numerically, graphically, or analytically. If the

---

*The plastic deformation consists of two kinds: useful deformation which is necessary in reducing the diameter from d_i to d_0 and redundant deformation which is due to the constriction imposed by the die. The redundant deformation can be as much as 50% of the total, but there is no simple technique of estimating the work done for the redundant deformation. Redundant deformation does not exist in a simple tensile specimen under uniaxial loading.

frictional work is negligible, Eqs. (IV.22) and (IV.24) may be equated to obtain $(\sigma_{zz})_0$ as

$$(\sigma_{zz})_0 = \int_0^{\bar{\epsilon}} \bar{\sigma} d\bar{\epsilon} \qquad (IV.25)$$

The equivalent strain is

$$\bar{\epsilon} = \epsilon_{zz} = \ln \frac{l}{l_0} = 2 \ln \frac{d_i}{d_0} \qquad (IV.26)$$

In deriving the above equation the constancy of volume is utilized. If the material is rigid-plastic, the expression for $(\sigma_{zz})_0$ becomes

$$(\sigma_{zz})_0 = 2 \ln \frac{d_i}{d_0} \bar{\sigma}_0 \qquad (IV.27)$$

The work done to overcome the interface friction may now be included. It turns out that there is no simple means of estimating the total frictional force, since the pressure acting on the die wall is not constant. If the average pressure acting on the die-work interface is P_{ave}, the work done to overcome the frictional force per unit time is

$$(\mu P_{\text{ave}} \cos \alpha) A_c V_{\text{ave}} = (\mu P_{\text{ave}} \cos \alpha)\left(\frac{\pi}{4} \frac{d_i^2 - d_0^2}{\sin \alpha}\right)\left(\frac{2d_0}{d_i + d_0}\right)^2 V \qquad (IV.28)$$

where A_c is the cone-surface area and V_{ave} is the velocity at the midpoint of the die, where the pressure is P_{ave}. Equating Eq. (IV.22) to the sum of Eqs. (IV.24) and (IV.28), we obtain

$$(\sigma_{zz})_0 = \bar{\sigma}_0 \ln\left(\frac{d_i}{d_0}\right)^2 + 4\mu P_{\text{ave}} \cot \alpha \left(\frac{d_i - d_0}{d_i + d_0}\right) \qquad (IV.29)$$

Here, P_{ave} may be estimated from the solution for p given by the slab method, i.e., Eq. (IV.38), or from the yield condition

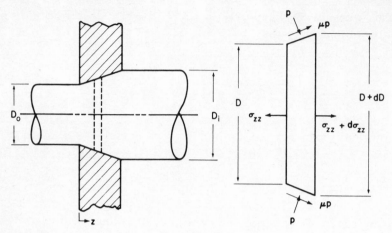

Fig. IV.20 Free-body diagram of an element of material in the wire drawing die.

$$p + \sigma_{zz} = 2k = \bar{\sigma}_0 \tag{IV.30}$$

At the inlet p is equal to $\bar{\sigma}_0$ and at the outlet it is equal to $\bar{\sigma}_0 - (\sigma_{zz})_0$. The pressure P_{ave} may be approximated as being the average of these two values.

Because of the many assumptions made in deriving Eq. (IV.29), the expression for $(\sigma_{zz})_0$ is only approximate. There are many other similar solutions available which can approximate the experimental results with varying degrees of success.

2) *Slab Method*

The wire drawing problem will now be analyzed from a different point of view. The equilibrium condition and the yield condition will be satisfied on the average at every cross section of the work material undergoing deformation. Although the true stress distribution is likely to vary across the cross section, it will be assumed to be constant throughout. All the forces acting on an isolated slab are shown in Fig. IV.20.

Since the inertia is negligible, the problem may be treated as a static problem and the equilibrium condition for the body may be written as

$$(\sigma_{zz} + d\sigma_{zz})\left(\frac{\pi}{4}\right)(D + dD)^2 - \frac{\pi}{4}D^2\sigma_{zz} + \frac{1}{2}\,p\pi DdD$$

$$+ \mu p \cos\alpha \left(\pi D\,\frac{dD}{2}\,\frac{1}{\sin\alpha}\right) = 0 \tag{IV.31}$$

The above equation may be simplified to be

$$\frac{d\sigma_{zz}}{p(1 + B) + \sigma_{zz}} = -2\frac{dD}{D} \tag{IV.32}$$

where $B = \mu \cot \alpha$. To solve Eq. (IV.32), p must be related to σ_{zz} using the yield condition. Since the angle α is small, $-\sigma_{rr} \simeq p$, and since σ_{rr} is assumed to be constant, radial equilibrium implies $\sigma_{rr} \simeq \sigma_{\theta\theta}$. Thus, the yield condition may be written as

$$\sigma_{zz} + p = \bar{\sigma}_0 = 2k \tag{IV.33}$$

Substituting Eq. (IV.33) into Eq. (IV.32) gives

$$\frac{d\sigma_{zz}}{B\sigma_{zz} - (1 + B)\bar{\sigma}_0} = 2\frac{dD}{D} \tag{IV.34}$$

The boundary condition may be written as

$$\sigma_{zz} = 0 \quad \text{at} \quad D = D_i \tag{IV.35}$$

Equation (IV.34) may be integrated to obtain

$$\frac{\sigma_{zz}}{\bar{\sigma}_0} = \frac{1 + B}{B}\left[1 - \left(\frac{D}{D_i}\right)^{2B}\right] \tag{IV.36}$$

The drawing stress $(\sigma_{zz})_0$ is

$$\frac{(\sigma_{zz})_0}{\bar{\sigma}_0} = \frac{1 + B}{B}\left[1 - \left(\frac{D_0}{D_i}\right)^{2B}\right] \tag{IV.37}$$

Using Eq. (IV.33) the pressure distribution may be obtained as

$$\frac{p}{\bar{\sigma}_0} = \left(\frac{D}{D_i}\right)^{2B}\left(1 + \frac{1}{B}\right) - \frac{1}{B} \tag{IV.38}$$

The maximum reduction of area is obtained when $(\sigma_{zz})_0/\bar{\sigma}_0 = 1$. In

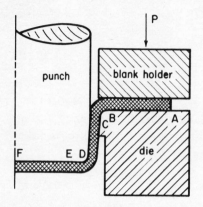

Fig. IV.21 Schematic diagram of a deep-drawing operation.

usual operations B is nearly equal to 1, and therefore the maximum reduction of area* is about 0.5. As stated in Sec. IV-7-a, this value for the drawing stress is a lower bound.

3) *Theoretical Predictions and Experimental Results*[†]

Solutions by the energy method have been found to predict the experimental results within approximately 20 to 50% over a range of die angles from 5° to 31°. The 20% accuracy is obtained when the redundant work as well as the frictional work is included in the energy method solution. Solutions obtained by the slab method are found to be accurate when the die angle is very small, but are not accurate when the die angle is large. The solutions by the slab method always predict values lower than the experimental results.

IV-7-c Deep Drawing of a Circular Cup

The problem to be considered now is the production of a cylindrical cup by deep drawing of a circular blank. The drawing arrangement is shown schematically in Fig. IV.21. The blank is initially held between the blank holder and the die. The punch with round edges pushes the blank down, bending the material between C and B, unbending it at C, and stretching it between C and D. The material FE under the head of the punch will be subject to biaxial tension, while the material ED will be subjected to plastic bending in addition to the biaxial tensile load. The material held between the blank

---

*The reduction in area $RA = (D_i^2 - D_0^2)/D_i^2$.

[†]For further details, see P. W. Whitton, "The Calculation of Drawing Force and Die Pressure in Wire Drawing," *The Wire Industry*, August, 1958, p. 735.

holder and the die is drawn radially inward, inducing compressive hoop stress as well as radial tensile stress in this region. As the flange material is drawn inward radially, the thickness of the sheet increases, the thickest section always being the rim. It has been found experimentally that the difference in thickness across the flange does not vary by more than 5% at any moment.

The blank holder is used to exert compressive stress on the flange so as to prevent wrinkling in the flange section AB. Unless the sheet is very thick, wrinkling occurs when the compressive hoop stress in the flange is sufficient to buckle the flange. The space between the die and the blank holder usually is provided with clearance in order to minimize interference with the free drawing process, otherwise "ironing" occurs. Therefore, the state of stress in the flange is approximately that of plane stress.

Most sheet metals are anisotropic since they are manufactured by cold rolling along one direction. Anistropy of the metal in the sheet plane requires a greater holding pressure. Anisotropy also causes the formation of ears, as shown in Fig. IV.23.

The drawing process is usually terminated when wrinkles form in the flange or when the material fractures at the thinnest section. This occurs near D under a combined loading of tensile and bending stresses.

The objectives of the following analysis are to determine the punch load as a function of the punch travel and to determine the maximum drawing ratio. The drawing ratio is defined as the ratio of the blank diameter to the throat diameter of the die. The exact analysis of the drawing process is difficult due to the complexity of the process. The analysis presented here is mainly concerned with the radial drawing of the flange, since it is the limiting process.

1) *Analysis*

Figure IV.22 shows an element isolated from the flange with all the forces acting on the element. Since the blank holder exerts a compressive force on the rim of the flange due to the greater thickness there, the normal load acting on the flange σ_{zz} and the frictional force $\mu\sigma_{zz}$ acting on the flange are not shown in the figure. The variation in the flange thickness will be disregarded, since the actual difference in thickness across the flange at any moment is small.

The equilibrium condition for the element may be written as

$$\left(\sigma_{rr} + \frac{d\sigma_{rr}}{dr}\right)(r + dr)\,t\,d\theta - \sigma_{rr}\,t\,rd\theta - \sigma_{\theta\theta}\,t\,dr\,d\theta = 0 \quad \text{(IV.39)}$$

Neglecting the higher-order terms, Eq. (IV.39) may be simplified to

$$\frac{d\sigma_{rr}}{\sigma_{rr} - \sigma_{\theta\theta}} + \frac{dr}{r} = 0 \quad \text{(IV.40)}$$

In writing Eq. (IV.40) the fact that, by symmetry, σ_{rr} is a function only of r is utilized. The Tresca yield criterion may be written as

$$\sigma_{rr} - \sigma_{\theta\theta} = \bar{\sigma} = 2k \quad \text{(IV.41)}$$

Equation (IV.41) may be substituted into Eq. (IV.40) yielding

$$d\sigma_{rr} = -\bar{\sigma}\,\frac{dr}{r} \quad \text{(IV.42)}$$

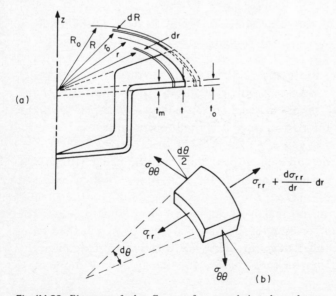

Fig. IV.22 Element of the flange of a cup being deep drawn, showing the displacements and the coordinate system.

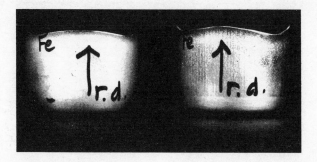

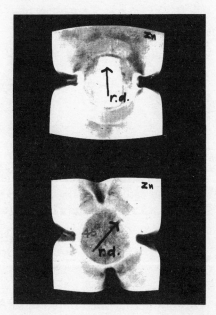

Fig. IV.23 Effect of plastic anisotropy on flange-wrinkling behavior during sheet metal forming; r.d. indicates the rolling direction of the sheet. (*From H. Naziri and R. Pearce, Intern. J. Mech. Sci., vol. 10, pp. 681–694, 1968.*)

For a rigid-plastic material, $\bar{\sigma}$ is equal to a constant σ_0 throughout the deformation process. For a work-hardening material, $\bar{\sigma}$ is a function of the plastic work done, i.e., $\bar{\sigma}$ depends on the deformation history.

1-a) *Rigid-Plastic Material*

The analysis is facilitated if the material may be characterized as being rigid-plastic. Equation (IV.42) may be integrated between the rim of the flange r_0 and any point on the flange r to yield

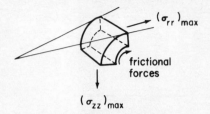

Fig. IV.24 Equilibrium of a section of the flange at the throat of the die, showing the effect of the frictional forces.

$$\sigma_{rr} - (\sigma_{rr})_{r=r_0} = \sigma_0 \ln \frac{r_0}{r} \qquad (IV.43)$$

where $(\sigma_{rr})_{r=r_0}$ is the radial stress at the rim. The stress $(\sigma_{rr})_{r=r_0}$ may be computed by noting that the frictional force at the rim due to the normal force P exerted by the blank holder is in equilibrium with the radial stress, i.e.,

$$(\sigma_{rr})_{r=r_0} = \frac{2\mu P}{2\pi r_0 t_0} \qquad (IV.44)$$

where μ is the coefficient of friction, t_0 is the flange thickness at the rim. The radial drawing stress is a maximum at $r = r_{min}$ according to Eq. (IV.43), which may be written as

$$(\sigma_{rr})_{r=r_i} = \sigma_0 \ln \frac{r_0}{r_i} + \frac{\mu P}{\pi r_0 t_0} \qquad (IV.45)$$

The load the punch has to exert to draw the blank can be obtained by considering the complete drawing process, such as bending of the sheet around the die profile. The "exact" solution of the problem is very complicated.* An approximate estimation of the axial stress in the straight section of the drawn cup suggested by Cook (Ref. 2) is to treat the problem as being the same as a rope around a capstan (see Fig. IV.24), i.e.,

$$\sigma_{zz} = \sigma_{rr} \exp \frac{\pi\mu}{2} \qquad (IV.46)$$

*See S. Y. Chung and H. W. Swift, "Cup-Drawing from a Flat Blank," *Proc. Inst. of Mech. Engrs.*, vol. 165, pp. 199–228, 1951.

When σ_{zz} exceeds the yield stress, the drawing operation fails. This usually occurs near the punch profile. The maximum drawing ratio is obtained when the coefficient of friction and the blank holding force are equal to zero. Then, from Eq. (IV.45), it is seen that the maximum radius of the blank holder is 2.72 times the throat diameter of the die.

1-b) *Work-Hardening Material*

The stress-strain relation of a work-hardening metal will be assumed to be represented by the so-called Ludwik power law,

$$\bar{\sigma} = A + C(\bar{\epsilon})^n \tag{IV.47}$$

Substituting Eq. (IV.47) into Eq. (IV.42), we obtain

$$d\sigma_{rr} = -(A + C\bar{\epsilon}^n)\frac{dr}{r} \tag{IV.48}$$

Equation (IV.48) can be integrated if the equivalent strain $\bar{\epsilon}$ can be expressed as a function of r. The strain $\bar{\epsilon}$ can be found by evaluating the strain components $d\epsilon_{\theta\theta}$, $d\epsilon_{rr}$, and $d\epsilon_{zz}$. The hoop strain $d\epsilon_{\theta\theta}$ is given by (u/r), which is equal to dr/r; the axial strain $d\epsilon_{zz}$ is equal to (dw/dz), which is equal to dt/t, assuming a uniform deformation throughout the thickness. The radial strain $d\epsilon_{rr}$ is (du/dr), which is equal to $-(dr/r + dt/t)$. In terms of these dimensional changes, i.e., dr and dt, Eq. (IV.48) cannot be evaluated. If the change in thickness may be neglected, $d\bar{\epsilon} = 1.15\ d\epsilon_{\theta\theta}$. Hill (Ref. 5) showed that, in reality, $\bar{\epsilon}$ is never more than 3% greater than the hoop strain. Thus, the equivalent strain may be approximated as being equal to the hoop strain, i.e.,

$$d\bar{\epsilon} = d\epsilon_{\theta\theta} = \frac{dr}{r} \tag{IV.49}$$

Integrating Eq. (IV.49) from the initial radius of the particle position R to the current particle position r, we obtain*

$$\bar{\epsilon} = \ln\frac{R}{r} \tag{IV.50}$$

*Note that $\bar{\epsilon}$ is always positive.

In order to write Eq. (IV.50) solely as a function of the current position r, the current mean thickness between the rim of the flange and r is denoted by t_m. The incompressibility condition yields

$$\pi(R_0^2 - R^2)t_0 = \pi(r_0^2 - r^2)t_m \qquad \text{(IV.51)}$$

which may be written as

$$\left(\frac{R}{r}\right)^2 = \left(\frac{c}{r^2} + 1\right)\frac{t_m}{t_0} \qquad \text{(IV.52)}$$

where the constant c is given by

$$c = R_0^2\left(\frac{t_0}{t_m}\right) - r_0^2 \qquad \text{(IV.53)}$$

Substituting Eqs. (IV.50) and (IV.52) into the equilibrium equation (IV.48), we obtain

$$d\sigma_{rr} = -\left(A + C\left\{\frac{1}{2}\ln\left[\left(\frac{c}{r^2} + 1\right)\frac{t_m}{t_0}\right]\right\}^n\right)\frac{dr}{r} \qquad \text{(IV.54)}$$

Equation (IV.54) may be integrated from the current rim radius r_0 to r, using the boundary condition given by Eq. (IV.44). The radial drawing stress then becomes

$$\sigma_{rr} = A\ln\frac{r_0}{r} - C\int_{r_0}^{r}\left\{\frac{1}{2}\ln\left[\left(\frac{c}{r^2} + 1\right)\frac{t_m}{t_0}\right]\right\}^n\frac{dr}{r} + \frac{\mu P}{\pi r_0 t_0} \qquad \text{(IV.55)}$$

In order to evaluate the above equation, the value of t_m and c must be known. They can be evaluated using an iterative scheme at a given r. The solution given by Chung and Swift is shown in Fig. IV.25. The curves shown in Fig. IV.25 give the radial stress distribution in the flange at various positions of the rim during a given drawing operation. The drawing ratio considered is two. The comparison of the work-hardening and the rigid-plastic solutions shown in Fig. IV.25 should be noted.

In order to evaluate the change in thickness, the stress-strain relations are considered next. The constitutive relations may be written as

$$d\epsilon_{rr} = \frac{1}{2} \frac{d\bar{\epsilon}}{\bar{\sigma}} (\sigma_{rr} + \bar{\sigma})$$

$$d\epsilon_{\theta\theta} = \frac{1}{2} \frac{d\bar{\epsilon}}{\bar{\sigma}} (\sigma_{rr} - 2\bar{\sigma}) = \frac{dr}{r} \qquad \text{(IV.56)}$$

$$d\epsilon_{zz} = \frac{1}{2} \frac{d\bar{\epsilon}}{\bar{\sigma}} (2\sigma_{rr} - \bar{\sigma}) = \frac{dt}{t}$$

Substituting Eq. (IV.55) into Eq. (IV.56) and integrating, the changes in the dimension of the blank may be computed numerically.

IV-7-d Plastic Instability under Tension (or Necking)

The necking of tensile specimens at the ultimate tensile strength is one of the most frequently observed phenomenon in materials testing laboratories. Less frequently observed necking occurs when tensile loads are applied to tubes, sheets, and spheres. When a specimen is loaded to the plastic state, plastic deformation does not occur uniformly throughout the specimen. It begins at the smallest or the weakest cross section, one of which must exist since no specimen has a perfectly uniform cross section or is perfectly homogeneous. With plastic deformation this deforming section becomes smaller in diameter and work hardens. When the increase in the flow stress due to work hardening is greater than the increase in stress due to the localized decrease in diameter, this section becomes stronger than other sections, and a new section deforms plastically. As long as this process continues, "uniform" plastic deformation of the specimen occurs. When the rate of work hardening cannot compensate for the localized increase in the applied stress due to the decrease in the cross-sectional area, deformation continues to occur in the same localized area. This is called necking. Plastic instability is controlled by the rate of work hardening and is important in many structural applications.

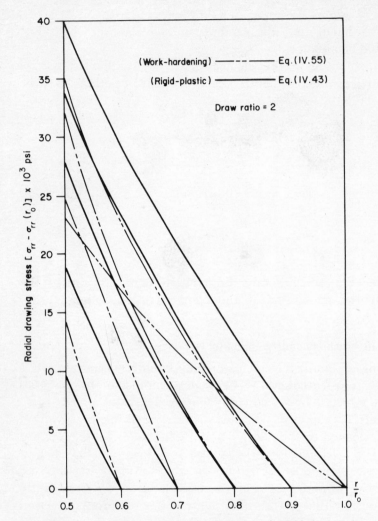

Fig. IV.25 Comparison of the radial drawing stress of a rigid plastic and a work-hardening material at a given radius during drawing. Constitutive relations: $\bar{\sigma} = 30{,}000 + 58{,}800\ \bar{\epsilon}^{0.49}$ psi for the work-hardening case; $\sigma_0 = 1/0.49 \int_0^{0.49} \bar{\sigma} d\bar{\epsilon} = 57{,}300$ psi for the rigid-plastic case. Draw ratio = 2. (*From S. Y. Chung and H. W. Swift, Cup Drawing from a Flat Blank*, Proc. Inst. Mech. Eng., *vol. 165, pp. 199–228, 1951.*)

In order to analyze this problem, consider a simple tensile specimen under the uniaxial load as shown in Fig. IV.26. Curve OY is the elastic region, YT is the plastic region of uniform elongation, and T is the necking point or the plastic instability point.

The point T can be found by determining the point at which the slope of F vs. ϵ_{11} is zero, i.e.,

$$\frac{dF}{d\epsilon_{11}} = 0 \tag{IV.57}$$

The stress-strain relationship is assumed to be

$$\sigma_{11} = C(\epsilon_{11})^n$$

or

$$\bar{\sigma} = C(\bar{\epsilon})^n \tag{IV.58}$$

The load is given by

$$F = A\sigma_{11} \tag{IV.59}$$

where A is the true cross-sectional area. The stress is assumed to be uniform throughout the cross section, which is a good assumption until necking occurs. Substituting Eq. (IV.59) into Eq. (IV.57), we obtain

$$\sigma_{11} \frac{dA}{d\epsilon_{11}} + A \frac{d\sigma_{11}}{d\epsilon_{11}} = 0 \tag{IV.60}$$

From the definition of true strain and the zero volume change requirement

$$\epsilon_{11} = \ln \frac{l}{l_0} = \ln \frac{A_0}{A} \tag{IV.61}$$

which may be rewritten as

$$A = A_0 e^{-\epsilon_{11}} \tag{IV.62}$$

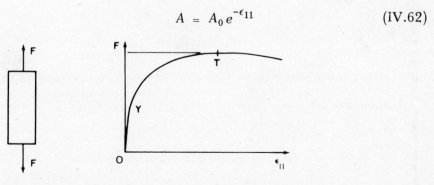

Fig. IV.26 Force-strain curve for a uniaxially loaded tensile specimen.

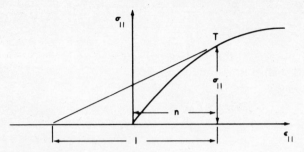

Fig. IV.27 True stress-true strain curve for a uniaxially loaded tensile specimen, showing the construction for determining the point where necking begins.

Equations (IV.58), (IV.60) and (IV.62) yield

$$\frac{d\sigma_{11}}{d\epsilon_{11}} = \sigma_{11} \qquad \text{(IV.63)}$$

or
$$\epsilon_{11} = n \qquad \text{(IV.64)}$$

On the true stress-strain curve, the ultimate tensile strength point T can be located as shown in Fig. IV.27. The results state that the strain at plastic instability is larger as the work-hardening exponent increases. The plastic instability criterion for other shapes, such as tubes, spheres, and sheets, may be obtained in a similar manner.

EXAMPLE IV.7—Plastic Instability of a Long
Thin-Walled Tube

Consider a long thin-walled tube with closed ends. The tube is made of soft aluminum, and the equivalent stress—equivalent strain curve is approximated by

$$\bar{\sigma} = 22,200\,(\bar{\epsilon})^{0.25}\,\text{psi}$$

The tube is loaded by internal pressure only. What is the value of $\bar{\epsilon}$ at which instability occurs? (Don't forget that this is a biaxial loading situation.) If the tube originally has a 4-in. diameter and a 1/16-in. wall thickness, at what pressure P will necking occur?

Solution:
The instability occurs when the applied load is a maximum, i.e.,

$$dP = 0 \qquad \text{(a)}$$

From the equilibrium condition (see the appendices to Chapter II),

$$\sigma_{rr} \simeq 0$$

$$\sigma_{zz} = \frac{PD}{4t}$$

$$\sigma_{\theta\theta} = \frac{PD}{2t} = 2\sigma_{zz}$$

(b)

where D is the diameter of the tube, and t is the wall thickness. Substituting the last equation of Eqs. (b) into Eq. (a) gives

$$\frac{dP}{P} = \frac{dt}{t} - \frac{dD}{D} + \frac{d\sigma_{\theta\theta}}{\sigma_{\theta\theta}} = 0$$

(c)

Since $dt/t = d\epsilon_{rr}$ and $dD/D = d\epsilon_{\theta\theta}$, the instability condition, Eq. (c), may be written as

$$d\epsilon_{rr} - d\epsilon_{\theta\theta} + \frac{d\sigma_{\theta\theta}}{\sigma_{\theta\theta}} = 0$$

(d)

Plastic instability occurs at large strains, and, therefore, elastic strains will be neglected.

Equation (d) can be solved if strains and stresses can be interrelated by using the Prandtl-Reuss equations. From Eq. (IV.20),

$$d\epsilon_{zz}{}^{p} = \frac{d\bar{\epsilon}^{p}}{\bar{\sigma}}\left(\sigma_{zz} - \frac{1}{2}\sigma_{\theta\theta}\right) = 0$$

$$d\epsilon_{rr}{}^{p} = -\frac{1}{2}\frac{d\bar{\epsilon}^{p}}{\bar{\sigma}}(\sigma_{\theta\theta} + \sigma_{zz}) = -\frac{3}{4}\frac{d\bar{\epsilon}^{p}}{\bar{\sigma}}\sigma_{\theta\theta}$$

(e)

$$d\epsilon_{\theta\theta}{}^{p} = \frac{d\bar{\epsilon}^{p}}{\bar{\sigma}}\left(\sigma_{\theta\theta} - \frac{1}{2}\sigma_{zz}\right) = \frac{3}{4}\frac{d\bar{\epsilon}^{p}}{\bar{\sigma}}\sigma_{\theta\theta}$$

From the definition of the equivalent stress, Eq. (IV.8), we have

$$\bar{\sigma} = \frac{\sqrt{3}}{2}\sigma_{\theta\theta}$$

(f)

The substitution of eqs. (e) and (f) into Eq. (d) yields

$$\frac{d\bar{\sigma}}{d\bar{\epsilon}^{p}} = \sqrt{3}\,\bar{\sigma}$$

(g)

Since the material behavior is such that $\bar{\sigma}$ is related to $\bar{\epsilon}^{p}$ by

$$\bar{\sigma} = C(\bar{\epsilon}^p)^n \tag{h}$$

The strain at necking is obtained as

$$\bar{\epsilon}^p = \frac{n}{\sqrt{3}} = \frac{0.25}{\sqrt{3}} \tag{i}$$

The pressure at necking is given by

$$P_{\text{necking}} = \frac{2t}{D}\,\sigma_{\theta\theta} \tag{j}$$

To evaluate Eq. (j) the hoop stress is found from Eq. (f) as

$$\bar{\sigma} = \frac{\sqrt{3}}{2}\,\sigma_{\theta\theta} = 22{,}200\,(\bar{\epsilon}^p)^{0.25} = 22{,}200\left(\frac{0.25}{\sqrt{3}}\right)^{0.25} \tag{k}$$

$$\bar{\sigma} = 13{,}680 \text{ psi} \qquad \sigma_{\theta\theta} = 15{,}800 \text{ psi}$$

The final dimensions of the tube at necking are obtained from Eqs. (e) and (f):

$$\epsilon_{rr} = \ln\frac{t}{t_0} = -\frac{\sqrt{3}}{2}\,\bar{\epsilon}$$

$$\epsilon_{\theta\theta} = \ln\frac{D}{D_0} = \frac{\sqrt{3}}{2}\,\bar{\epsilon} \tag{l}$$

which give

$$t = 0.555 \text{ in.} \quad D = 4.54 \text{ in.} \tag{m}$$

Substituting Eqs. (k) and (m) into Eq. (j), the load at necking is obtained as

$$P_{\text{necking}} = 383 \text{ psi} \tag{n}$$

EXAMPLE IV.8—Explosive Setting of
Hollow Rivets

A company in Boston is interested in developing rivets which can be set on construction sites with minimum use of energy. The potential applications of the rivets are said to lie mostly in underwater constructions, the potential for on-ground applications depending a great deal on their cost. The solution suggested by their chief engineer is to use hollow rivets filled with solid propellants which can be ignited electrically using batteries. The schematic diagram of the suggested rivet is shown in Fig. IV.28. Suitable metals for the rivets are to be chosen, together with appropriate annealing and cold-working treatments.

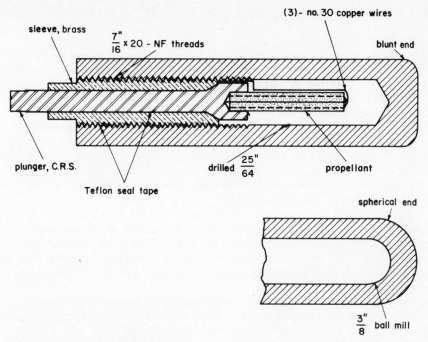

Fig. IV.28   Design of an explosive rivet.

Solution:

Before undertaking the selection process, the criteria for acceptable rivet material must be established. The metal must be plastically stable so as to prevent necking and should not fracture under actual operating conditions. The operating conditions are characterized by varying temperatures, perhaps from $-30°C$ to $30°C$, and high strain rates associated with the use of solid propellants. The strain rate may be in the range of 1.6×10^2 sec^{-1}, if a 3/4-in. diameter rivet is to be expanded to 7/8-in. in diameter during the propellant deflagration time of 10^{-3} sec. The flow stress of metals increases slightly with strain rate, which in turn affects the ductility of some and the ductile-brittle transition of others, such as steel. When necking occurs, the local strain rate may be much higher than the given value.

The metals to be considered are brass, commercially pure aluminum (1100-0), Monel 400, nickel 200, AISI 1141 steel, and AISI 1018 steel. These metals are chosen somewhat arbitrarily in order to establish the criteria for acceptable metals. From the commercial point of view, steel rivets are most desirable because of their low cost and high strength. In the application cited, the corrosion resistance may be equally important, in which case nickel, stainless steel, aluminum and the like may be more suitable.

The state of stress in the tube wall may be approximated to be that of a thin-walled tube under internal pressure, i.e.,

$$\sigma_{rr} = 0, \sigma_{\theta\theta} = \frac{pr}{t}, \sigma_{zz} = \frac{pr}{2t} \tag{a}$$

The strains at necking under this type of loading condition are linear functions of the work-hardening-rate exponent n. (See Example IV.7.) It has been shown that instability develops in a pressurized tube as a result of the hoop stress.

As shown in Figs. IV.29(a) and (b), which were obtained at low strain rates, the work-hardening-rate exponents of annealed metals are larger than those of cold-worked metals, indicating that annealed metals are better suited for rivet material, since plastic instability occurs at larger strains. The work-hardening rates of Monel and nickel are about the same, generally being higher than that of annealed aluminum. Stress-strain measurements of aluminum indicate that the work-hardening rate increases with strain rate [Fig. IV.29(c)]. Table IV.2 shows the values of n for various metals.

Figure IV.29(a) shows the stress-strain curves of AISI 1018 and AISI 1141 steel. In this case, also, the annealed metals are superior to the cold-worked material, and, from the plastic instability point of view, annealed medium-carbon AISI 1141 steel would appear to be quite satisfactory. Actually, however, at high strain rates it undergoes brittle fracture and is therefore totally unacceptable.

Experimental results obtained at room temperature showed that, in general, all rivets made of annealed metals expanded satisfactorily, except those of medium-carbon steel. Annealed brass rivets expanded uniformly and symmetrically, whereas those made of cold-worked brass reached plastic instability at small strains (see Fig. IV.30), with the necked region undergoing ductile fracture.* The strain rate at the necked region was much higher than estimated, since the flow of metals was limited to this region. The increase in flow stress due to the high strain rate must have been the cause of the ductile fracture at small strains. Aluminum rivets expanded satisfactorily, although a rivet overloaded with propellant ruptured. Note that in both of these cases the failure was caused by the hoop stress, as the theory predicts. Although annealed low-carbon AISI 1018 steel performed well, a rivet made of cold-worked AISI 1018 steel fractured in a brittle manner with little expansion. However, in the case of medium-carbon steel (AISI 1141), even the annealed rivets failed by brittle fracture. This is because, due to the relatively high carbon content, the steel becomes brittle at high strain rates even at room temperature.

Tests performed at the three low temperatures of $-15°F$, $-20°F$ and $-109°F$ showed that steel rivets are seriously liable to brittle fracture. Other metals, all of which had face-centered cubic structure, performed satisfactorily. Thus, it can be speculated that austenitic stainless steel, which has an f.c.c. structure, should also be a satisfactory rivet material. Stainless steels are particularly promising, since they generally have high work hardening rates.

*Fracture and rupture of metals are discussed in Chapters VIII and IX.

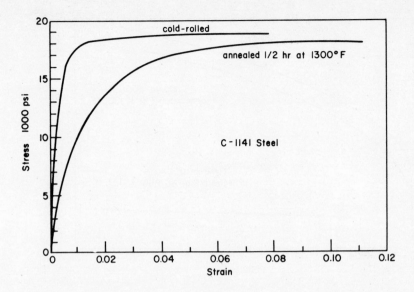

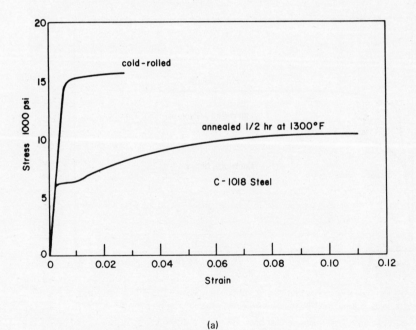

(a)

Fig. IV.29 Stress-strain curves for various materials considered for use in explosive rivets. (a) C-1141 and C-1018 steel.

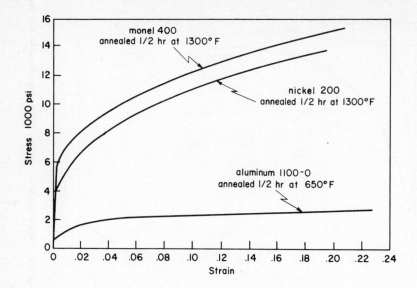

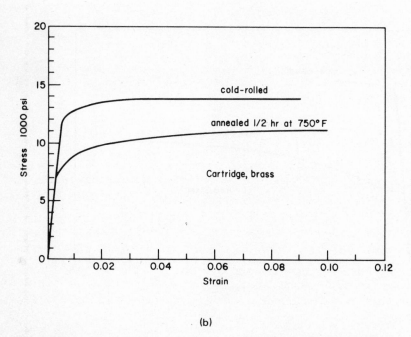

(b)

Fig. IV.29 (*Continued*)   Stress-strain curves for various materials considered for use in explosive rivets. (b) Various f.c.c. metals. Monel 400, nickel 200, aluminum 1100-0, and cartridge brass.

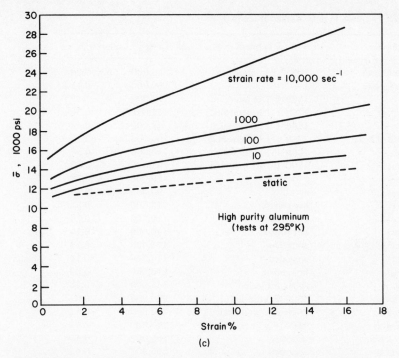

Fig. IV.29 (*Continued*) Stress-strain curves for various materials considered for use in explosive rivets. (c) The effect of strain rate on the work-hardening rate of high-purity aluminum. (*Data for Fig. IV.29(c) from F. E. Hauser et al.*, Rept. to Convair, *Series 133, No. 3, University of California, Berkley, 1960.*)

IV-7-e Buckling of Columns—Plastic Instability under Compression

In a previous section, the instability of materials under tensile loading conditions was considered. The instability was manifested in the form of localized deformation in the necked region. In this section, the instability under compressive loading conditions, normally called buckling, will be analyzed. In particular, the plastic buckling of a column will be considered.* The analysis of buckling problems for plates, shells, and other geometric shapes is much more complicated, but the general concept discussed in this section still holds true.

The analysis of an elastic column in Chapter III [Eq. (III.39)] shows that if the column is very short, the buckling load will be greater than the yield strength of the material. Furthermore, if the

*The elastic buckling of a column was treated in Chapter III, Sec. III-10.

TABLE IV.2—Some Typical Values of n and σ_0 *

Material	Treatment	σ_0	n
1100 aluminum	900°F 1 hr annealed	26,000	0.20
2024 aluminum	T-4	100,000	0.15
Copper	1000°F 1 hr annealed	78,000	1.19
Copper	1250°F 1 hr annealed	72,000	0.50
Copper	1500°F 1 hr annealed	68,000	0.48
70-30 leaded brass	1250°F 1 hr annealed	105,000	0.50
70-30 brass	1000°F 1 hr annealed	110,000	0.56
70-30 brass	1200°F 1 hr annealed	105,000	0.52
1002 steel	annealed	80,000	0.32
1018 steel	annealed	90,000	0.25
1020 steel	hot rolled	115,000	0.22
1212 steel	hot rolled	110,000	0.24
1045 steel	hot rolled	140,000	0.14
1144 steel	annealed	144,000	0.14
1144 steel	annealed	144,000	0.14
4340 steel	hot rolled	210,000	0.09
52100 steel	spher. annealed	165,000	0.18
52100 steel	1500°F annealed	210,000	0.07
18-8 stainless	1600°F 1 hr annealed	210,000	0.51
18-8 stainless	1800°F 1 hr annealed	230,000	0.53
304 stainless	annealed	185,000	0.45
303 stainless	annealed	205,000	0.51
202 stainless	1900°F 1 hr annealed	195,000	0.30
17-4 PH stainless	1100°F aged	260,000	0.01
17-4 PH stainless	annealed	173,000	0.05
Molybdenum	ext. annealed	105,000	0.13
Cobalt base alloy	solution H. T.	300,000	0.50
Cobalt base alloy	solution H. T.	300,000	0.50
Vanadium	annealed	112,000	0.35

*From G. Datsko, *Material Properties and Manufacturing Processes*, Wiley, 1966. n and σ_0 are constants in the equation $\bar{\sigma} = \sigma_0 \, (\bar{\epsilon})^n$.

material is elastic-perfectly plastic, i.e., nonworkhardening, the column will collapse once the yield stress is reached. This is illustrated in Fig. IV.31, which is a plot of the buckling (and collapse) stress vs. the slenderness ratio (L/k) based on Eq. (III.39). From the figure, it is seen that in region (A) elastic buckling is the failure mode, whereas in regions (B) and (C) the column will fail by yielding. Region (S) is the safe region, regardless of whether or not the material work hardens.

If the design considerations require the determination of the "exact" failure envelope for a work hardening material, plastic buckling must be analyzed. Plastic deformation can develop after the

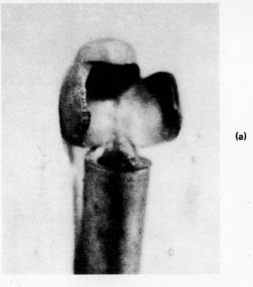

(a)

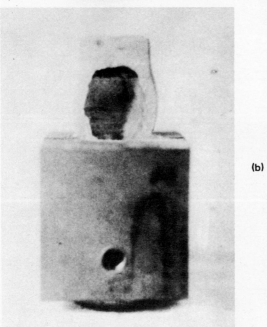

(b)

Fig. IV.30 Results of experiments on explosive rivets of various materials. (a) Brittle fracture of cold-drawn cartridge brass at room temperature [0.995 gm of double-base (d.b.) propellant]. Note the plastic instability before brittle fracture. (b) Uniform deformation of annealed cartridge brass at room temperature (0.795 gm of d.b. propellant). (*From N. P. Suh, Feasibility Study of Explosive Rivets, Rept. to USM Corp., University of South Carolina, September 1966.*)

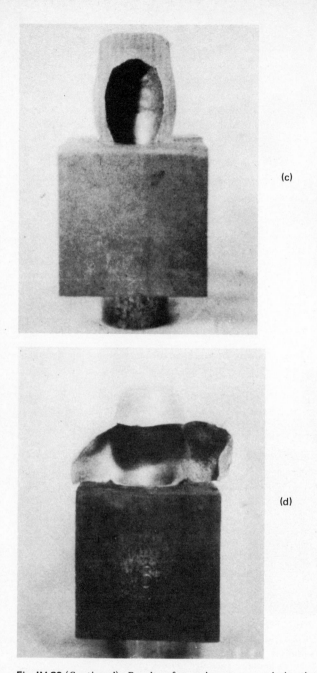

(c)

(d)

Fig. IV.30 (*Continued*) Results of experiments on explosive rivets of various materials. (c) Uniform deformation of annealed 1100-0 aluminum at room temperature (0.298 gm d.b. propellant). (d) Rupture failure of annealed 1100-0 aluminum at room temperature (0.472 gm d.b. propellant).

(e)

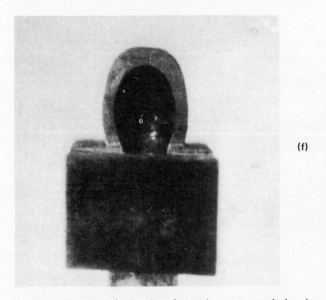

(f)

Fig. IV.30 (*Continued*)   Results of experiments on explosive rivets of various materials. (e) Brittle failure of C-1018 steel at room temperature (1.095 gm d.b. propellant). (f) Uniform deformation of spherical end rivet of annealed cartridge brass at room temperature (0.794 gm d.b. propellant).

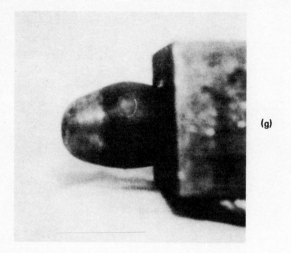

(g)

(h)

Fig. IV.30 (*Continued*) Results of experiments on explosive rivets of various materials. (g) Unsymmetrical deformation of annealed C-1018 steel at room temperature (0.995 gm d.b. propellant). (h) Brittle fracture of annealed C-1141 steel at room temperature (0.997 gm d.b. propellant). Note negligible expansion before fracture.

(i)

(j)

Fig. IV.30 (*Continued*) Results of experiments on explosive rivets of various materials. (i) Unsymmetrical expansion of annealed C-1018 steel at $-15°$F (0.998 gm d.b. propellant). (j) Brittle fracture of annealed C-1018 steel at $-109°$F (0.797 gm d.b. propellant).

elastic buckling has taken place, but it is of no concern, since, for engineering purposes, failure has already occurred. The plastic buckling failure of interest lies in region (C), where the stiffness of the material is less than in the elastic case, as a result of plastic deformation, and yet has a higher flow stress, as a result of work hardening.

In order to find the plastic buckling load, the applied load that can exist in equilibrium with the plastically bent column must be determined. If we isolate the column between the plastically buckled region and one of its ends, as shown in Fig. IV.32(a), and assume that the plastic deformation is confined to a particular cross section, the equilibrium conditions may be written as

$$P = P_1 \qquad Pu_2 = M \qquad (IV.65)$$

The expression for the bending moment has to be determined in terms of the stress distribution of the buckled cross section and is not as simple as in the elastic case because of the involvement of plastic deformation. The stress distribution may be represented as the sum of the stresses due to the axial load and those due to the bending moment, as shown in Fig. IV.32(b) and (c).

There are two possible loading cases: If the applied load remains constant during the bending process [Fig. IV.32(b)], one side of the column cross section will unload elastically, as shown, while the other side will deform further, plastically, due to the additional stress generated by the bending stress $\sigma_m{}^p$. If, however, the applied load continuously increases, so that the increase in the compressive strain due to the increased applied load is larger than the maximum tensile strain caused by bending, the entire cross section will remain plastically deformed without any elastic unloading [see Fig. IV.32(c)]. Since the stress-strain relationship is then the same on

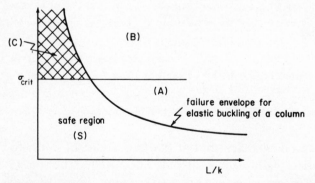

Fig. IV.31 Effect of the slenderness ratio L/k on the critical buckling load of a column.

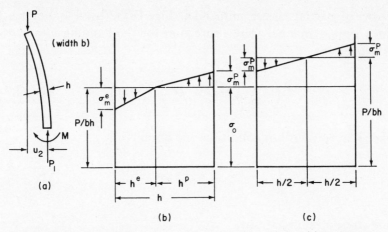

Fig. IV.32 Stress distribution in a plastically buckling column. (a) Equilibrium of a section of a column. (b) Stress distribution in a column when buckling causes part of the column to unload. (c) Stress distribution when the whole cross section undergoes plastic deformation.

both sides of the neutral axis, the latter remains at the same position as in the elastic case. The expression for the bending moment may be written as

$$M_0 = 2\left(\frac{2}{3}\right)\frac{h}{2}\left(\frac{1}{2}\sigma_m^{\,p}\,b\,\frac{h}{2}\right) = -\frac{E_t I}{R} = -\frac{d^2 u_2}{dx_1^{\,2}}E_t I \qquad \text{(IV.66)}$$

where I is the moment of inertia, R is the radius of curvature of the bent column, (assuming small deflections) and E_t, the slope of the stress-strain curve in the plastic range, is what is normally called the tangent modulus. Obviously, if the work hardening rate is a function of strain, E_t will vary throughout the cross section, and Eq. (IV.66) is not correct. However, since only small deformations are considered, E_t may be assumed to be constant for this material and may be determined either from a true stress-true strain curve or from an engineering stress-strain curve. Substituting Eq. (IV.66) into Eq. (IV.65) and solving for the critical buckling load in a manner similar to that for the elastic case, the buckling load is found to be

$$\sigma^p_{\text{buckling}} = \pi^2\,\frac{E_t}{(L/k)^2} \qquad \text{(IV.67)}$$

where k is the radius of gyration. These expressions are the same as for the elastic case, except that E is replaced by E_t.

The case of partial elastic unloading, Fig. IV.32(b), can be analyzed in a similar manner. The expression for the bending moment may be written as

$$M_0 = \frac{2}{3}\left(h^p \frac{1}{2} \sigma_m{}^p bh^p + h^e \frac{1}{2} \sigma_m{}^e bh^e\right) \qquad (IV.68)$$

From the axial equilibrium condition one obtains

$$\frac{1}{2}\sigma_m{}^p bh^p = \frac{1}{2}\sigma_m{}^e bh^e \qquad (IV.69)$$

From the stress-strain relations,

$$\sigma_m{}^p = E_t \epsilon_m{}^p = E_t\left(\frac{h^p}{R}\right)$$

$$\sigma_m{}^e = E\epsilon_m{}^e = E\left(\frac{h^e}{R}\right) \qquad (IV.70)$$

Substituting Eqs. (IV.68) and (IV.70) into Eq. (IV.65) and noting that $h = h^e + h^p$, the following is derived:

$$Pu_2 = \bar{E}\frac{I}{R} = \bar{E}I\frac{d^2 u_2}{dx_1{}^2}$$

$$(IV.71)$$

where
$$\bar{E} = \frac{4EE_t}{(\sqrt{E} + \sqrt{E_t})^2}$$

The parameter $\bar{E}$ is commonly referred to as the double modulus. Therefore, the buckling load becomes

$$P = \pi^2 \frac{\bar{E}I}{L^2} \qquad (IV.72)$$

Since $\bar{E}$ is greater than E_t, the double modulus theory predicts a higher buckling load than the tangent modulus theory. Experimental results correlate better with the latter.

IV-8 APPLICATIONS—ELASTOPLASTIC PROBLEMS

IV-8-a Introduction

When the extent of plastic deformation is small, both elastic and plastic deformations may have to be considered simultaneously in the analysis. Some of the numerous elastoplastic problems considered in the literature are the bending of sheets and beams, the deformation of vessels, the propagation of cracks in metals, the torsion of bars, and the rolling of sheets. The mathematical difficulties multiply rapidly in comparison with the fully plastic case discussed in the preceding section, and the available solutions are generally limited to the simpler problems.

Most cold-worked metals have residual stresses. This is caused by the elastoplastic nature of all postyield deformation processes. In general, metals are not uniformly deformed during processing, and the stress-strain relationship for the unloading cycle is different from that of the loading cycle. Under fatigue and impact loading conditions, compressive residual stresses at the surface usually play a beneficial role by preventing crack propagation. In this section, the bending and unloading of a wide plate will be considered to illustrate how residual stresses are generated and how "springback" occurs.

IV-8-b Bending and Unloading of a Wide Plate

The problem to be considered here is the bending of a wide plate of nonhardening material. The right-hand coordinate system (x_1, x_2, x_3) is chosen such that the x_1- and x_3-axes lie on the centroidal plane of the plate, the x_3-axis being directed along the width of the plate, as shown in Fig. IV.33. The x_2-axis is perpendicular to the plane of the plate. The plate is subjected to an externally applied bending moment $\mathbf{M} = -M_3\mathbf{k}$.

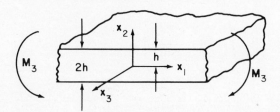

Fig. IV.33 Uniform bending of a wide plate.

At the onset of loading, the deformation is everywhere elastic, and the magnitude of strain is greatest at the outer surfaces. By symmetry, the neutral axis coincides with the centroid throughout the deformation process. As the bending moment continues to increase, the surface first reaches the yield condition. As the load is increased, the plastically deformed zone thickens and the elastic-plastic boundary moves in toward the netural axis. It will be assumed that the strain along the x_3-direction, ϵ_{33}, is equal to zero due to its large width, i.e., plane strain conditions prevail. This is a reasonable assumption, since transverse curvature (anticlastic curvature) is confined to the edges.* The strain along the x_1-direction is

$$-\epsilon_{11} = \frac{x_2}{R} \tag{IV.73}$$

By limiting attention to the case where $R \gg 2h$, the induced transverse stress along the x_2-direction may be neglected. From the stress-strain relation in the elastic region, we obtain

$$\sigma_{33} = \nu\sigma_{11} \quad (\text{from } \epsilon_{33} = 0)$$

$$\sigma_{11} = -\frac{Ex_2}{(1 - \nu^2)R} \tag{IV.74}$$

The Tresca yield criterion may be applied at $x_2 = h$ in order to find the incipient yield condition, i.e.,

$$\sigma_0 = (\sigma_{11})_{x_2 = h} = \frac{Eh}{(1 - \nu^2)R} \tag{IV.75}$$

At this point the applied bending moment per unit width is

$$M_3 = -\int_{-h}^{h} \sigma_{11}x_2 dx_2 = \int_{-h}^{h} \frac{Ex_2^2}{(1 - \nu^2)R} dx_2 = \frac{2}{3}\sigma_0 h^2 \tag{IV.76}$$

---

*If the plate were a narrow beam, the anticlastic curvature would be induced uniformly because of the Poisson effect.

As the bending moment is increased, the plastic zone at the surface grows inward toward the neutral axis. The distance from the x_1–x_3-plane to the elastoplastic interface will be denoted by c. The curvature is still governed by the elastic core, which may be determined from the Tresca yield criterion as

$$\sigma_0 = \frac{Ec}{(1 - \nu^2)R_c} \tag{IV.77}$$

The radius R_c is related to the radius of curvature R_y at the incipient yielding by the expression

$$\frac{R_c}{R_y} = \frac{c}{h} \tag{IV.78}$$

It should be noted at this time that σ_{33} in the plastic zone is

$$\sigma_{33} = \frac{1}{2}\sigma_{11} \tag{IV.79}$$

Therefore, unless $\nu = 1/2$, the stresses on the elastic and the plastic side may be discontinuous across the elastic-plastic boundary. The exact solution which takes this into account is not yet available. However, for this problem, the discontinuity does not influence the Tresca yield condition, if it is assumed that σ_{11} is not affected (i.e., σ_{33} is unchanged) at the elastoplastic transition.

The bending moment per unit width now becomes

$$M_3 = \int_{-c}^{c} \frac{Ex_2^2}{(1 - \nu^2)R_c} dx_2 + 2\int_{c}^{h} \sigma_0 x_2 dx_2 = \sigma_0\left(h^2 - \frac{1}{3}c^2\right) \tag{IV.80}$$

The maximum bending moment is obtained when $c = 0$, i.e.,

$$M_{max} = \frac{3}{2}M_Y \tag{IV.81}$$

which states that the maximum bending moment is 50% higher than the bending moment at the incipient yielding. However, the accuracy of Eq. (IV.81) is not clearly known, since the assumption that $R \gg x_2$ is violated at this state of bending. It should be noted that

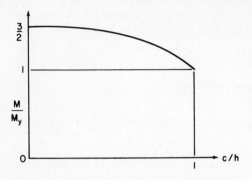

Fig. IV.34 Dimensionless bending moment M/M_y as a function of the thickness of the plastic zone.

these equations are only valid if the metal behaves in the same manner both under tension and compression. Figure IV.34 shows how the bending moment varies as a function of the thickness of the plastic zone, indicating that the plate can be bent far beyond the incipient yield point without suffering total collapse.

Let us assume that the plate was bent so that $c = h/2$. Then, the stress distribution takes the form of the light solid line in Fig. IV.35. At unloading, the process may be thought of as the superposition of a bending moment having a magnitude equal but opposite in direction to the original bending moment. The unloading moment imposes a purely elastic stress field, as shown by the dotted line in Fig. IV.35, with the peak unloading stress σ_u given by Eqs. (IV.80) and (IV.76), where σ_0 in Eq. (IV.76) is replaced by σ_u. The peak unloading stress is $11/8$ (σ_0) for the case where half the plate thickness is plastically deformed (i.e., $c = h/2$). The residual stress

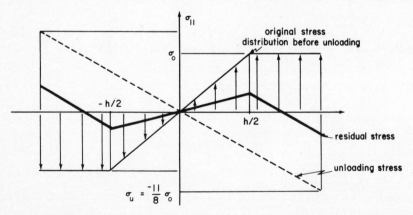

Fig. IV.35 Stress distribution in a wide plate subjected to bending before and after elastic unloading.

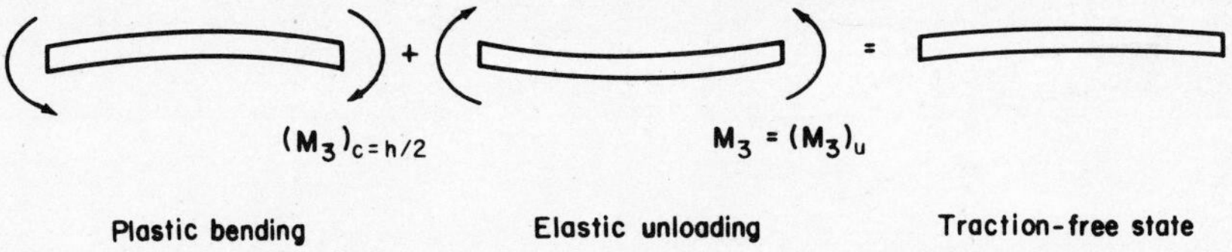

Plastic bending   Elastic unloading   Traction-free state

Fig. IV.36 A method of finding the distribution of residual stress in a plate which has been bent into the plastic state with a moment M_3 by superimposing an elastic stress distribution with resultant moment $-M_3$ on the original elastic-plastic stress distribution. This gives the stress in the moment-free unloaded state.

field which is obtained by adding these two stress fields is shown by the heavy solid line in the same figure. The process of imposing an elastic unloading moment is shown in Fig. IV.36.

The curvature of the plate after unloading can be determined by adding the curvatures associated with plastic bending and elastic unloading. This fact can readily be demonstrated by considering a position vector from the origin of a fixed coordinate system to the surface of a plate. Let **r** be this position vector. The first derivative of **r** with respect to the distance along the plate surface s is the tangent vector **t**, that is (see Fig. IV.37),

$$\frac{d\mathbf{r}}{ds} = \mathbf{t} \quad \text{where} \quad |\mathbf{t}| = 1 \tag{IV.82}$$

The second derivative of **r** with respect to s may be written as

$$\frac{d^2\mathbf{r}}{ds^2} = \frac{1}{R}\mathbf{n} \tag{IV.83}$$

where **n** is a unit vector normal to the tangent vector and the surface,* while R is the radius of curvature. As a result, curvatures in the same direction can simply be added algebraically. The curvature of the plate after unloading, $1/R$, is thus

$$\frac{1}{R} = \frac{1}{R_u} + \frac{1}{R_p} \tag{IV.84}$$

where R_u is the radius of curvature associated with elastic unloading, and R_p is the radius of curvature associated with plastic bending.

---

*This fact can easily be shown by differentiating $\mathbf{t} \cdot \mathbf{t} = 1$ with respect to s and realizing that the dot product of two orthogonal vectors is zero.

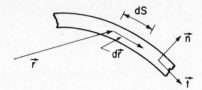

Fig. IV.37 Position, tangent, and normal vectors which define the shape of a bent plate.

IV-9 PLASTIC LIMIT ANALYSIS

To obtain an exact solution in continuum mechanics, three basic conditions must be satisfied: equilibrium; geometric compatibility; and the constitutive relationships of the continuum. The solutions given as applications in Sec. IV.7-b and IV.7-c were obtained without satisfying all these conditions and are, therefore, *approximate* solutions. As frequently happens, the complexity of the case compelled their use. Furthermore, in many practical applications, such as wire and cup drawing, the complete history of deformation is not required, and only the limiting load is of interest. Therefore, it will be very desirable to have a technique which approximates limiting solutions in a simple manner and at the same time sets bounds on the degree of approximation involved.

In this section, *limit theorems* will be presented, which will provide the basis for determining limiting solutions. The limiting solutions can be classified into *upper-bound* and *lower-bound* solutions. An upper-bound solution overestimates the true load while a lower-bound solution underestimates it. If both the upper and lower bounds can be established, the range of the true load is determined. When the difference between the upper- and lower-bound solutions is insignificant, the approximation provides answers which are very close to the true solutions. Although it is not always possible to determine both the upper- and lower-bound solutions, any limiting solution can be useful in a proper context. For example, in the materials processing field, the upper-bound solution can provide answers for stress (and power) which will be more than sufficient to perform a given task. On the other hand, in structural designs, the lower-bound solutions can provide a conservative estimate of the load-carrying capacity of a structure, and is thus much more desirable than the overestimated upper bound.

The theorems of limit analysis are developed for *rigid-plastic* materials. These theorems are similar to the theorem of minimum potential energy and the theorem of minimum complementary energy in elasticity. The upper- and lower-bound theorems will be simply stated here, and example problems will be discussed. The derivation of these theorems are given in Appendix IV-B for those interested. The limit theorems provide a powerful tool to engineers, and the importance of this technique cannot be overemphasized.

The Lower-Bound Theorem—The load corresponding to an assumed stress field which satisfies:

1) the equilibrium condition everywhere in the continuum
2) the yield condition
3) the stress boundary conditions

is always less than that corresponding to the true stress field. Therefore, the lower-bound to the limit load can be obtained by assuming a stress field which satisfies the above three conditions, totally neglecting the geometric compatibility condition.

The Upper-Bound Theorem—The actual work done in deforming a rigid-plastic continuum is always less than the work done by an assumed displacement which:

1) satisfies the displacement boundary condition
2) yields a strain field which satisfies the incompressibility condition.

Therefore, the upper bound to the limit load can be obtained by assuming a displacement field which satisfies the above two conditions and by equating the work done by the external agent with that done deforming the material along the assumed displacement field. The equilibrium condition is totally neglected in obtaining the upper-bound solution.

EXAMPLE IV.9—Simple Punch Identation Problem
(Idealized Hardness Test)

A semi-infinite solid is indented by a flat punch, as shown in Fig. IV.38, in a manner similar to the hardness test. The material may be assumed to be rigid-plastic with the shear yield stress k. Determine the upper- and lower-bound indentation loads. Assume that plane strain conditions prevail and that there is no friction between the indenter and the work piece.

Solution:

A) *Lower-Bound Solution*

In order to obtain the lower-bound solution, note that the material in region AB must yield and be displaced for the punch to move downward. Therefore, the state of stress existing in region AB may be represented by a compressive

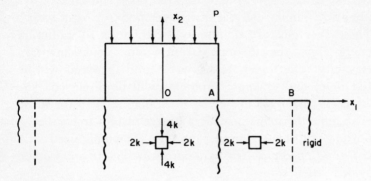

Fig. IV.38 Stress distribution for the lower-bound solution for the punch indentation problem.

stress with magnitude $\sigma_{11} = -2k$, which also satisfies the yield condition. Then, the state of stress in region OA must be such that it is in equilibrium with the state of stress in region AB and also satisfies the yield condition. This is the case when the state of stress is as shown in Fig. IV.38, i.e., $\sigma_{11} = -2k$ and $\sigma_{22} = -4k$. Therefore, the punch must exert pressure p given by

$$p \geq 4k \tag{a}$$

B) *Upper-Bound Solution*

The assumed displacement field is as shown in Fig. IV.39. The material shears only along the shear displacement lines shown in the figure. The shear stress that needs to be overcome along these displacement lines is the critical shear stress k. As the punch P moves downward, three blocks (A, B, and C) of the work material are displaced, always in contact with each other, but sliding along the rigid boundaries shown in Fig. IV.39. The assumed displacement field satisfies the condition of incompressibility.

If the relative displacement per unit time between the rigid portion of the work material and block A is denoted by q_{AR}, etc., and its interface area by

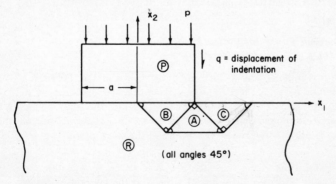

Fig. IV.39 Displacement field for the upper-bound solution of the punch indentation problem.

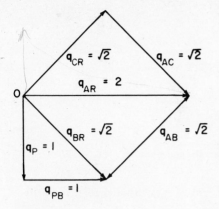

Fig. IV.40 Hodograph of the displacement field for the punch indentation problem shown in Fig. IV.39.

A_{AR}, etc., the work done by the punch can be related to the work done in deforming the work material. The work performed by the punch is

$$pqa = W_p \qquad \text{(b)}$$

where a is one-half of the cross-sectional area of the punch. The work done in deforming the material is

$$k(A_{AR}q_{AR} + A_{AB}q_{AB} + A_{BR}q_{BR} + A_{AC}q_{AC} + A_{CR}q_{CR}) = W_d \qquad \text{(c)}$$

Equating Eqs. (b) and (c), the pressure p can be determined. The areas are simply determined from the geometric relations.

The major task now is the determination of the relative velocity between the different blocks themselves, between the blocks and the rigid portion R of the work material, and between the blocks and the punch. The velocities, i.e., the displacements per unit time, are shown in Fig. IV.40 in the form of a vector diagram commonly known as a hodograph. The velocities must satisfy the requirements of compatibility and consequently the velocity vectors always form a closed loop. Compatibility requires that rigid blocks adjacent to the rigid region must slide parallel to the slip lines between the blocks and the rigid material. Similarly, the relative velocities between blocks themselves are always parallel to the slip lines between blocks.

In the construction of a hodograph, all the absolute velocities are shown to emanate from a single stationary point, i.e., point O in Fig. IV.40, which represents the entire rigid section. Since the punch is moving downward, a vector representing q_p is drawn downward. Then, recognizing that block B has to move parallel to the slip line between B and R, the absolute velocity of B is drawn from the stationary point O along the direction shown. Similarly, the absolute velocities of blocks A and C are drawn from O. The magnitudes of these velocities are determined by setting the magnitude of q_p arbitrarily to unity. The absolute velocity of block B, q_{BR}, is determined by drawing a horizontal line from the end of the vector q_p until it intersects q_{BR}, because the relative velocity between the punch and block B has to be horizontal. The magnitude of the absolute velocity q_{AR} is similarly determined by drawing a

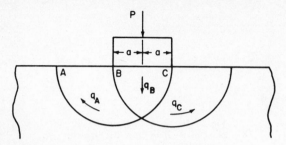

Fig. IV.41 Alternative displacement field for an upper-bound solution of the punch indentation problem.

relative velocity vector q_{AB} from the end of q_{BR} parallel to the slip line between blocks A and B.

Substituting these values into Eq. (c) and setting the latter equal to Eq. (b), we obtain the upper-bound solution

$$p \le 6k \tag{d}$$

Therefore, the true indentation pressure lies between the two limiting values given by Eqs. (a) and (d):

$$4k \le p \le 6k \tag{e}$$

The exact solution, obtained using the slip line method, gives $P = 5.14k$.

The upper-bound solution could have been obtained much more readily by assuming the displacement field to be a semicircle, as shown in Fig. IV.41. Using this displacement field, the solution is then found to be $p \le 6.28k$.

The indenters used in the standard hardness tests are not flat but shaped in the form of balls, pyramids, and cones. In some of the tests, e.g., the Rockwell hardness test, the depth rather than the area of indentation is measured. In the Brinell, Vickers, and Knoop hardness tests the hardness is given in terms of the average load per unit area, where the unit is typically in kg/mm^2. In all of these tests, plane strain conditions are inoperative. However, the indentation loads for the plane strain case are not substantially different from, although always less than, the axisymmetric case. Experimentally, the indentation hardness in a typical test is found to range from 6.2 to 6.4k.[*]

EXAMPLE IV.10—Extrusion of Strips Through a Die

The extrusion of a strip through a die is considered. Since the axisymmetric problems are difficult to handle, the extrusion of a wide sheet is once more considered so that plane strain approximations can be made, i.e., $\epsilon_{zz} = 0$. Only the upper-bound solution is desired.

Solution:

The assumed displacement field is shown in Fig. IV.42, and the hodograph is shown in Fig. IV.43. Note again that the absolute velocities always emanate from the stationary point O in the hodograph along directions dictated by the stationary surfaces. The material in a corner A cannot move, i.e., it is "dead

*For a more detailed discussion of hardness tests, see Ref. 1.

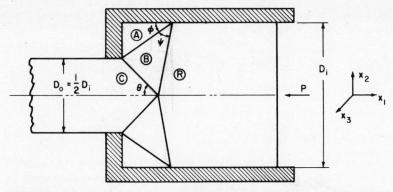

Fig. IV.42 Displacement field for an upper-bound solution of the strip extrusion problem.

metal." The exit velocity of the material, q_C, is determined from the continuity and incompressibility conditions. It can be seen that, depending on the values assumed for θ and ϕ, the value of P will vary. The optimum values of θ and ϕ are those that give minimum P, since the upper-bound solution is the one being sought. These optimums could be found by defining P in terms of the unknowns θ and ϕ and determining the stationary values of P at which P is a minimum. However, in order to simplify the algebra, we set $\theta = \phi = \pi/4$. The magnitude of q_R may be taken to be any arbitrary value, say 1.

From the hodograph, q_C is seen to be related to q_{BA} and q_{BC} through the expressions

$$q_C = 2 = q_{BA} \cos\phi + q_{BC} \cos\theta$$

and
$$q_{BA} = q_{BC} \tag{a}$$

The two equations above yield

$$q_{BA} = q_{BC} = \sqrt{2} \tag{b}$$

Similarly, q_{BR} is obtained as

$$q_{BR} = 1 \tag{c}$$

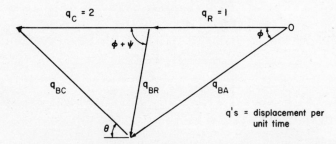

Fig. IV.43 Hodograph of the displacement field for the strip extrusion problem shown in Fig. IV.42.

The work done by the external agent may now be equated to the work done during plastic deformation as follows:

$$P \cdot 1 \cdot 1 = k\left(\frac{\sqrt{2}}{2} q_{BA} + \frac{\sqrt{2}}{2} q_{BC} + 1 q_{BR}\right)$$

or $P \leq 3k$ (d)

EXAMPLE IV.11—Forging of Metals

During the forging of a metal in a closed die, the metal is squeezed a number of times to bring the work to the desired shape. The final operation squeezes the excess metal of the rough forging into a flash through the space between the closed dies. The pressure necessary to squeeze the flash through the die opening is a good estimate of the pressure required to perform the entire final step in the forging operation. That final step of the closed die forging is illustrated in Fig. IV.44. Assuming a frictionless die, determine the forging load F by finding an upper-bound solution. The work material behaves like a rigid-perfectly plastic solid. A possible deformation field is shown.

Solution:

The hodograph can be constructed as shown in Fig. IV.45, noting that the block B moves downward as well as horizontally. Again, all of the absolute-velocity vectors emanate from point O, and the relative velocity vectors have the same direction as the slip lines. After the magnitude and direction of q_P are arbitrarily drawn at O, the directions of q_A and q_D can be established. The magnitude of q_A is determined by drawing q_{AC}. The

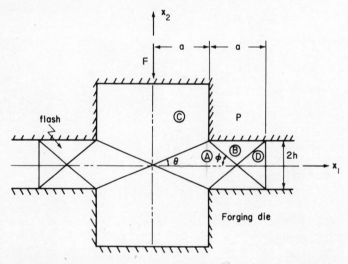

Fig. IV.44 Displacement field for an upper-bound solution of a forging operation.

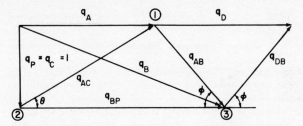

Fig.IV.45 Hodograph corresponding to the displacement field for the forging operation shown in Fig. IV.44.

magnitude of q_B is found by drawing q_{AB} from point 1 and q_{BP} from point 2. Vector q_D is determined by drawing q_{DB} from point 3.

By symmetry, only the first quadrant of the $x_1 - x_2$-space needs to be examined. The condition of incompressibility gives

$$q_D = \frac{2a}{h} q_P \tag{a}$$

which is also derivable from the hodograph. The rate of work done by the external force F on the first quadrant is

$$\frac{F}{2} \cdot q_P = W_P \tag{b}$$

The rate of deformation work done is

$$k(q_{AC} A_{AC} + q_{AB} A_{AB} + q_{BD} A_{BD}) = W_D \tag{c}$$

where

$$q_{AC} = \frac{q_P}{\sin \theta} = \frac{q_P}{h/\sqrt{a^2 + h^2}} \qquad A_{AC} = \sqrt{a^2 + h^2}$$

$$q_{AB} = q_{BD} = \frac{q_P}{h/\sqrt{(a/2)^2 + h^2}} \qquad A_{AB} = A_{BD} = \sqrt{\left(\frac{a}{2}\right)^2 + h^2} \tag{d}$$

Equating Eqs. (b) and (c), we obtain

$$F \leq 2kh\left\{\left[\left(\frac{a}{h}\right)^2 + 1\right] + 2\left[\left(\frac{a}{2h}\right)^2 + 1\right]\right\}$$

$$\leq 6kh\left[\frac{1}{2}\left(\frac{a}{h}\right)^2 + 1\right] \tag{e}$$

Note that the forging force is a sensitive function of the ratio (a/h).

REFERENCES

1. McClintock, F. A., and A. S. Argon: "Mechanical Behavior of Materials," Addison-Wesley, Reading, Mass., 1966.
2. Cook, N. H.: "Manufacturing Analysis," Addison-Wesley, Reading, Mass., 1966.
3. Thomsen, E. G., C. T. Yang, and S. Kobayashi: "Mechanics of Plastic Deformation in Metal Processing," Macmillan, New York, 1965.
4. Mendelson, A.: "Plasticity: Theory and Application," Macmillan, New York, 1968.
5. Hill, R.: "Mathematical Theory of Plasticity," Oxford at Clarendon Press, London, 1960.
6. Johnson, W., and P. B. Mellor: "Plasticity for Mechanical Engineers," Van Nostrand, London, 1962.
7. Drucker, D. C.: "Introduction to the Mechanics of Deformable Solids," McGraw-Hill, New York, 1967.
8. Naghdi, P. M.: Stress-Strain Relations and Thermoplasticity, *Plasticity, Proc. 2d Sympos. Naval Struct. Mech., ONR Structural Mechanics Series*, Pergamon Press, New York, 1960.
9. Eirich, F. R.: "Rheology, Theory and Application," vol. 1, Academic Press, New York, 1956.
10. Shanley, F. R.: "Strength of Materials," McGraw-Hill, New York, 1957.

PROBLEMS

IV.1   Show that the equivalent strain is equal to the uniaxial strain in a tensile test.

IV.2   Derive the expressions for true strains given in Eqs. (IV.2), (IV.3), and (IV.4).

IV.3   A cylindrical bar, 6 in. in diameter, is uniformly deformed plastically to a final shape of 4 in. in diameter and 6 ft long. Determine: a) the original dimension of the bar; and b) the true strains and the equivalent strain, under the assumption that the strain ratios remain constant during the entire deformation process.

IV.4   A thin-walled hollow aluminum tube, shown on the next page, is deformed plastically as follows:

    a) It is first deformed axially to make $L = 21$ in. and $d = 10$ in.

    b) It is then deformed radially to make $d = 11$ in. while L remains 21 in.

    c) It is compressed to make $L = 20$ in. while the diameter remains 11 in.

    d) Finally, the diameter is reduced to 10 in. while L is held at 20 in.

Determine the wall thickness at the end of each incremental loading, and the total equivalent strain. Do you think that the equivalent strain can be used to indicate the degree of cold working? Why? Is the state of the metal the same before and after the deformation?

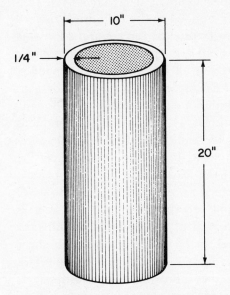

IV.5 Suppose a state of stress is given by the following stress tensor:

$$\sigma_{ij} = \begin{bmatrix} \sigma_{11} & \sigma_{12} & 0 \\ \sigma_{21} & \sigma_{22} & 0 \\ 0 & 0 & \sigma_{33} \end{bmatrix} = \begin{bmatrix} 5{,}000 & 2{,}000 & 0 \\ 2{,}000 & 3{,}000 & 0 \\ 0 & 0 & 10{,}000 \end{bmatrix}$$

Determine the maximum shear stress and the plane on which the maximum shear stress lies. If the critical shear stress for yielding is 20,000 psi, find the value of σ_{33} which will cause yielding when the other stress components remain the same.

IV.6 Consider the state of plane stress given by the following stress tensor:

$$\sigma_{ij} = \begin{bmatrix} 0 & 0 & 0 \\ 0 & 8{,}000 & 5{,}000 \\ 0 & 5{,}000 & 15{,}000 \end{bmatrix}$$

If the critical shear stress for plastic deformation is 8,000 psi, is the given state of stress sufficient to cause plastic deformation?

IV.7   A 10-in.-long, 2-in.-diam steel cylinder is covered by a thin layer of lead which is firmly bonded to the cylindrical surface

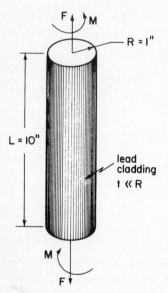

of the rod. An axial force F and a twisting moment M are applied to the cylinder such that, at all times during the loading, $M = G_s FR/2E_s$, where R is the radius of the cylinder. You may make the following assumptions: a) The force and moment carried by the lead are negligible compared to the force and moment in the steel; b) the lead is rigid perfectly plastic, i.e., there is no elastic strain and no work hardening; c) the strain in the lead is uniform; and d) the Young's modulus for the steel is $E_s = 3 \times 10^7$ psi, the shear modulus $G_s = 1.1 \times 10^7$ psi, and $\nu = 1/3$. Answer the following two questions: a) As the force is increased from F to $F + dF$ (and the moment is increased from $G_s FR/2E$ to $G_s(F + dF)R/2E$), what are the plastic strain increments in the lead? b) What is the stress distribution in the lead at this time?

IV.8   A thin-walled tube made from 1100 aluminum with a stress-strain curve given by

$$\bar{\sigma} = 26,000 \, (\bar{\epsilon}^p)^{0.2} \, \text{psi}$$

is to be work hardened by alternately stretching and compressing it. While it is being deformed it is placed over a mandrel which keeps its diameter constant at all times. If it is stretched until its length is increased by 10% and then compressed to its original length on each cycle, how many cycles are required to raise the flow stress to 20,000 psi? Assume uniform strain throughout the cold-working process.

IV.9   The aluminum of Problem IV. 4 has an experimentally determined stress-strain curve approximated by

$$\bar{\sigma} = 15,000 + 20,000\,(\bar{\epsilon})^{0.25}$$

If the tube of Problem IV.4 is loaded by twisting after it undergoes the deformation described in the statement of the problem, determine the twisting moment at which further plastic deformation will occur.

IV.10 The yield stress of most metals depends on the grain size. The relationship between the yield stress σ_Y and the grain size D is given by

$$\sigma_Y = \sigma_i + K/\sqrt{D}$$

where σ_i is commonly known as the friction stress, and K is a proportionality constant. One way of reducing the grain size is to cold work the metal and then recrystallize it. To increase the grain size, the metal is simply annealed.

A customer of Franklin Steel Corp. orders annealed 316 stainless steel tube, specifying a yield strength of 45,000 psi. However, the 316 stainless steel tube in stock at F. S. C. has a yield strength of only 35,000 psi. One of the engineers suggests that they can cold work the material by simply twisting the tubes over a mandrel and then by recrystallizing them, reduce the grain size and obtain an increased yield strength. Is this feasible?

The tubing is 12 in. long with a 3-in. ID and a 1/16-in.-thick wall. The stress-strain relationship for the steel is given by

$$\bar{\sigma} = 28,000 + 220,000\,(\bar{\epsilon})^{1/2}\,\text{psi}$$

If the tube has to be twisted by one complete turn, what is the maximum twisting moment required to do the job?

IV.11 A precision pressure vessel, shown below, is built of hot-rolled 1020 steel. During final inspection, it is discovered that the two lines from the chamber are misaligned by 5°, and it is decided that a torque will be applied to the ends in order to deform the vessel into alignment.

a) How much torque must be applied to the ends in order to correct this problem?

b) If the vessel is accidentally overloaded under pressure, at what pressure will it begin to deform plastically?

Regard the vessel as an ideal thin-walled tube with closed ends and neglect elastic strains (you might begin to think about how to handle elastic strains in such a problem). The experimental stress-strain curve for the steel is approximated by

$$\bar{\sigma} = 40,000 + 200,000 (\bar{\epsilon})^{1.0} \text{psi}$$

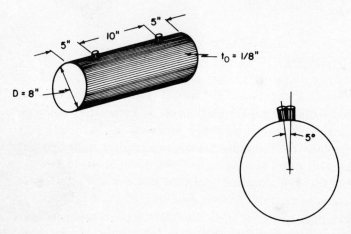

misalignment (looking along the axis)

IV.12 A thin-walled tube with *open* ends is made of soft aluminum with a stress-strain curve approximated by

$$\bar{\sigma} = 22,000 (\bar{\epsilon})^{0.25}$$

The tube is initially pressurized to 200 psi. An axial force of 9,000 lb is then applied to the pressurized tube.
a) What are the dimensions after the tube is pressurized?
b) What is the state of loading as F is applied? How does further deformation affect this state of loading?
c) What happens as the force increases from 0 to 9,000 lb? Does the tube load further into the plastic regime or does it begin to unload at first?
d) As F increases, make an approximation which will allow you to directly integrate the governing Prandtl-Reuss equations. Discuss why this approximation is necessary.

e) Carefully outline your solution to this problem and discuss the effects of the approximations which you have made. (If you wish, find the final dimensions of the tube. Again, neglect elastic deformation.)

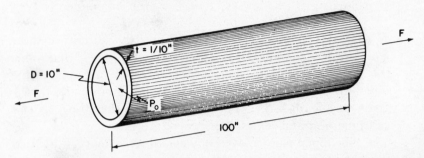

IV.13 Consider a thin-walled tube made of hot-rolled AISI 1020 steel. We approximate the equivalent stress-equivalent strain curve by two moduli. The elastic modulus E is 30×10^6 psi and the tangent modulus in the plastic regime, E_t, is approximated to be 200,000 psi; the yield stress is 40,000 psi. Poisson's ratio is 0.3, and the shear modulus is 11.5×10^6 psi. The tube has an initial OD of 10″, length of 100″, and wall thickness of 1/8″. The initial state of loading consists of a compressive axial force of 120,000 lb and an internal pressure of 750 psi. A twisting moment in 800,000 in.-lb is then applied to the ends of the tube.

a) Calculate the initial increments of the plastic strain as the yield point is reached, and determine the magnitude of the twisting moment which will initiate yield.

b) Calculate the final dimensions of the tube after the loads are relaxed, making whatever approximations are convenient and valid.

c) Determine the final angle of twist of the tube.

d) If a tensile axial load is applied to the tube after the other loads have been relaxed, determine the maximum load (lb) which can be applied without the occurrence of further yielding.

IV.14 A flat plate of soft brass 60 in. long, 20 in. wide, and 3 in. thick, is pulled with uniform tension across its width. The material is constrained to maintain its original width while it is stretched to a final length of 70-in. Assuming the stress-strain curve to be

$$\bar{\sigma} = 100{,}000\,(\bar{\epsilon})^{0.5} \text{ psi}$$

a) What is the work required to perform this operation?
b) If the deformation takes place with the end stretching at a velocity of 1 in./sec, what is the maximum power requirement? (Neglect elastic deformation.)

IV.15 A disk of a rigid-perfectly plastic material with critical shear strength k is compressed between rigid flat plates. Determine the force required to deform the material in terms of the dimensions h and r. Assume that Coulomb friction acts on the faces of the disk, i.e., $F_r = \mu F_n$. Use the slab (ring, in this case) method of analysis.

   If the upper plate moves downward at velocity V, discuss the horsepower requirement to compress the disk in terms of $V, k, r, h,$ and μ.

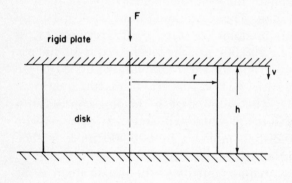

IV.16 Consider a long thin-walled tube with closed ends. The tube is made of soft aluminum, and the equivalent stress-equivalent strain curve is approximated by

$$\bar{\sigma} = 22{,}200\,(\bar{\epsilon})^{0.25} \text{ psi}$$

The tube is loaded by internal pressure only. What is the value of $\bar{\epsilon}$ at which instability occurs? (Don't forget that this is a biaxially loaded situation.) If the tube originally has a 4-in. diameter with a 1/16-in. wall thickness, at what pressure will necking occur? (Instability occurs when the applied *load* is a maximum.)

IV.17 Determine the necking condition of a specimen subjected to uniaxial tension with a superimposed hydrostatic pressure P.

IV.18 A strip of aluminum is to be drawn through a die in order to reduce its thickness from 1 in. to 7/16 in. The width of the strip is 12 in. The dimensions of the die are shown below. Estimate the drawing force required, if the yield stress of the metal is 10,000 psi. The coefficient of friction is 0.05.

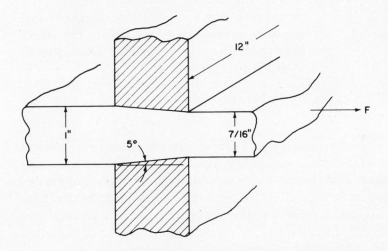

IV.19 It is proposed to roll square threads on the circumference of a large-diameter shaft of normalized 1045 steel. As a step toward making an estimate of the forces and torques required, determine the force per unit area required to press a square thread of the shape shown in the figure below. Assume plane strain. Assume that the yield strength of the steel at large strains is 130,000 psi.

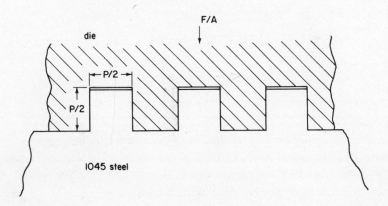

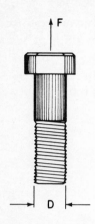

IV.20 Consider the bolt shown at left. Its root diameter is given by D, and it has n threads per inch. The material of the bolt is assumed to be sufficiently ductile that brittle fracture will not occur as a result of the stress concentration of the threads. How many threads must be engaged into a tapped hole to ensure that the threads will strip off before the shank fails under the applied load F? (Use approximate limit solutions.)

IV.21 A plane strain model of a die for extruding tubing is shown in the sketch below. The die is well lubricated so you may assume that there is no friction between the work and the die. The tensile yield stress for the work material is 5,000 psi. Give an upper bound for the force required to extrude the material. Identify any necessary angles in your deformation field and give the answer in terms of these angles.

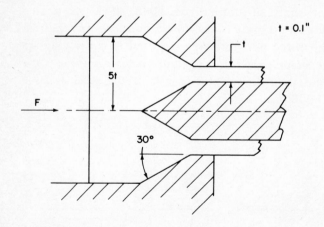

IV.22 What is an upper bound to the force necessary to compress an ingot of a material with critical shear stress k? Assume that this is a problem of plane strain, i.e., $\epsilon_{33} = 0$. Give your answer in terms of the height h and the width w.

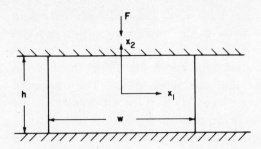

IV.23 A V-groove is to be rolled into a sheet of material with critical shear strength k. As a first approximation to the work required to perform this operation, find the load required to drive a pointed indenter with included angle 2θ into the material as a function of the depth of indentation x. Propose two or three slip line fields and comment briefly about the merits of each. Choose one and solve for the required load.

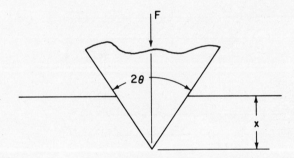

IV.24 A notched specimen is subjected to a bending moment as shown. Determine the limiting bending moments if the yield stress $(Y = 2k)$ is 30,000 psi. (Find both an upper and a lower bound.)

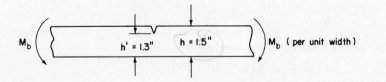

IV.25 Determine a lower-bound solution for the extrusion problem of Example IV.10 of the text.

IV.26   Figure a illustrates the backward (or indirect) extrusion process, while Fig. b illustrates the forward extrusion process. The major difference between the two processes is that the forward extrusion process requires overcoming the friction force between the die and the work piece. The backward extrusion process is free of the friction force between the die and the work piece, but it requires a long, hollow ram.

The backward extrusion process is to be analyzed. Assuming that both the work piece and the ram behave as rigid-perfectly plastic metals, and that the flow stress of the ram material is three times the flow stress of the work piece, determine the

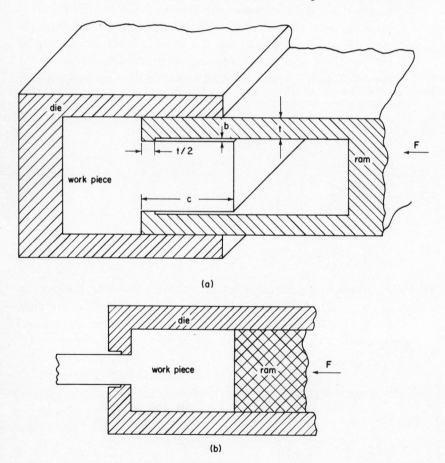

(a)

(b)

maximum extrusion ratio at a given wall thickness of the ram. In order to do this, find the appropriate upper- or lower-bound solution for the ram and for the work piece at the stage shown in Fig.a. Assume that a sticking condition prevails between the die and the work piece.

IV.27 A wide flat plate is to be bent over a template to a constant curvature, as shown below. Determine the maximum bending moment that has to be applied to the plate. You may neglect the elastic strain during the loading.

The stress-strain relationship of the plate material is given by

$$\bar{\sigma} = 40{,}000 + 10^5 (\bar{\epsilon})^{1.0} \text{ psi}$$

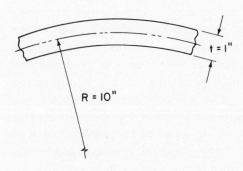

Determine the residual stress remaining in the plate after it is unloaded. (Elastic modulus $E = 30 \times 10^6$ and Poisson's ratio $\nu = 0.29$.)

IV.28 In a coining operation, a die is pressed into a blank of material so that material is forced up into the cavities in the die to give the profile required. Some material is also forced out of the open edges of the die. Consider the simple plane-strain model of the coining process shown in the figure and estimate, in terms of the shear strength k of the work material, the pressure P which must be applied to the die to perform the first increment of the deformation. A sticking condition exists at the die-blank interface.

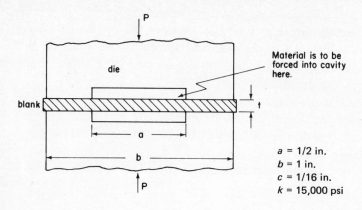

a = 1/2 in.
b = 1 in.
c = 1/16 in.
k = 15,000 psi

IV.29 In the backward extrusion process shown in the figure below, the pointed punch is forced into a block of material contained within a rigid die. The material displaced by the advancing punch is forced backward along the sides of the punch. At the stage of the process shown, find an upper bound for the force F which must be applied to the punch to continue the extrusion. Due to the high pressures developed during the extrusion, the real junction area where the work contacts the wall of the die or punch is the same as the apparent area of contact. Express your answer in terms of the shear strength of the material, k. The dimension along the direction perpendicular to the plane of the figure is much larger than a or b.

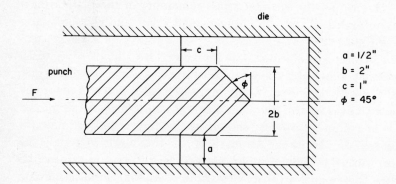

a = 1/2"
b = 2"
c = 1"
ϕ = 45°

IV.30 Extension and twisting are to be imparted to a thin-walled tube by a device which causes the tube to extend such that the change in length is proportional to the change in angle of twist $(dl = C\,d\theta_T)$. The tube is constrained by a mandrel coated with "frictionless" Teflon, such that the radius of the tube remains constant during the operation. The tube is strained until the angle of twist is 45°. The proportionality constant C is such that the tube extends 4 in. for each revolution of twist. Write an expression for the state of stress in the tube as a function of the twist angle θ_T, if the stress-strain curve of the material is

$$\bar{\sigma} = (4 \times 10^4 + 10^5 \bar{\epsilon}^p)\,\text{psi}$$

What are the final values of the stresses? The tube is 10″ long × 2″ in diameter × 0.040″ thick.

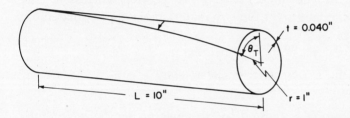

IV.31 A student designed a safety device which relieves the pressure in a utility line when the pressure exceeds the rated critical value. The device consists of a thin-walled tube which is inserted between two rigid walls and held in place by means of rubber gaskets. The rubber gaskets are compressed from their original thickness of 0.060 in. to 0.030 in. during the installation of the tube. The device relieves the pressure when the gaskets come loose. Assuming that the stiffness of the rubber is small and the axial force is negligible, determine the maximum pressure that the safety device can withstand. You may assume that the tube will not reach plastic instability. The tube is made of aluminum, whose stress-strain relationship is given by

$$\bar{\sigma} = 15{,}000 + 12{,}500\,\bar{\epsilon}^p\,\text{psi}$$

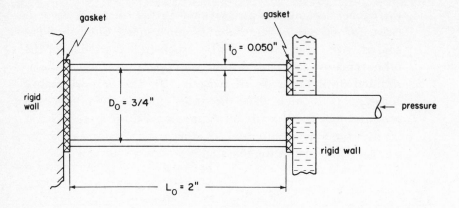

IV.32 A composite chemical transport pipe is made of a killed steel outer layer and a stainless steel inner layer. The metals have been bonded together during a drawing operation. The purpose of the inner core is to prevent corrosion. The stress-strain curves for the two materials are approximated by:

$$\bar{\sigma}_1 = 90,000\,\bar{\epsilon}_1^{0.2} \text{ psi (steel)}$$

$$\bar{\sigma}_2 = 220,000\,\bar{\epsilon}_2^{0.5} \text{ psi (stainless steel)}$$

1) If the tube is twisted by an angle ψ per 10 in. of tube length, give an expression for the moment required as a function of ψ. Neglect elastic deformation during plastic loading.
2) If the tube is twisted to a final value of 45°/10-in. length, what is the maximum twisting moment M_t?
3) If the tube is unloaded elastically, sketch the residual stress distribution. Neglect deformation changes during elastic deformation.

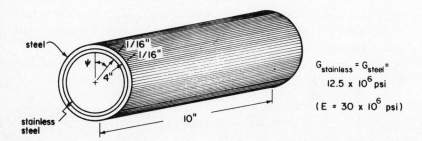

$G_{stainless} = G_{steel} = 12.5 \times 10^6 \text{ psi}$

$(E = 30 \times 10^6 \text{ psi})$

4) If the tube is loaded by internal pressure (the tube is modeled as having closed ends), at what pressure will it begin to yield and where will this occur? Include prior deformation in this calculation.

IV.33 A support device for a large impact forging machine is shown below. A cost and weight conscious designer has suggested that material be eliminated from the original casting in the manner shown.

1) Your job is to tell him how much material he can remove and still be *certain* that the device's load carrying ability is not changed. Use limit analysis and assume that, initially, the possible failure modes are "mushrooming" of the loading joint (A) or shearing of the flange ears (B).

2) You may as well tell him which of the two "normal" failure modes is more probable by comparing the limit loads calculated above.

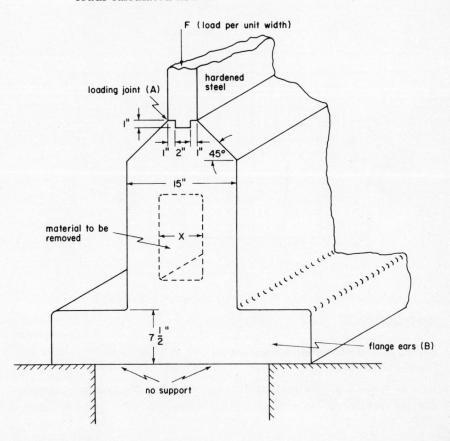

Neglect friction and assume that the metal is not likely to fail by fatigue or brittle fracture. The support member is made of mild steel with critical shear stress k, and the loaded arm is of a much higher strength steel (i.e., it will not fail).

IV.34 A new mechanical bonding system has been devised. It incorporates casting of loose-fitting T-slots into the two pieces to be joined. A punch is applied to the surface of the "male" piece in order to set the joint (close the gaps shown).

1) In terms of the critical shear stress of the material, what is the load necessary to apply to the punch in order to cause the necessary deformation?

2) Recommend a critical shear stress for the punch to ensure no failure during the setting of the joint.

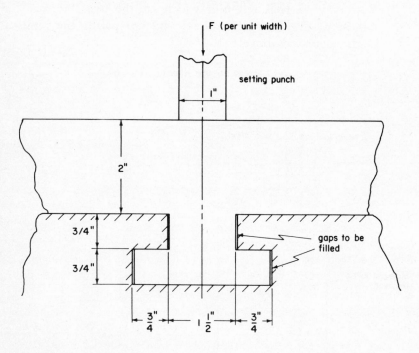

APPENDIX IV-A

Derivation of the Constitutive Relations for Plastic Deformation

In this appendix, the constitutive relations for the plastic deformation of metals will be derived from the *principle of maximum plastic work*. From this derivation, the necessary convexity of the yield surface and normality of the plastic strain on the yield surface will be demonstrated. At the base of all constitutive relations is empirical observation, e.g., Hooke's law, and the same holds true for plastic deformation. Thus, although our discussion begins with the principle of maximum plastic work, which is based on a postulate, it could well be developed from the Mises criterion or the Prandtl-Reuss stress-strain relationships, which have their origin in empirical observations.

1) *Principle of Maximum Plastic Work*

The principle will be derived from Drucker's postulate:[†] "If a load increment is slowly applied to, and removed from, an element which is in some state of stress, the work done by the added stress, during the application and removal of the added stress, is positive, if plastic deformation has occurred during the cycle." If the initial state of stress is denoted by σ_{ij}^*, which is inside the yield curve, and the final state of stress, corresponding to the plastic strain rate $\dot{\epsilon}_{ij}^p$, is denoted by σ_{ij}^f, the principle of maximum plastic work may be stated as

$$W^p = \int_{t_1}^{t_2} \sum_{\substack{i=1 \\ j=1}}^{3} (\sigma_{ij}^f - \sigma_{ij}^*)\dot{\epsilon}_{ij}^p \, dt \geq 0 \tag{A1}$$

where t_1 is the moment at which the state of stress satisfies the yield condition for the first time, $(t_2 - t_1)$ is the time increment during which the load is continuously applied, and $t = t_2$ is the point at which the load is finally removed. At $t > t_2$, the state of stress is finally brought back to σ_{ij}^*. Since unloading takes place elastically and since all the elastic work done is recovered, Eq. (A1) represents only the work done during plastic deformation by the added increment of stress. Equation (A1) is satisfied if

[†]Drucker, D. C., "Stress-Strain Relations in Plastic Range of Metals—Experiments and Basic Concepts," *Rheology, Theory and Applications*, Vol. 1, Ed. by F. R. Eirich, Academic Press, 1956. Also see P. M. Naghdi, "Stress-Strain Relations and Thermoplasticity," in *Plasticity*, Proc. of Second Symposium on Naval Structural Mechanics, ONR Structural Mechanics Series, Pergamon Press, 1960.

$$\sum_{\substack{i=1 \\ j=1}}^{3} (\sigma_{ij}{}^{f} - \sigma_{ij}^{*})\dot{\epsilon}_{ij}{}^{p} \geq 0 \qquad\qquad\text{(A2)}$$

2) *Convexity of the Yield Surface and Normality of the Plastic Strain Vector Upon It*

Consider a yield surface, given by $f(\sigma_{ij}) = k$, which is regular (i.e., without any discontinuity) in the principal-stress space $(\sigma_{\mathrm{I}}, \sigma_{\mathrm{II}}, \sigma_{\mathrm{III}})$. On a plane perpendicular to the direction $(1/\sqrt{3}, 1/\sqrt{3}, 1/\sqrt{3})$ in the principal-stress space the normal stress is equal to the first stress invariant, and the shear stress (or deviator stress), which lies on this plane, is proportional to the distance from this direction vector. If the principal-strain space $(\epsilon_{\mathrm{I}}, \epsilon_{\mathrm{II}}, \epsilon_{\mathrm{III}})$ is superimposed upon the principal-stress space $(\sigma_{\mathrm{I}}, \sigma_{\mathrm{II}}, \sigma_{\mathrm{III}})$ such that the corresponding axes, i.e., σ_{I} and ϵ_{I}, σ_{II} and ϵ_{II}, coincide along the direction vector $(1/\sqrt{3}, 1/\sqrt{3}, 1/\sqrt{3})$, then the first invariant is equal to the elastic volume change, and the shear (or deviator) strain again lies on the plane perpendicular to the direction vector at a distance from the direction vector proportional to the maximum resultant shear strain.

Consider an element which is loaded such that the state of stress satisfies the yield condition, i.e., it lies on the surface $f = k_0$. If an incremental load $d\sigma_{ij}$ is applied to the element at equilibrium, the element will deform at an incremental plastic strain rate of $\dot{\epsilon}_{ij}{}^{p}$, as shown in Fig. IV.A1. The strain rate vector $\dot{\epsilon}_{ij}{}^{p}$ will point outward

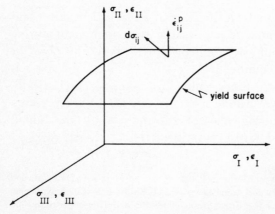

Fig. IV.A1 Plastic strain rate increment $\dot{\epsilon}_{ij}{}^{p}$ perpendicular to the yield surface, which results from an increment of load $d\sigma_{ij}$.

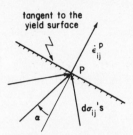

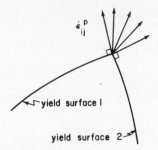

Fig. IV.A2 Plane passed through a point P on the yield surface perpendicular to the strain rate vector, separating stress increments which give positive plastic work increments from those which give negative ones.

Fig. IV.A3 Yield surface containing a corner. The strain rate vector at the corner can lie anywhere in the angle defined by the normals of the adjacent portions of the surface.

from the surface, if σ_{ij} is such that $f(\sigma_{ij} + d\sigma_{ij}) > k_0$. Note that there are many different combinations of $d\sigma_{ij}$ that can initiate the plastic deformation, since in the elastic regime the final state of stress does not depend on the path of loading. If a plane is passed perpendicular to the strain rate vector at its end (point P in Fig. IV.A2), all possible incremental stresses must lie on one side of the plane, as shown in Fig. IV.A2, since the angle between $d\sigma_{ij}$ and $\epsilon_{ij}{}^p$ must be acute, if the plastic work done is to be positive; otherwise, the principle of maximum plastic work would be violated. Furthermore, the plane must be tangent to the yield surface $f = k_0$, as otherwise, some of the stress vectors would be on the wrong side of the tangent plane. The plane can be tangent to the yield surface only if the yield surface is convex.

If the yield surface is not regular but has discontinuities, e.g., the Tresca yield surface, at the point of discontinuity the strain rate vector* (or the incremental strain) must lie between two vectors normal to the yield surfaces meeting at the point as shown in Fig. IV.A3.

3) *Associated Flow Rule (Normality Flow Rule)*

It was shown in the preceding section that the strain rate vector is always normal to the yield surface. Based on this fact the stress strain relationship will be developed.

---

*Strain and stress are not vectors but second-order tensors. However, on a given plane one may treat them as vectors for the purpose of the discussion here, provided no attempt is made to transform them to some other coordinate system using the rules of vector transformation (i.e., the rules for first-order tensor transformation).

Consider a yield surface given by

$$f\left[(\sigma_{ij})_1, (\sigma_{ij})_2, (\sigma_{ij})_3\right] = k = \text{constant} \qquad (A3)$$

in a Cartesian coordinate system defined by the axes $(\sigma_{ij})_1$, $(\sigma_{ij})_2$ and $(\sigma_{ij})_3$. The function f is a scalar function. The gradient of the scalar function f is

$$\nabla f = \frac{\partial f}{\partial (\sigma_{ij})_1}\, i + \frac{\partial f}{\partial (\sigma_{ij})_2}\, j + \frac{\partial f}{\partial (\sigma_{ij})_3}\, k \qquad (A4)$$

where i, j, and k are unit vectors along the $(\sigma_{ij})_1$-, $(\sigma_{ij})_2$-, and $(\sigma_{ij})_3$-axes. If a vector $[\sigma_{ij}]$ is drawn from the origin to the point P on the yield surface at which the strain rate vector is located (Fig. IV.A4), it may be represented as

$$[\sigma_{ij}] = (\sigma_{ij})_1 i + (\sigma_{ij})_2 j + (\sigma_{ij})_3 k \qquad (A5)$$

The change of $[\sigma_{ij}]$ along any direction is given by

$$d[\sigma_{ij}] = d(\sigma_{ij})_1 i + d(\sigma_{ij})_2 j + d(\sigma_{ij})_3 k \qquad (A6)$$

Taking the dot product of f with $d[\sigma_{ij}]$ yields

$$\nabla f \cdot d[\sigma_{ij}] = \frac{\partial f}{\partial (\sigma_{ij})_1}\, d(\sigma_{ij})_1 + \frac{\partial f}{\partial (\sigma_{ij})_2}\, d(\sigma_{ij})_2 + \frac{\partial f}{\partial (\sigma_{ij})_3}\, d(\sigma_{ij})_3$$

$$(A7)$$

Equation (A7) is identical with df. Since on a yield surface $df = 0$, it follows that Eq. (A7) is equal to zero. Equation (A7) states that ∇f is

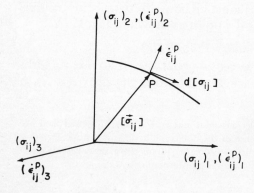

Fig. IV.A4 Orientation of the yield surface and the stress and strain increment vectors in stress-strain space.

orthogonal to $d[\sigma_{ij}]$. Since $d[\sigma_{ij}]$ lies on the plane $f = k$, then ∇f must be perpendicular to the yield surface at P. Since the strain rate vector $\dot{\epsilon}_{ij}{}^p$ is also perpendicular to the yield surface, $\dot{\epsilon}_{ij}{}^p$ and ∇f must be parallel to each other. The strain rate vector also has three components along the corresponding axes, i.e.,

$$[\dot{\epsilon}_{ij}{}^p] = \left(\dot{\epsilon}_{ij}{}^p\right)_1 \mathbf{i} + \left(\dot{\epsilon}_{ij}{}^p\right)_2 \mathbf{j} + \left(\dot{\epsilon}_{ij}{}^p\right)_3 \mathbf{k} \tag{A8}$$

Therefore, comparing Eqs. (A4) and (A8), it is seen that the corresponding strain and $\partial f/\partial(\sigma_{ij})$ components must be proportional to each other, if the condition of normality of the strain rate vector to the yield surface is to be satisfied. That is,

$$\dot{\epsilon}_{ij}{}^p = \dot{\lambda} \frac{\partial f}{\partial \sigma_{ij}} \tag{A9}$$

where $\dot{\lambda}$ is a proportionality constant which is always positive and which may depend on stress, strain, loading history, and strain rate. Integrating Eq. (A9) with respect to time, we can write

$$d\epsilon_{ij}{}^p = d\lambda \frac{\partial f}{\partial \sigma_{ij}} \tag{A10}$$

Equation (A10) cannot readily be integrated to obtain the total strain, since λ depends on stress, strain, loading history, and strain rate. It should also be noted that the equation $f = k$ is valid for the whole loading curve (which is a locus of successive yield points) beyond the incipient yield point, as k increases with work hardening. Therefore, Eqs. (A7) and (A10) are valid for a work-hardening material. Equations (A9) and (A10) are alternate statements of the *associated flow rule*.

Equation (A10) indicates that the choice of a yield criterion (loading function) dictates the exact form of the stress-strain relationship. Consider the Mises criterion, i.e.,

$$f = \bar{\sigma} = -\sqrt{-3J_2'} = \left(\frac{3}{2} \sum_{\substack{i=1 \\ j=1}}^{3} \sigma_{ij}' \sigma_{ij}' \right)^{1/2} \tag{A11}$$

Partial differentiation of f with respect to σ_{ij} yields

$$\frac{\partial f}{\partial \sigma_{ij}} = \frac{3}{2} \frac{\sigma'_{ij}}{\bar{\sigma}} \tag{A12}$$

Substituting Eq. (A12) into Eq. (A10), one obtains

$$d\epsilon_{ij}{}^p = \frac{3}{2} \frac{d\lambda}{\bar{\sigma}} \sigma'_{ij} \tag{A13}$$

Along the x_1-axis, Eq. (A13) takes the form

$$d\epsilon_{11}{}^p = \frac{3}{2} \frac{d\lambda}{\bar{\sigma}} \left[\frac{2}{3} \sigma_{11} - \frac{1}{3} (\sigma_{22} + \sigma_{33}) \right]$$

$$d\sigma_{12}{}^p = \frac{3}{2} \frac{d\lambda}{\bar{\sigma}} \sigma_{12} \tag{A14}$$

$$d\epsilon_{13}{}^p = \frac{3}{2} \frac{d\lambda}{\bar{\sigma}} \sigma_{13}$$

Under the uniaxial loading condition along the x_1-axis, Eq. (A14) becomes

$$\bar{\sigma} = \sigma_{11}$$
$$d\epsilon_{11}{}^p = d\lambda = d\bar{\epsilon}^p \tag{A15}$$

Along the x_1-, x_2-, and x_3-axes, Eq. (A13) may then be written as

$$d\epsilon_{11}{}^p = \frac{d\bar{\epsilon}^p}{\bar{\sigma}} \left[\sigma_{11} - \frac{1}{2} (\sigma_{22} + \sigma_{33}) \right]$$

$$d\epsilon_{22}{}^p = \frac{d\bar{\epsilon}^p}{\bar{\sigma}} \left[\sigma_{22} - \frac{1}{2} (\sigma_{11} + \sigma_{33}) \right]$$

$$d\epsilon_{33}{}^p = \frac{d\bar{\epsilon}^p}{\bar{\sigma}} \left[\sigma_{33} - \frac{1}{2} (\sigma_{22} + \sigma_{11}) \right]$$

$$d\epsilon_{12}{}^p = \frac{3}{2} \frac{d\bar{\epsilon}^p}{\bar{\sigma}} \sigma_{12} \tag{A16}$$

$$d\epsilon_{13}{}^p = \frac{3}{2} \frac{d\bar{\epsilon}^p}{\bar{\sigma}} \sigma_{13}$$

$$d\epsilon_{23}{}^p = \frac{3}{2} \frac{d\bar{\epsilon}^p}{\bar{\sigma}} \sigma_{23}$$

Equations (A16), which are consequences of the normality flow rule, are almost identical to Hooke's law, except that the proportionality constant $(d\bar{\epsilon}^p/\bar{\sigma})$ replaces $1/E$, and Poisson's ratio is replaced by 1/2 for the plastic case. The most significant difference is that in the plastic case the stress can be related only to the incremental strains, whereas in the elastic case it is related to the total strain. Equations (A16) are known as the Prandtl-Reuss equations, or the proportionality flow rule. If the strain is very large, or the material is rigid-plastic, the plastic strains are nearly equal to the total strains. In this case, it may be assumed that

$$d\epsilon_{ij}{}^p = d\epsilon_{ij} \tag{A17}$$

When Eq. (A17) is assumed to be valid, Eq. (A16) are then referred to as the Levy-Mises equations.* If, in those cases where only the principal stresses are involved, the Tresca yield criterion is used in place of the Mises yield criterion, the resulting stress-strain relationship is much simplified. However, the Prandtl-Reuss equations are more versatile in that they can be used in all cases.

APPENDIX IV-B

Proof of Theorems of Limit Analysis[†]

1) *The Lower-Bound Theorem (Statically Admissible Solution)*

Consider an element enclosed by a surface S. The boundary conditions on the surface are specified in terms of the velocity components on a portion of the surface, S_V and in terms of the surface stress vector $\mathbf{F}_i$ on the remainder of the surface, S_F. The surface stress vector $\mathbf{F}_i$ is defined as the sum of the three components of $\sigma_{ij}\nu_j$, where ν_j is the direction vector normal to the surface and pointing outward.

The actual stress field will be denoted by σ_{ij}, and any arbitrary stress field that satisfies the equilibrium condition and the boundary condition on S_F is denoted by σ_{ij}^*. The rate of work done by the external load is

---

*It should be noted that the simultaneous use of the Levy-Mises equations and the Mises yield criterion in solving a given problem is inconsistent, according to the principle of maximum plastic work. However, simplicity recommends it.

[†]For an alternate proof of these theorems, see Ref. 1, p. 364.

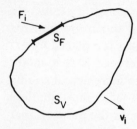

Fig. IV.B1 Boundary conditions on a body, expressed in terms of the displacements of the boundary on the surface S_v and the tractions applied to the boundary on the surface S_F.

$$\dot{W} = \int_S \sum_{i=1}^{3} F_i v_i \, ds = \int_S \sum_{i=1}^{3} \sum_{j=1}^{3} \sigma_{ij} \nu_j v_i \, ds$$

(B1)

where v_i is the actual velocity field associated with the stress field. Application of the divergence theorem to Eq. (B1) yields*

$$\dot{W} = \int_V \sum_{i=1}^{3} \sum_{j=1}^{3} (\sigma_{ij} v_{i,j} + \sigma_{ij,j} v_j) \, dV \quad (B2)$$

Since the equilibrium condition requires that $\sigma_{ij,j} = 0$, Eq. (B2) may be simplified to

$$\dot{W} = \int_V \sum_{i=1}^{3} \sum_{j=1}^{3} \sigma_{ij} v_{i,j} \, dV$$

(B3)

In terms of the deviator and hydrostatic stress components, Eq. (B3) may be rewritten as

$$\dot{W} = \int_V \sum_{i=1}^{3} \sum_{j=1}^{3} \left(\sigma'_{ij} + \frac{1}{3} \delta_{ij} \sum_{k=1}^{3} \sigma_{kk} \right) v_{i,j} \, dV$$

$$= \int_V \sum_{i=1}^{3} \sum_{j=1}^{3} \left(\sigma'_{ij} v_{i,j} + \frac{1}{3} \sum_{k=1}^{3} \sigma_{kk} v_{i,i} \right) dV$$

(B4)

The conservation of mass condition requires that $v_{i,j} = 0$, since the material is incompressible. The velocity field $v_{i,j}$ may be written in terms of the strain rate tensor $\dot{\epsilon}_{ij}$ and rotation rate tensors $\dot{\omega}_{ij}$ as

$$v_{i,j} = \frac{1}{2}(v_{i,j} + v_{j,i}) + \frac{1}{2}(v_{i,j} - v_{j,i})$$

$$= \dot{\epsilon}_{ij} + \dot{\omega}_{ij}$$

(B5)

Note that $\dot{\omega}_{ij}$ is a skew-symmetric tensor, where $\dot{\epsilon}_{ij}$ and σ_{ij} are symmetric tensors. Note also that the product of a symmetric tensor

---

*The notation $\sigma_{ij,j} = \partial\sigma_{ij}/\partial x_j$, while $v_{i,j} = \partial v_i/\partial x_j$. Commas denote partial differentiation with respect to the coordinate axis indicated by the subscript following the comma.

and a skew-symmetric tensor equals zero. Upon the substitution of Eq. (B5), Eq. (B4) may be rewritten as

$$\dot{W} = \int_V \sum_{i=1}^{3} \sum_{j=1}^{3} \sigma'_{ij} \dot{\epsilon}_{ij} \, dV = \int_S \sum_{i=1}^{3} F_i v_i \, ds \tag{B6}$$

A similar expression may be obtained using an assumed stress:

$$\dot{W}^* = \int_V \sum_{i=1}^{3} \sum_{j=1}^{3} \sigma'^*_{ij} \dot{\epsilon}_{ij} \, dV = \int_S \sum_{i=1}^{3} F_i^* v_i \, ds \tag{B7}$$

where F_i^* is the unknown boundary condition in equilibrium with the assumed stress field σ_{ij}^*. Since F_i is known on S_F, Eq. (B7) may be written as

$$\dot{W}^* = \int_{S_v} \sum_{i=1}^{3} F_i^* v_i \, ds + \int_{S_F} \sum_{i=1}^{3} F_i v_i \, ds = \int_V \sum_{i=1}^{3} \sum_{j=1}^{3} \sigma'^*_{ij} \dot{\epsilon}_{ij} \, dV \tag{B8}$$

Similarly, Eq. (B6) may be written as

$$\dot{W} = \int_{S_v} \sum_{i=1}^{3} F_i v_i \, ds + \int_{S_F} \sum_{i=1}^{3} F_i v_i \, ds = \int_V \sum_{i=1}^{3} \sum_{j=1}^{3} \sigma'_{ij} \dot{\epsilon}_{ij} \, dV \tag{B9}$$

Subtracting Eq. (B8) from Eq. (B9), we obtain

$$\dot{W} - \dot{W}^* = \int_{S_v} \sum_{i=1}^{3} (F_i - F_i^*) v_i \, ds = \int_V \sum_{i=1}^{3} \sum_{j=1}^{3} (\sigma'_{ij} - \sigma'^*_{ij}) \dot{\epsilon}_{ij} \, dV$$
$$\tag{B10}$$

Applying the principle of maximum plastic work, we have

$$W \geq W^* \tag{B11}$$

Equation (B11) states that the actual work done is always greater than that produced by any assumed stress field. It should be noted that the statically admissible lower-bound solution must satisfy: a) *equilibrium conditions*; b) *boundary conditions*; and c) the *yield condition*.

2) *The Upper-Bound Theorem (Kinematically Admissible Solution)*

Consider again the same element discussed in the previous section. This time the equilibrium condition will be disregarded and consideration given only to the deformation, including slip along an internal slip surface S_D. Let v_i be the true velocity field and v_i^* the assumed velocity field.

The rate of work done by the true stress and velocity field is derived as before:

$$\dot{W} = \int_S \sum_{i=1}^3 F_i v_i \, ds = \int_S \sum_{i=1}^3 \sum_{j=1}^3 \sigma_{ij} \nu_j v_i \, ds$$

$$= \int_V \sigma'_{ij} \dot{\epsilon}_{ij} \, dV + \int_{S_D} k \, |v_D| \, dS_D \tag{B12}$$

where v_D is the absolute value of the relative velocity on S_D, and k is the critical shear stress. The above expression is similar to Eq. (B6), except in the very last term, which incorporates the work done on internal slip planes. Similarly the rate of work done by the assumed velocity field is

$$\dot{W}^* = \int_S \sum_{i=1}^3 F_i v_i^* \, ds = \int_V \sum_{i=1}^3 \sum_{j=1}^3 \sigma'_{ij} \dot{\epsilon}_{ij}^* \, dV + \int_{S_D} k \, |v_D^*| \, dS_D \tag{B13}$$

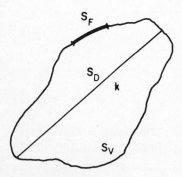

Fig. IV.B2 Boundary conditions on a body, specified in terms of the displacements of the boundary on the surface S_v and the tractions applied to the boundary on the surface S_F. Slip is allowed to occur along some internal surface S_D with a shear stress k acting across the slip surface.

The principle of maximum plastic work states that if the deviator stress field, corresponding to an assumed strain rate field $\dot{\epsilon}_{ij}^*$ (for the assumed velocity field v_i^*), is $\sigma_{ij}'^*$, then

$$\int_V \sum_{i=1}^3 \sum_{j=1}^3 (\sigma_{ij}'^* - \sigma_{ij}') \dot{\epsilon}_{ij}^* \, dV \geq 0 \tag{B14}$$

Substituting Eq. (B14) into Eq. (B13), we obtain

$$\dot{W}^* = \int_S \sum_{i=1}^3 F_i v_i^* \, ds \leq \int_V \sum_{i=1}^3 \sum_{j=1}^3 \sigma_{ij}^* \dot{\epsilon}_{ij}^* \, dV + \int_{S_D} k \, |v_D^*| \, dS_D \tag{B15}$$

Since $v_i^* = v_i$ on S , Eq. (B15) may be rewritten as

$$\int_{S_V} \sum_{i=1}^3 F_i v_i \, ds + \int_{S_F} \sum_{i=1}^3 F_i v_i^* \, ds \leq \int_V \sum_{i=1}^3 \sum_{j=1}^3 \sigma_{ij}^* \dot{\epsilon}_{ij}^* \, dV$$
$$+ \int_{S_D} k \, |v_D^*| \, dS_D \tag{B16}$$

Equation (B16) can be transformed to

$$\int_{S_V} \sum_{i=1}^3 F_i v_i \, ds \leq \int_V \sum_{i=1}^3 \sum_{j=1}^3 \sigma_{ij}^* \dot{\epsilon}_{ij}^* \, dV + \int_{S_D} k \, |v_D^*| \, dS_D - \int_{S_F} \sum_{i=1}^3 F_i v_i^* \, ds$$
$$= \int_{S_V} \sum_{i=1}^3 F_i^* v_i^* \, ds = \int_{S_V} \sum_{i=1}^3 F_i^* v_i \, ds \tag{B17}$$

The right-hand side of the equation represents the work done by the unknown stress field which corresponds to the assumed velocity field on S_V. Equation (B17) states that the actual work done on S_V by the true stress and velocity fields is always less than that done by the assumed velocity field. In the limiting case of $S_F = 0$, it is clear from

Eq. (B17) that the actual work done over the entire surface is also always less than the work done by the assumed velocity field and its corresponding stress field.

The upper-bound solutions can be obtained if the assumed displacement field satisfies the velocity boundary conditions and the condition of incompressibility. The equilibrium conditions and the yield condition need not be satisfied.

CHAPTER FIVE

Microscopic Basis of Plastic Behavior

V-1 INTRODUCTION

In Chapters III and IV a number of assumptions were made regarding the nature of elastic and plastic deformations in order to develop a continuum treatment of the deformation processes for isotropic materials. In the present chapter the validity of these assumptions will be examined in terms of the microscopic behavior of metals based on dislocation theory. First, the basic aspects of dislocation theory are reviewed. Then, the validity of ideal elasticity and plasticity are examined. The relationship between dislocation motion and the macroscopic stress-strain curve is discussed, and the effects of metallurgical variables such as grain size, impurity content, and phase structure on the dislocation motion and on the stress-strain curve are considered.

The purpose of these discussions is to give the reader an understanding of the underlying physical mechanisms which are

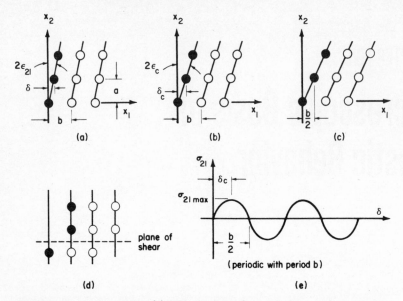

Fig. V.1 Shear of a crystal. (a) Shear is elastic for small shear strains such that the displacement $\delta \ll b$ or a. (b) Stress reaches a maximum for some critical shear strain $\epsilon_c (\delta < b/2)$. (c) The lattice is in an unstable equilibrium for $\delta = b/2$. (d) This configuration is again a zero-stress equilibrium, but the first and second planes have been displaced by a distance b. (e) The stress-displacement diagram must be periodic with period b as shown, but the function need not be sinusoidal.

responsible for plastic behavior. The materials presented in this chapter should enable the reader to understand the limitations of the assumptions made in deriving the phenomenological constitutive relations presented in Chapters III and IV.

V-2 HISTORY OF DISLOCATIONS

The discovery that x-rays could be diffracted by metals established the crystalline nature of their structure. Once the structure was established, it was possible to attempt theoretical calculations of the shear strength of metals and compare this with the shear strengths measured experimentally. The most significant work in this area is that of Frenkel [J. Frenkel, Z. *Phys.*, 37:752 (1926)]. As Frenkel proposed, if two planes of atoms are sheared relative to each other, the atoms are first displaced from their equilibrium positions elastically. At some point, such as that shown in Fig. V.1(b), the resistance to displacement reaches a maximum. At the displacement shown at (c) the resistance drops again to zero. Finally, at (d) the atoms are again in equilibrium but displaced by a distance b from their initial positions. The force resisting this shearing process must be periodic with b as shown in Fig. V.1(e). Since the force displacement curve for small displacements must give the shear modulus, the maximum force can be calculated if the exact shape of the force displacement curve is known. Calculations using various forms of the force displacement law give theoretical values for the maximum shear stress, σ_{max}, ranging from G/30 to G/5.

EXAMPLE V.1—Theoretical Shear Strength for a Sinusoidal
Force Rule

Assuming that the resistance to shearing is sinusoidal with period b, calculate the shear strength of a crystal with shear modulus G and spacing between shear planes a.

Solution:

The assumed sinusoidal nature of the shear strength may be expressed as

$$\sigma_{21} = \sigma_{max} \sin \frac{2\pi\delta}{b} \tag{a}$$

where σ_{max} is the maximum shear strength shown in Fig. V.1. By definition the shear strain ϵ_{21} is equal to $\delta/2a$ as shown in Fig. V.1. Therefore (a) may be written as

$$\sigma_{21} = \sigma_{max} \sin \frac{4\pi\epsilon_{21}a}{b} \tag{b}$$

Since the shear strain is related to the shear stress by $2G$ for small strains

$$\epsilon_{21} = \frac{\sigma_{21}}{2G} \tag{c}$$

For small displacements, Eq. (b) can be linearized,

$$\sigma_{21} \simeq \sigma_{max} \frac{4\pi\epsilon_{21}a}{b} = 2G\epsilon_{21} \tag{d}$$

so that

$$\sigma_{max} = G \frac{b}{2\pi a} \simeq \frac{G}{6} \tag{e}$$

Therefore, the ideal shear strength of metals is about one-tenth of the shear modulus.

At the time of Frenkel's work, the best experimental values of crystal shear strengths ranged from 10^{-3} to 10^{-4} G. This discrepancy was clearly too large to be explained away by the choice of the force-displacement relationship. It was therefore clear that a new model was required to explain the extremely low shear strengths of real metals in comparison with the theoretical strength. Obviously, the reason for the large strength calculated from the Frenkel model results from the necessity of displacing all of the atoms on a plane simultaneously. Recognition of the possibility for a lattice defect which would allow the displacement to occur over only part of the plane at a time led three men, Orowan, Polanyi, and Taylor (1934), to propose the existence of the lattice defect now called the edge dislocation. Finally, in 1939, Burgers added a description of what is known as the screw dislocation.

V-3 BASIC DISLOCATION THEORY

V-3-a Dislocations, Burgers Vectors, and Glide Planes

When slip occurs over an entire plane of a crystal simultaneously, the perfect crystal structure is restored after each increment b of sliding has occurred. However, when slip occurs over only part of a plane in a crystal, the crystal structure in the vicinity of the boundary between the slipped and unslipped regions of the plane cannot be perfect. Therefore, this boundary between slipped and unslipped regions must be some type of crystal defect and for convenience it is called a dislocation. The nature of dislocations can be illustrated pictorially by considering partially sheared single crystals as shown in Fig. V.2. When the boundary between slipped and unslipped portions of a crystal plane is perpendicular to the direction of slip, as in Fig. V.2(a) an extra half-plane of atoms is created perpendicular to the slip plane above or below the boundary of the slipped region. This is called an edge dislocation. The arrangement of atoms around an edge dislocation is shown in Fig. V.3. When the boundary of the slipped region of a plane is parallel to the direction of shear, a screw dislocation exists at the boundary as illustrated in Fig. V.2(b).

It should be noted that these dislocations can only be created under the influence of the shear stress. The motion of the dislocations is primarily controlled by shear stresses and, therefore,

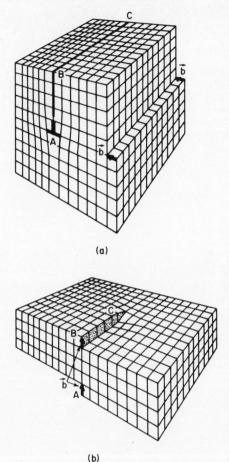

(a)

(b)

Fig. V.2 Partially slipped crystals containing line defects which separate the slipped portion of the slip plane from the unslipped portion. These defects are called dislocations. (a) This type of dislocation is called an edge dislocation. The dislocation is at A. An extra half-plane of atoms ABC has been created by the slip. b is the slip vector. (b) This type of dislocation is called a screw dislocation. The dislocation here is at C. b is the slip vector. Note that the slip vector of an edge dislocation is perpendicular to the dislocation line, and that the slip vector of a screw dislocation is parallel to the dislocation line.

the earlier assumption that the plastic deformation is caused only by shear stress is reasonable. The motion of dislocations may be compared to moving a large rug by introducing a wrinkle at one end and then accomplishing the displacement by propagating the wrinkle across the rug, rather than moving the rug all at once by pulling on an edge.

Motion of either type of dislocation produces a relative shear across the plane of motion. However, it is in the nature of the defects that only atoms immediately adjacent to the dislocation are displaced large distances from their equilibrium positions. Therefore, motion of dislocations provides a mechanism for producing permanent shear by sequential motion of rows of atoms.

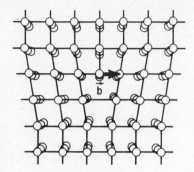

Fig. V.3 Arrangement of atoms around an edge dislocation, showing the extra half-plane of atoms above the dislocation.

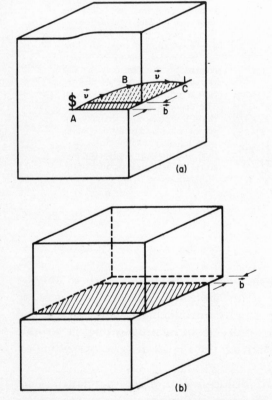

Fig. V.4 Burgers vector **b**, which describes the relative slip across the slip plane, is a constant property of the dislocation. The curved dislocation in this partially slipped crystal in (a) is a screw dislocation at A and an edge dislocation at C, but **b** is the same all along the dislocation. At B the dislocation is of mixed character. If the dislocation loop grows to cover the entire slip plane, the top block of material is displaced by **b** relative to the bottom block as shown in (b).

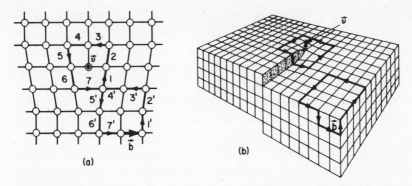

Fig. V.5 Examples of Burgers circuits for (a) edge and (b) screw dislocations. When the same atom-by-atom circuits are taken in good material away from the dislocation, there is a closure failure which defines the Burgers vector **b**.

The concept of a dislocation can be further generalized by noticing that an edge dislocation becomes a screw dislocation by turning through an angle of ninety degrees, as shown in Fig. V.4. Such a situation exists when the slip is confined to only a corner of a crystal. The dislocation part of the way around the corner is neither edge nor screw, but some of each. Such a dislocation is called mixed. The one property which is common to the entire length of the dislocation in Fig. V.4 is its slip or *Burgers vector*. Fundamentally, this vector describes the direction of slip which has occurred in the slipped region bounded by the dislocation. In Fig. V.2(a) this vector is from right to left, as shown, and in Fig. V.2(b) it is directed upward, as also shown.

A method for defining the Burgers vector makes use of the concept of a Burgers circuit. Consider first a region of the crystal containing a dislocation, such as that shown in Fig. V.5(a) or (b). A positive direction along the dislocation line must first be arbitrarily assigned. Since such an assignment is arbitrary, the directions indicated by the vectors **v** are chosen. A Burgers circuit which encircles the dislocation in the right-hand screw sense with respect to the positive direction of the dislocation is now chosen. A Burgers circuit is *any* path around the dislocation which returns to the same point where it starts. Examples of Burgers circuits are indicated in Figs. V.5(a) and (b). The same circuit is now traced in material away from the dislocation, following the original circuit exactly on an atom-by-atom basis. These circuits are also shown in the figure. The circuits in good material do not end at the same place that they started but fail to close by some amount indicated by the vectors **b**. *This closure* *failure is defined as the Burgers vector of the dislocation enclosed by the initial Burgers circuit.* The Burgers vector of a dislocation is an unvarying property of any particular dislocation. Although the dislocation may wind through the crystal, changing its character from edge to screw or mixed, its Burgers vector will be the same at all points. An additional corollary of this is that a dislocation cannot end within the crystal but must always extend to the outside surface of the crystal, unless it is in the form of a closed loop. Once again this statement is obvious if a dislocation is considered to be the boundary between slipped and unslipped regions of a plane. Since a boundary must be continuous, a dislocation cannot end inside a crystal.

For an edge dislocation, the direction of the dislocation line and the Burgers vector define the glide plane of the dislocation. Note from Fig. V.5(b) that motion of the dislocation in a vertical direction would require that atoms be removed from the crystal (upward motion) or that atoms be added to the crystal (downward motion). Such motion can only occur in a material if diffusion takes place to remove or provide the required atoms. Such motion of a dislocation is called climb and is thought to be an important part of the time-dependent plastic deformation of metals at elevated temperatures. At ambient temperature, diffusion in most metals is too slow to allow much deformation from dislocation *climb*. Notice, however, that motion of the dislocation in the horizontal direction can take place without changing the number of atoms in the crystal. Such motion is called *glide*. The horizontal plane, which contains the Burgers vector and the dislocation line, is called the glide plane of the dislocation.

For the screw dislocation, the glide plane is not unique, since the dislocation vector and the Burgers vector are in the same direction. Notice that the screw dislocation in Fig. V.6 has moved into the block of crystal in the direction indicated by u_1. The motion of the dislocation has resulted in a relative displacement of the two parts of the crystal by b. It should be further noted that if the dislocation were now to change its direction and move in a direction u_2, the motion of the dislocation could still cause a relative displacement b without necessitating addition or removal of material. In fact, the screw dislocation of Fig. V.6 can move in any plane which contains the vertical dislocation line. Therefore, all vertical planes are glide planes for this screw dislocation and no process equivalent to climb

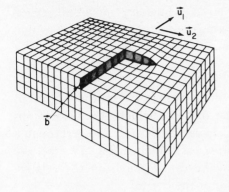

Fig. V.6 Screw dislocation which can move on any plane containing the dislocation line. Here a screw dislocation has moved into the crystal along one glide plane and then turned onto another glide plane. Note that no material had to be added to or removed from the crystal to allow the dislocation to move on the different slip plane.

of an edge dislocation exists for a screw dislocation. This is consistent with the definition of the glide plane as the plane that contains the dislocation line and the Burgers vector, since these two directions are parallel for the screw dislocation and no longer define a unique plane. This would imply that a dislocation with any degree of edge character such that the Burgers vector and the dislocation line define a plane is constrained to move in this slip plane. Motion of a dislocation in its slip plane is called conservative motion, while motion perpendicular to the glide plane is called nonconservative motion.

In a homogeneous, isotropic continuum where all directions and all planes are equivalent, it would appear that dislocations could be formed on any plane in the material and that the Burgers vector could be any direction in the body. A screw dislocation in an isotropic continuum would seemingly be able to glide in any direction. However, in real materials which are crystalline, this situation does not exist. It can be shown that Burgers vectors will be limited to a few particular lattice vectors and that glide of dislocations will take place on only a few particular types of planes in the crystal structure.* For example, Burgers vectors in a face-centered cubic (f.c.c.) crystal are almost always the vectors from an atom at the corner of the cube to an atom at the center of the face of the cube. The preferred glide planes in f.c.c. are the four planes which are perpendicular to the cube diagonals. It should be noted that a Burgers vector corresponds to the vector between nearest-neighbor atoms and that the preferred glide planes are the planes of closest atomic packing. A pure screw dislocation in most crystals of high symmetry usually lies on several intersecting

*Common structures of metallic elements are discussed in Appendix V-A.

preferred glide planes so that it can, in fact, move in several directions, but its freedom of motion is not in general as great as that implied by the isotropic continuum.

V-3-b Strain Produced by Dislocation Motion

The average strain produced in a block of material by the motion of dislocations can be related to the distance which the dislocations move. If N dislocations with Burgers vector $\mathbf{b} = b\mathbf{e}_1$ move through a crystal of dimensions dx_1, dx_2, dx_3, as shown in Fig. IV.7(a), the average plastic strain $(\epsilon_{13}{}^p)_{\text{ave}}$ can be calculated from the relative displacement of the top and bottom of the crystal.

$$\left(\epsilon_{13}{}^p\right)_{\text{ave}} = \frac{1}{2}\frac{Nb}{dx_3} = \frac{1}{2}\rho dx_1 b \tag{V.1}$$

where ρ is the dislocation density $N/dx_1 dx_3$. Since dx_1 is also equal to d, the distance the dislocations have moved, the equation can be written in the standard form

$$\left(\epsilon_{13}{}^p\right)_{\text{ave}} = \frac{1}{2}\rho bd \tag{V.2}$$

When the dislocations do not move entirely through the crystal, the displacement of the outside surfaces of the crystal are not constant and averaging the strain becomes more difficult. A reasonable approach for making this average is to equate the work done by the external stresses with the local work done on the sliding slip plane. Figure V.7 shows a small crystal under a uniform shear stress σ_{13} which has been partially traversed by a dislocation which has moved a distance d into the crystal. Since the stress on the slip plane is σ_{13}, the work done by the stress on the plane during the slip is simply the product of the stress times the displacement times the area of the plane which has slipped.

$$W = \sigma_{13}\, bd\, dx_2 \tag{V.3}$$

The work done at the external surfaces of the crystal is given by

$$W = 2\sigma_{13}\left(\epsilon_{13}\right)_{\text{ave}} dx_1 dx_2 dx_3 \tag{V.4}$$

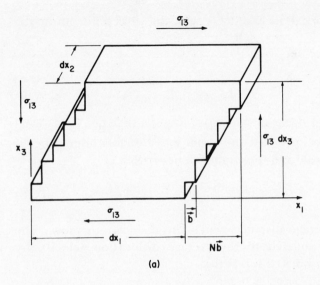

(a)

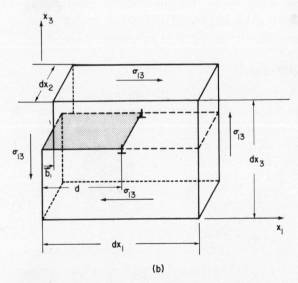

(b)

Fig. V.7 (a) If N dislocations with Burgers vector $\mathbf{b} = b\,\mathbf{e}_1$ move through a crystal on a slip plane perpendicular to the x_3-axis, the top of the crystal is displaced by $N\mathbf{b}$ relative to the bottom. (b) If a dislocation moves only part of the way through the crystal, the average strain is proportional to the slipped area (shaded region of the slip plane).

By equating the work done externally with the work done on the slip plane, one finds

$$\left(\epsilon_{13}\right)_{\text{ave}} = \frac{1}{2}\frac{bd}{dx_1 dx_3} = \frac{1}{2}\rho bd \tag{V.5}$$

since $1/dx_1 dx_3$ is in this case the dislocation density. In a real crystal where many dislocations exist the strain is reasonably homogeneous on a macroscopic scale and the strain can be written as

$$\epsilon_{13}{}^{p} = \frac{1}{2}\rho b\tilde{d} \tag{V.6}$$

Where $\tilde{d}$ is the average distance which the dislocations move. The strain rate can be related to the average dislocation velocity by differentiating both sides of Eq. (V.6).

$$\dot{\epsilon}_{13}{}^{p} = \frac{1}{2}\rho b\tilde{v} \quad * \tag{V.7}$$

where $\tilde{v}$ is the average velocity of the dislocations measured in a direction perpendicular to their length. (The Burgers vector is taken to be in the e_1-direction and the slip plane is perpendicular to e_3.)

V-3-c Strain Fields of Dislocations[†]

Dislocations can be introduced into a dislocation-free body by cutting the body along a plane, displacing the surfaces on either side of the cut by **b** relative to each other and then rejoining them. In Example II.6 it was seen that such a slitting operation was necessary to produce the displacements

$$u_r = u_\theta = 0$$
$$u_z = \frac{b\theta}{2\pi} \tag{V.8}$$

---

*The density of dislocations, ρ, generally increases with strain, so that ρ is not a constant. Since the distance moved, d, for any newly created dislocation is zero, the term $1/2\,\dot{\rho}bd$ is zero.

[†]The stress and strain distributions discussed in this section assume that the material containing the dislocation is isotropic. This is never the case. Since dislocations are crystal defects, their true elastic behavior can only be calculated using the anisotropic properties of the crystal. However, the results presented here are qualitatively correct and allow discussion in much more simple terms than when anisotropic calculations are used.

corresponding to the strains

$$\epsilon_{z\theta} = \frac{b}{4\pi r} \tag{V.9}$$

These displacements in a cylinder correspond to a screw dislocation along the axis of the cylinder with Burgers vector [Fig. V.8(a)]

$$\mathbf{b} = b\mathbf{e}_z \tag{V.10}$$

The stresses are then derived from Hooke's law with only $\sigma_{z\theta}$ nonzero

$$\sigma_{z\theta} = \frac{Gb}{2\pi r} \tag{V.11}$$

This satisfies the equilibrium equations and can be shown to approach the proper boundary conditions as the radius of the cylinder increases. Equations (V.8), (V.9), and (V.11) then give a complete description of the internal stresses and strains associated with a screw dislocation.

An edge dislocation can be formed in a cylinder by cutting it as shown in Fig. V.8(b) and displacing the surfaces above and below the cuts by $\mathbf{b}$ relative to each other parallel to the x_1-direction as shown. The distribution of stress which results from this operation is

$$\sigma_{rr} = \sigma_{\theta\theta} = -\frac{Gb \sin\theta}{2\pi(1-\nu)} \frac{1}{r}$$

$$\sigma_{r\theta} = \frac{Gb \cos\theta}{2\pi(1-\nu)} \frac{1}{r} \tag{V.12}$$

$$\sigma_{zz} = -\frac{Gb\nu \sin\theta}{\pi(1-\nu)r}$$

$$\sigma_{rz} = \sigma_{\theta z} = 0$$

These equations form a description of the internal stresses associated with an edge dislocation in a large body.

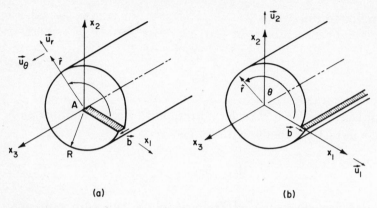

Fig. V.8 Introduction of screw and edge dislocations into the cylinders by cutting along certain planes. If the cut surfaces are displaced parallel to the cylinder axis (a), a screw dislocation is formed. If the displacement is perpendicular to the cylinder axis (b), an edge dislocation is formed.

V-3-d Dislocation Strain Energy and Line Tension

Knowing the stress and strain distributions around a dislocation, it is possible to calculate the strain energy of the dislocation. Considering a single dislocation lying along the axis of a circular cylinder of radius R and ignoring the portion of the body inside a radius r_0 where the strain is too large to be considered elastic, one obtains for the strain energy of an edge dislocation

$$\frac{U}{L} = \frac{Gb^2}{4\pi(1 - \nu)} \ln \frac{R}{r_0} \qquad (V.13)$$

and for the screw dislocation

$$\frac{U}{L} = \frac{Gb^2}{4\pi} \ln \frac{R}{r_0} \qquad (V.14)$$

where L is the length of the dislocation and the cylinder. For this simple case of a body containing a single dislocation, the energy of the dislocation depends on the size of the body. This is a reflection of the fact that the strain field of the dislocation decreases quite slowly (proportional to $1/r$) with increasing distance from the dislocation.

In a real body there are large numbers of dislocations. Therefore, each length of dislocation with Burgers vector **b** can be assumed to be associated with another dislocation segment with Burgers vector -**b** at a distance r_1 from it. If these two lengths of dislocation are considered together, one finds that their combined strain fields decrease more rapidly with increasing distance from the dislocation pair (proportional to $1/r^2$). The strain energy of such a combination of dislocations is given by

$$\frac{U}{L} = \frac{Gb^2}{2\pi(1-\nu)}\left(\ln\frac{r_1}{r_0} - C\right) \qquad (V.15)$$

where the constant C is between zero and $1/2$, depending on how the dislocations are separated with respect to their glide plane. For a screw dislocation pair the energy is

$$\frac{U}{L} = \frac{Gb^2}{2\pi}\ln\frac{r_1}{r_0} \qquad (V.16)$$

For typical values of the dislocation density (10^6 to $10^{12}/\text{cm}^2$), r_1 is 10^{-3} to 10^{-6} cm. Here r_0 is of the order of $2b \simeq 10^{-7}$ cm. Therefore, $(\ln r_1/r_0)/2\pi$ is of the order of one and the energy per unit length of dislocation is given by the approximate expression

$$\frac{U}{L} \simeq Gb^2 \qquad (V.17)$$

For metals this energy is of the order of 5–10 ev/atom length.

Since the energy of a dislocation increases with increasing length, the principle of minimum energy implies that the dislocation will attempt to be as short as possible. In this respect, a dislocation approximates the behavior of a stretched string with constant line tension. In the case of the string, the work required to increase its length by an amount dL is TdL. For the dislocation, the work necessary to increase its length is equal to the increase in strain energy

$$dU = Gb^2dL \qquad (V.18)$$

By comparison, the dislocation behaves like a stretched string with line tension

$$T = Gb^2 \qquad (V.19)$$

V-3-e Force on a Dislocation

In Sec. V-3-b it was shown that motion of a dislocation by an amount d resulted in an average strain in the body. If a stress is applied to a body and a strain results, the stresses do work on the body. Therefore, it is to be expected that stress will tend to make a dislocation move such as to allow work to be done. Consider the volume element containing a dislocation shown in Fig. V.7. As the dislocation moves forward by a distance d, the work done by the shear stress is

$$W = \tau \cdot b \cdot L \cdot d \qquad (V.20)$$

where τ is the shear stress on a plane parallel to the slip plane in the direction parallel to b. In Fig. V.7, $\tau = \sigma_{13}$. It may be assumed that the effect of the stress is to move the dislocation just as if a force F were acting on the dislocation line directly. Then, F can be evaluated by letting the work done by the force equal the work done by the stress, where F/L is the force per unit length acting on the dislocation in a direction perpendicular to the dislocation line. Thus, $W = Fd$ and $F/L = \tau b$. The parameter τ is called the *resolved shear stress* acting on the dislocation (i.e., the component of stress which acts on the dislocation slip plane parallel to the dislocation Burgers vector). This is the only component of stress which tends to move the dislocation by conservative glide. Other components of stress produce a climb force on the dislocations. However, for consideration of room-temperature plasticity of metals, climb of dislocations can be ignored.

V-3-f Interaction Between Parallel Dislocations

In Sec. V-3-d, the concept of paired segments of dislocations of opposite sign was used to determine the approximate energy per unit length of a dislocation. In that discussion it was implicit that there is an interaction energy between two parallel dislocations which varies with the spacing between the two dislocations. This means that the two dislocations must exert forces on each other.

This force can be determined by considering the effect of the internal stress field of one dislocation on the other dislocation. For example, consider a dislocation with Burgers vector be_1 lying along the x_3-axis and another dislocation with Burgers vector $-be_1$, lying parallel to the x_3-axis through the point (x_1, x_2) as shown in Fig. V.9. The stress at the point x_1, x_2 from the dislocation at the origin is given in Sec. V-3-c. As stated in Sec. V-3-e the only component of stress which produces a glide force on a dislocation is the resolved shear stress. On the second dislocation the resolved shear stress is the stress component σ_{12} which is given by

$$\sigma_{12}(x_1, x_2) = \frac{Gb}{2\pi(1 - \nu)} \frac{x_1(x_1^2 - x_2^2)}{(x_1^2 + x_2^2)^2} \tag{V.21}$$

The force per unit length which dislocation 1 exerts on dislocation 2 (or vice-versa since the force must be equal and opposite) is

$$\frac{F}{L} = -\frac{Gb^2}{2\pi(1 - \nu)} \frac{x_1(x_1^2 - x_2^2)}{(x_1^2 + x_2^2)^2} \tag{V.22}$$

which is attractive for the region $|x_1| > |x_2|$. The expression for the force between two edge dislocations of the same sign is the negative of Eq. (V.22). Therefore, two edge dislocations with the same sign for the Burgers vector repel each other when $|x_1| > |x_2|$.

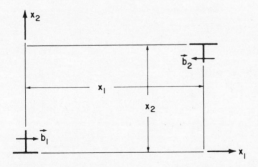

Fig. V.9 Interaction of the stress fields of two dislocations to produce a force between them.

A similar calculation can be made for the case of two screw dislocations. In this case any force in the x_1-x_2-plane is a glide force, so that the latter is simply

$$\frac{F}{L} = \pm\sigma_{z\theta}b = \frac{Gb_1 \cdot b_2}{2\pi r} \tag{V.23}$$

where the force is directed along the line between the two dislocations, and r is the distance between them.

Similar arguments can be used to determine that a screw dislocation exerts no force on a parallel edge dislocation. Therefore, the interaction between two parallel mixed dislocations is the sum of the interaction force between the two edge components plus the interaction between the two screw components. This is illustrated by the following example.

EXAMPLE V.2—Interaction Between Two Parallel
Dislocations of Mixed Character

Consider a dislocation with Burgers vector $b_1 = b_{1e}e_1 + b_{1s}e_3$ lying along the line $x_1 = x_2 = 0$ and another dislocation with Burgers vector $b_2 = b_{2e}e_1 + b_{2s}e_3$ lying along the line $x_1 = r_1$, $x_2 = 0$. What is the glide force which dislocation b_1 produces on dislocation b_2.

Solution:

The stress field of dislocation b_1 is obtained by superimposing the stress distribution for the edge component b_{1e} upon the stress distribution of the screw component b_{1s}. Using Eq. (V.11) and Eq. (V.12), we obtain

$$\sigma_{rr} = \sigma_{\theta\theta} = -\frac{Gb_{1e}\sin\theta}{2\pi(1-\nu)}\frac{1}{r}$$

$$\sigma_{r\theta} = \frac{Gb_{1e}\cos\theta}{2\pi(1-\nu)}\frac{1}{r}$$

$$\sigma_{zz} = -\frac{Gb_{1e}\nu\sin\theta}{2\pi(1-\nu)}\frac{1}{r} \tag{a}$$

$$\sigma_{z\theta} = \frac{Gb_{1s}}{2\pi r}; \qquad \sigma_{rz} = 0$$

Evaluating this stress at $\theta = 0$, $r = r_1$ gives

$$\sigma_{rr} = \sigma_{\theta\theta} = \sigma_{zz} = 0$$

$$\sigma_{r\theta} = \frac{Gb_{1e}}{2\pi(1 - \nu)} \frac{1}{r} \tag{b}$$

$$\sigma_{z\theta} = \frac{Gb_{1s}}{2\pi} \frac{1}{r}$$

Since the plane perpendicular to the θ-direction is the glide plane of the dislocation b_2, the glide force $F/L = \tau b$ becomes

$$\frac{F}{L} = \tau b = (\sigma_{r\theta}e_1 + \sigma_{z\theta}e_3) \cdot b_2$$

$$\frac{F}{L} = \frac{Gb_{1e} b_{2e}}{2\pi(1 - \nu)} \frac{1}{r} + \frac{Gb_{1s} b_{2s}}{2\pi} \frac{1}{r} \tag{c}$$

Note that the total interaction glide force is simply the sum of the interaction between the edge components and the interaction between the screw components. There is no interaction between the edge component of one dislocation and the screw component of the other.

V-3-g Interaction Between a Dislocation and a Stress-Free Surface

If a dislocation exists in a body at a finite distance d from a stress free surface (Fig. V.10), the stress fields given in Sec. V-3-c cannot apply because they violate the boundary conditions on the surface. Additional stress terms must be added in order to cancel the stresses on the surface. These stresses will produce a force on the dislocation. A simple procedure for calculating the necessary additional stresses makes use of an image dislocation reflected through the free surface. In other words, if two screw dislocations* of opposite sign $(b_1 = -b_2)$ are spaced a distance $2d$ apart in an infinite body, the plane perpendicular to the line between them and passing at a distance d from each will be stress free (i.e., in Fig. V.10 the stresses σ_{11}, σ_{12} and σ_{13} are zero on the plane indicated). Therefore, the

---

*The case of the edge dislocation is more complex since the use of an image dislocation alone does not produce a stress free surface. However, it can be shown that the attractive force between the dislocation and the surface is given by the image dislocation since the other stresses which arise from the boundary condition do not produce a force on the dislocation.

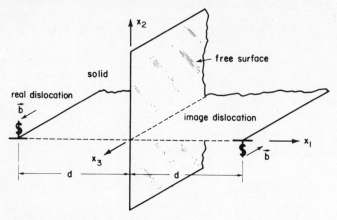

Fig. V.10 Movement of screw dislocation toward a free surface. The attraction to the surface is as if there were a dislocation of opposite sign located a distance d outside of the free surface and $2d$ away from the real dislocation. The imaginary dislocation outside the surface is called on image dislocation.

effect of a stress-free surface on a screw dislocation at a distance d below the surface is the same as the effect of having a dislocation of opposite sign at a distance $2d$ away from the dislocation. Thus, dislocations are always attracted to a free surface by an image force which varies inversely with the distance from the dislocation to the surface. The image force is

$$F_1 = \sigma_{12}b = \frac{Gb^2}{4\pi d} \tag{V.24}$$

The image force also exists on a dislocation near the interface between materials with different elastic constants. In this case the dislocations can be either attracted to or repelled from the interface depending on the ratio of the elastic constants.

V-3-h Generation of Dislocations

Consider a single crystal of metal which is a 1-cm cube. If this is a well annealed crystal, it would have a dislocation density between 10^3 and 10^6 cm^{-2}. The dimension b is typically 2–5 Å, and the dislocations which are present in the crystal can only move distances on the order of 1 cm without reaching a surface of the crystal. From Sec. V-3-b one can calculate the maximum strain which can be achieved before all of these dislocations are pushed out of the crystal

$$\epsilon_{ij}{}^{p} = \frac{1}{2}\rho b\tilde{d} = \frac{10^6}{2}(5 \times 10^{-8}) = 2.5 \times 10^{-2} = 2.5 \text{ percent}$$

$$(V.25)$$

Strains much larger than this are observed in experiments. It is also observed that dislocation densities increase during deformation. These observations imply that dislocations must be generated during deformation.

Generation at the free surfaces is impossible because of the large attractive force between dislocations and the free surface. Therefore, the dislocations must be generated internally. A mechanism by which this might occur is called the *Frank-Read dislocation source*. The stages of such a source are shown in Fig. V.11. The starting point is a segment of dislocation line lying in one plane with the adjacent parts

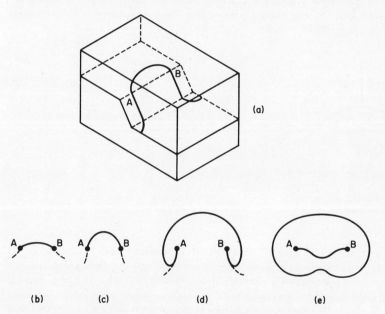

Fig. V.11 Operation of a Frank-Read source after cross slip. (a) A segment of dislocation cross slips from one plane to another. (b) At low resolved shear stress, the dislocation bows out to take the shape of the arc of a circle. (c) Increasing the resolved shear stress reduces the radius of curvature of the dislocation segment. At the critical stress, the segment has the shape of a semicircle. (d) Further increase of the resolved shear stress causes the dislocation segment to become unstable and expand around the pinning points at A and B. (e) When the two sections of the unstable loop come together, they annihilate each other to form a complete dislocation loop and a line segment joining A and B. Dashed lines indicate the remainder of the dislocation which is on a different plane as indicated in (a).

lying out of that plane. (Sometimes this type of dislocation source is called a Koehler source). In the example, this situation has resulted from the cross slip of a segment of screw dislocation on an alternative glide plane. The operation of the Frank-Read source from this segment is illustrated in the following example.

EXAMPLE V.3—Stress Required to Operate a
Frank-Read Dislocation Source

Consider a dislocation segment which is pinned at points A and B which are a distance L apart. Find the equilibrium configuration of the dislocation as a function of the applied resolved shear stress, τ. Is there a maximum stress τ above which no stable configuration exists?

Solution:

Consider the equilibrium of a small segment of dislocation dL acted on by a resolved shear stress τ giving a uniform force per unit length F/L along the dislocation. The force in the x_1- and x_2-directions must be zero if the element is in equilibrium. Therefore,

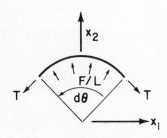

$$\sum F_1 = T \cos \frac{d\theta}{2} - T \cos \frac{d\theta}{2} = 0$$

$$\sum F_2 = \frac{F}{L} R \, d\theta - 2T \sin \frac{d\theta}{2} = 0$$

$$\frac{FR}{L} = T$$

(a)

Fig. V.12 Forces acting on a segment of dislocation line.

$$R = \frac{T}{F/L} = \frac{Gb^2}{\tau b} = \frac{Gb}{\tau}$$

If the resolved shear stress τ is constant between A and B, the radius of curvature of the segment of dislocation is constant and the dislocation assumes the shape of a segment of a circle as shown in Fig. V.11(b). If τ is increased, the radius of curvature decreases until the loop becomes a semicircle as shown in Fig. V.11(c). The radius of curvature at this point is $L/2$ corresponding to a resolved shear stress

$$\tau = \frac{2Gb}{L}$$

(b)

Since no circle with radius less than $L/2$ can pass through both A and B, a stable configuration does not exist for larger resolved shear stresses.

As shown in Example V.3, $\tau = 2Gb/L$ is the largest stress which yields a stable configuration for a dislocation segment of length L. At that stress, the shape of the segment is a semicircle. At this point the loop becomes unstable since further motion requires the radius of curvature to increase. The loop therefore grows bigger and it bends back on itself about the anchor points. Eventually the two recurved sections meet and are of such a relationship to each other that they cancel each other leaving a complete loop of dislocation plus the original segment which began the process. This segment can again initiate the generation process, generating loops of dislocation without limit.

Since L in Eq. (V.26) can easily be of the order of 10^2 to 10^4 b, generation can occur at resolved shear stress given by Eq. (V.25) as $\tau \simeq 2 \times 10^{-2}$ G to 2×10^{-4} G. This is a range which includes the yield stresses of most metals, so that the stress at yield is usually more than enough to allow dislocation generation.

V-3-i Lattice Resistance to Dislocation Motion

The preceding discussion of the dislocation theory showed that a finite strain energy is stored around a dislocation, that the dislocation experiences a force in a stress field created either by other dislocations or externally, and that dislocations can be generated by external forces. The discussion on the basic aspects of the dislocation theory will now be completed by considering the question of whether or not a dislocation experiences a resistance to its motion in a pure single crystal.

In a homogeneous, infinite body, the energy of a dislocation is independent of its position. Therefore, any resolved shear stress should be able to cause the dislocation to move without encountering any resistance to its motion. However, dislocations encounter finite resistance even in a pure single crystal due to the periodic, discrete nature of the crystal lattice. The periodicity of the crystal structure implies that all positions of the dislocation are not equivalent. Because of the differences in energy of the various configurations, a periodic force with period b acts on the dislocation. This force is known as the Peierls force or Peierls stress. This force must be overcome if the dislocation is going to move. The Peierls stress is therefore the minimum stress which must be applied to a crystal containing a dislocation in order to cause plastic deformation, in the absence of thermal activation. The magnitude of the Peierls stress is generally dependent on the slip plane and the Burgers vector. It is a minimum when b is minimum and when the spacing between slip planes is maximum. The Peierls stress is therefore one of the factors which cause crystals to deform on specific types of crystal planes. The Peierls stress is generally thought to be negligibly small for close-packed metals such as copper and aluminum when slipping occurs on the close-packed planes. It may be larger for body-centered cubic metals such as iron and tungsten (perhaps as great as a few thousand psi) and may be responsible for much of the low temperature strength of these metals. However, controversy still exists about the size of the Peierls stress in b.c.c. metals. The Peierls stress is probably greater for ionic and covalent crystals.

V-4 STRESS-STRAIN RELATIONS OF METALS IN TERMS OF DISLOCATION THEORY

In Chapter IV various types of experimentally measured stress-strain curves for metals were presented. It was shown that some metals exhibited a sharp discontinuous yielding phenomenon with a drop in the stress from the upper to the lower yield point occurring at the "beginning" of plastic deformation. In other metals it is found that deviation from elastic behavior occurs much more gradually, but even in these cases it was assumed that the boundary between elastic and plastic deformation could be described by a yield strength for the material. Since it is known that dislocation motion is the basic mechanism of plastic deformation, it should be possible to explain the empirical stress-strain curves in terms of the behavior of dislocations. This ideal is far from being achieved on a quantitative basis because of the extreme complexity of the problem. However, a great many features of the plastic stress-strain curve can be qualitatively related to aspects of dislocation behavior.

V-4-a Stress-Strain Relations of Metals Which Exhibit Gradual Yield

The simplest explanation for the elastic-plastic transition is that the yield strength of the material represents the stress necessary to cause dislocations to move. This picture can be shown to be incorrect in most cases for a number of reasons. One reason is that careful experiments have shown that small amounts of plastic deformation occur at stresses well below the yield strength of the material.

In a tensile stress-strain test the total strain rate, $\dot{\epsilon}_{11} = \dot{\epsilon}_{11}^{\ e} + \dot{\epsilon}_{11}^{\ p}$ is held constant. The elastic strain is proportional to the stress, $\dot{\epsilon}_{11}^{\ e} = \dot{\sigma}_{11}/E$. The plastic strain rate was shown in Sec. V.3-c to be of the form

$$\dot{\epsilon}_{11}^{\ p} = C\rho b\tilde{v} \tag{V.26}$$

where ρ is the mobile dislocation density, b is the magnitude of the Burgers vector, $\tilde{v}$ is the average dislocation velocity, and C is a constant which arises because, in a tensile stress-strain test on a polycrystal, the strain rate must be averaged over the contributions of a number of slip systems. Therefore, the strain rate equation can be written as

$$\dot{\epsilon}_{11} = \dot{\epsilon}_{11}^{\ e} + \dot{\epsilon}_{11}^{\ p} = \frac{\dot{\sigma}_{11}}{E} + C\rho b\tilde{v} \tag{V.27}$$

During the early portion of the stress-strain test, on an annealed material, the density of mobile dislocations, ρ, is small and the average velocity, $\tilde{v}$, is small compared to the imposed strain rate. Therefore, the majority of the strain occurs elastically and the stress increases nearly linearly with total strain. As the stress increases, two things occur which cause the plastic strain rate to increase. First, the average velocity of the dislocations increases as the stress increases and second, the dislocation motion results in the generation of new dislocations, so that ρ increases. As the yield strength is approached, the velocity and numbers of dislocations have increased to the point where the plastic strain rate is comparable to the elastic strain rate, and deviation from linear stress-strain behavior becomes obvious. It is usually found that a very rapid increase in the dislocation density occurs as the material passes through the macroscopic yield.

As straining continues beyond yield, the increase in the density of dislocations results in more frequent interactions between dislocations. This means that the stress required to maintain a given average dislocation velocity increases with increasing strain and the stress continues to rise throughout the work-hardening portion of the stress-strain curve.

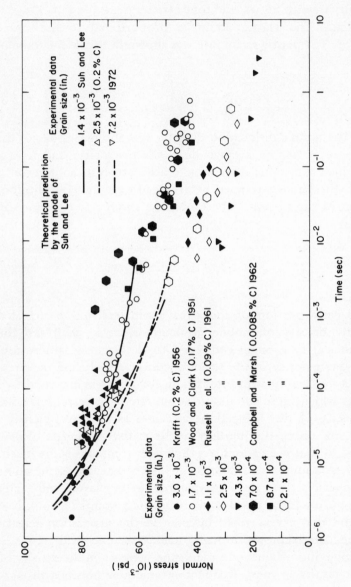

Fig. V.13 Experimental and theoretical delayed yielding time as a function of the normal stress for various grain sizes. *(From N. P. Suh and R. S. Lee, A Dislocation Model for the Delayed Yielding Phenomenon, Materials Sci. & Eng., vol. 10, pp. 269–278, 1972.)*

V-4-b Discontinuous Yield Points

Certain metals, notably annealed low-carbon steel, exhibit the yield point phenomenon at the transition between the elastic and plastic deformation. When such a material is strained at a constant strain rate, the stress rises nearly elastically to the upper yield stress of the material then as plastic strain commences the stress drops suddenly to the lower yield stress. A closely related phenomenon is delayed yielding, which occurs when the stress is raised rapidly to a stress above the lower yield stress and then held constant. The loading occurs elastically, then after a definite delay, appreciable plastic deformation occurs. The delay time before yielding occurs depends on the magnitude of the stress as shown in Fig. V.13. Therefore, when the duration of loading is very short, a much higher stress level must be applied to induce plastic deformation than when the duration of loading is long.

The exact origin of the yield point instability and the delayed yielding phenomenon is still debated. An early theory proposed by Cottrell and Bilby was that the instability resulted because in materials which exhibit unstable yield, dislocations would be pinned by atmospheres of impurity atoms in the manner described in Sec. V-6-a. The upper yield stress would then represent the stress necessary to break the dislocations loose from their pinned positions, and the lower yield stress would represent the stress necessary to move the dislocations after they had been pulled loose from their pinned positions. By invoking the idea that the unpinning process can be aided by thermal activation, which would require a waiting time before it occurred, this explanation could also be made to explain delayed yielding.

More recent work has shown that some plastic strain of magnitude less than the elastic strain precedes macroscopic yielding. This indicates that all dislocations are not pinned prior to the onset of yielding. More recent theories therefore tend to favor a rapid rate of generation of new dislocations, rather than release of existing dislocations, as being the source of the stress drop following the upper yield point. In these theories, the role of impurities is simply to keep the initial *mobile* dislocation density well below the total dislocation density, since a low initial mobile dislocation density is required to produce a sharp discontinuous yield. In the case of polycrystalline materials a major portion of the yield process is probably associated with the spread of deformation from one grain to another. Since grain boundaries present a very strong obstacle to dislocation motion, yielding of a single grain in a polycrystalline aggregate does not result in macroscopic yield. There is ample experimental evidence to indicate that some grains in a polycrystalline material yield at lower stresses or at shorter times than the rest of the material. General macroscopic yield of the test sample cannot occur until all of the grains in a cross section of the body have yielded. In this model, the delay time is associated with the time required for the dislocation motion in the yielded grains to produce sufficient stress concentrations where slip bands are blocked at grain boundaries to initiate yield in adjacent grains. Since the stress concentration factor which can be produced at a grain boundary by a slip band depends on the grain size, this model gives rise to a grain size effect on the delay time or the upper yield stress; the delay time becomes longer and the upper yield stress becomes larger as the grain size becomes smaller.

A similar grain size effect is observed in the lower yield stress. Experiments show that the lower yield stress σ_{ol} observed in low strain rate experiments is related to the grain sized by the Hall-Petch relation

$$\sigma_{ol} = \sigma_f + kd^{-1/2} \qquad (V.28)$$

where σ_f and k are constants of the material. The stress σ_f is normally called the dislocation "friction stress." This relation is in good agreement with experimental results when the mean grain diameter is greater than 1 micron (1 μm). For submicron-size grains the lower yield stress seems to be inversely proportional to d. The explanation for the grain size effect on the lower yield stress is essentially the same as for the grain size effect on the delay time. In materials which exhibit unstable yield, plastic deformation occurs first at some particular cross section of the sample forming a band of plastically deformed material called Lüders band. Additional plastic extension at the lower yield stress occurs by growth of the band into essentially undeformed material. The lower yield stress is therefore the stress required to propagate deformation from deformed grains into undeformed grains. Since the stress concentrations at grain boundaries play an important part in this process, the grain size effect again arises.

V-4-c Serrated Yielding (Portevin-Le Chatelier Effect)

Certain metals, such as mild steel, aluminum, and zinc exhibit serrated or multiple yielding behavior when they are elongated at low strain rates in certain temperature ranges as shown in Fig. V.14. This phenomenon in steel may be explained by the pinning effect of impurity atoms on the dislocations. At room temperature, the dislocations which are either released from their atmospheres at yield or generated by subsequent dislocation motion, remain free of pinning impurity atmospheres. At somewhat elevated temperatures, the rates of diffusion of the impurities become sufficiently high so that dislocations can collect new atmospheres of impurities very rapidly. Thus, when dislocations become temporarily immobilized at obstacles or by interaction with other dislocations they can be quickly pinned so as to become permanently immobile. Therefore, the high density of mobile dislocations produced during each discontinuous yield event becomes rapidly depleted, requiring the yield process to be essentially repeated periodically as plastic flow continues. This phenomenon is often called the Portevin-Le Chatelier Effect. In steel the serrated yielding arises in the temperature range where a blue oxide forms on the surface of the metal and the reduced ability of the metal to flow plastically leads to increased embrittlement. Because of this, the phenomenon in steel is often called "blue brittleness."

V-5 WORK HARDENING MECHANISMS

Interactions between dislocations are the hardening mechanisms which change with strain and account for work hardening. Dislocation–dislocation interactions are of three general types: junction reactions, jog formation, and stress field interactions.

Junction reactions occur when two dislocations intersect. Since plastic deformation in polycrystalline materials requires dislocation motion on several different sets of crystallographic planes, dislocations must continually intersect and pass through one another. When two dislocations approach intersection, they may attract each other and combine along some length to form a dislocation with a Burgers vector equal to the sum of the two individual Burgers vectors, or the two intersecting dislocations may repel each other. In the first

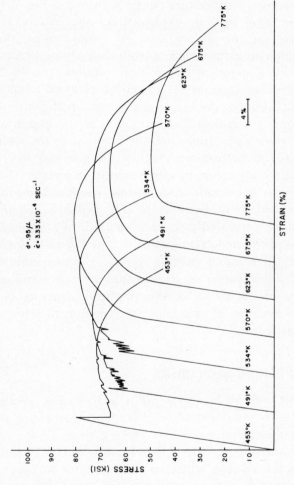

Fig. V.14 Stress-strain curves of a low-carbon steel (composition: 0.17 wt% C, 0.58 wt% Mn, 0.50 wt% W, and 0.24 wt% Zr) from 453°K to 775°K. Grain size = 0.95 μm. (*From V. Ramachandran and E. P. Abrahamson, II, Plastic Instability in an Ultrafine Grain Steel, Scripta Metallurgica, vol. 6, p. 187, 1972.*)

instance the two dislocations are said to form an attractive junction and, in the second instance, to form a repulsive junction. A general rule for the nature of the interaction is derived from an energy balance argument. If the two dislocations have Burgers vectors $\mathbf{b}_1$ and $\mathbf{b}_2$, they have specific energies of approximately $G|\mathbf{b}_1|^2$ and $G|\mathbf{b}_2|^2$. The combined dislocation has Burgers vector $(\mathbf{b}_1 + \mathbf{b}_2)$ and energy $G|\mathbf{b}_1 + \mathbf{b}_2|^2$. If the combined energy is less than the sum of the separate energies, an attractive junction occurs. If the combined energy is greater, a repulsive junction occurs. Geometrically this can be stated that attractive junctions occur when $\mathbf{b}_1$ and $\mathbf{b}_2$ make an angle less than 90 degrees, and repulsive junctions occur when they make an angle of more than 90 degrees. Since both types of interaction produce a retarding force on the dislocation during some portion of the cutting process, both resist dislocation motion. However, the attractive junction is stronger and resists motion over longer distances.

EXAMPLE V.5—Junction Reactions for
        Dislocations

In the face-centered cubic crystal system with coordinates along the cube edges, the preferred Burgers vectors are given by*

$$\pm \mathbf{b}_1 = \frac{a}{2}\mathbf{e}_1 + \frac{a}{2}\mathbf{e}_2$$

$$\pm \mathbf{b}_2 = \frac{a}{2}\mathbf{e}_1 - \frac{a}{2}\mathbf{e}_2$$

$$\pm \mathbf{b}_3 = \frac{a}{2}\mathbf{e}_2 + \frac{a}{2}\mathbf{e}_3$$

$$\pm \mathbf{b}_4 = \frac{a}{2}\mathbf{e}_2 + \frac{a}{2}\mathbf{e}_3$$ (a)

$$\pm \mathbf{b}_5 = \frac{a}{2}\mathbf{e}_1 + \frac{a}{2}\mathbf{e}_3$$

$$\pm \mathbf{b}_6 = \frac{a}{2}\mathbf{e}_1 - \frac{a}{2}\mathbf{e}_3$$

where a is the length of a cube edge, and the $\pm$ sign indicates that dislocations can have either the positive or the negative Burgers vector for a given direction. In what cases are interactions between pairs of dislocations with these Burgers vectors favorable for forming junctions and in which cases are they not favorable? (See Fig. V.15 for the coordinate geometry.)

*In terms of the Miller indices $\mathbf{b}_1$ is expressed as $a/2\,[110]$ (see Appendix V-A).

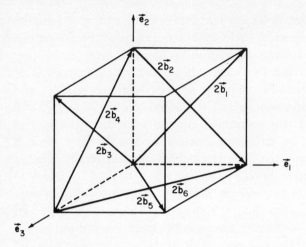

Fig. V.15 Burgers vectors in the f.c.c. crystal system.

Solution:
    The junction is favorable when the energy per unit length of the junction dislocation is less than the sum of the energies of the two separate dislocations. The energy is given approximately by Eq. (V.17).

$$\frac{U}{L} = Gb^2 \tag{b}$$

Therefore, the reaction is favorable if

$$|\mathbf{b}_i|^2 + |\mathbf{b}_j|^2 > |(\mathbf{b}_i + \mathbf{b}_j)|^2 \tag{c}$$

where i and j range from 1 to 6. Considering the possible combinations, one finds that the pairs which are favorable for junction formation are

$$\begin{array}{lll}
\mathbf{b}_1 - \mathbf{b}_3 = \mathbf{b}_6 \;; & \mathbf{b}_1 - \mathbf{b}_4 = \mathbf{b}_5 \;; & \mathbf{b}_1 - \mathbf{b}_5 = \mathbf{b}_4 \;; \\
\mathbf{b}_1 - \mathbf{b}_6 = \mathbf{b}_3 \;; & \mathbf{b}_2 + \mathbf{b}_3 = \mathbf{b}_5 \;; & \mathbf{b}_2 + \mathbf{b}_4 = \mathbf{b}_6 \;; \\
\mathbf{b}_2 - \mathbf{b}_5 = -\mathbf{b}_3 \;; & \mathbf{b}_2 - \mathbf{b}_6 = -\mathbf{b}_4 \;; & \mathbf{b}_3 - \mathbf{b}_5 = -\mathbf{b}_2 \;; \\
\mathbf{b}_3 + \mathbf{b}_6 = \mathbf{b}_1 \;; & \mathbf{b}_4 + \mathbf{b}_5 = \mathbf{b}_1 \;; & \mathbf{b}_4 - \mathbf{b}_6 = -\mathbf{b}_2
\end{array} \tag{d}$$

The unfavorable combinations are

$$\begin{array}{llll}
\mathbf{b}_1 + \mathbf{b}_3 \;; & \mathbf{b}_1 + \mathbf{b}_4 \;; & \mathbf{b}_1 + \mathbf{b}_5 \;; & \mathbf{b}_1 + \mathbf{b}_6 \;; \\
\mathbf{b}_2 - \mathbf{b}_3 \;; & \mathbf{b}_2 - \mathbf{b}_4 \;; & \mathbf{b}_2 + \mathbf{b}_5 \;; & \mathbf{b}_2 + \mathbf{b}_6 \;; \\
\mathbf{b}_3 + \mathbf{b}_5 \;; & \mathbf{b}_3 - \mathbf{b}_6 \;; & \mathbf{b}_4 - \mathbf{b}_5 \;; & \mathbf{b}_4 + \mathbf{b}_6
\end{array} \tag{e}$$

The combinations

$$b_1 \pm b_2 \; ; \quad b_3 \pm b_4 \; ; \quad b_5 \pm b_6 \qquad \text{(f)}$$

are neutral reactions.

Note that for all of the favorable reactions the resulting Burgers vector is one of the normal f.c.c. Burgers vectors. The energy relationship for these combinations is given by

$$\frac{a^2}{2} + \frac{a^2}{2} > \frac{a^2}{2} \qquad \text{(g)}$$

For the unfavorable combinations the relationship is

$$\frac{a^2}{2} + \frac{a^2}{2} < \frac{3a^2}{2} \qquad \text{(h)}$$

The neutral reaction occurs between the Burgers vectors which are the two diagonals of the same face of the cube and therefore are perpendicular to each other.

Jog formation also occurs during the dislocation intersection and cutting process. Whenever a dislocation cuts through another dislocation which is partially screw, a jog is formed on the cutting dislocation, as shown in Fig. V.16. If the cutting dislocation is an edge dislocation, the glide plane of the jog will be parallel to the direction of motion of the dislocation, and such a jog will be able to move along with the dislocation. However, if the cutting dislocation is a screw dislocation, the glide plane of the jog will be perpendicular to the direction of the dislocation motion. The jog is thus unable to move by glide. In order to move along with the rest of the dislocation, the jog must move by climb. This means that the jog will produce vacancies or interstitial atoms for each atomic distance moved. The energy required to form these point defects in a crystal is quite high, so the presence of a sessile* jog will result in a large drag on the dislocation. In some cases the energy required to produce the vacancies or interstitials is so high that less energy is required to simply leave a trail of a dislocation dipole behind the moving dislocation. This trail is composed of the two parallel segments of

---

*In some circumstances such as by interaction of two dislocations, a dislocation can be formed which does not lie on a glide plane of the crystal. Such a dislocation cannot move and is called sessile. A dislocation which lies on a slip plane and can move by glide is called glissile.

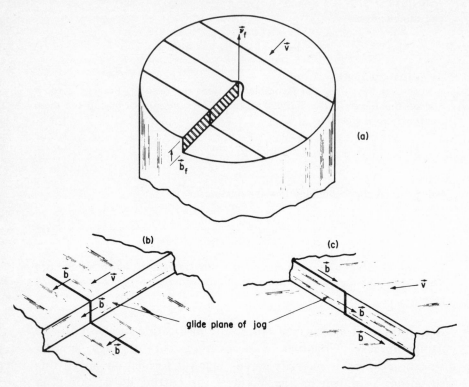

Fig. V.16 Formation of a jog on a gliding dislocation by intersection with a forest dislocation. (a) A jog is formed on a glide dislocation when it passes through a screw dislocation which intersects its slip plane. (b) If the gliding dislocation is edge, the jog can move by glide. (c) If the gliding dislocation is screw, the jog cannot move by glide in the direction of motion of the dislocation. ($\bar{v}$ indicates the direction of motion of the dislocation.)

dislocation which have passed on opposite sides of the obstacle dislocation, as shown in Fig. V.17. After encircling the obstacle these two segments are on different atomic planes. Since these segments of dislocation are edge in character they cannot move out of their planes and the dipole arrangement is stable until diffusion can cause climb to bring the two parts of the dipole together.

Interactions between dislocations caused by the forces which the stress field of one dislocation produces on the other dislocation probably account for a large part of work hardening. As described in Sec. V.3-f, dislocations of opposite sign attract each other. If, during deformation, two such dislocations approach sufficiently closely, the applied stress will not be sufficient to pull the two dislocations apart. The combination of dislocations formed is again a dislocation dipole.

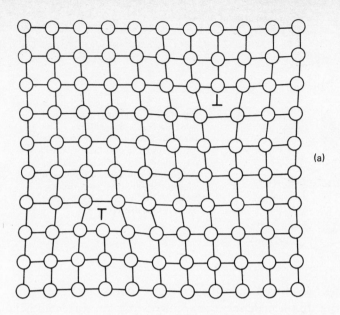

(a)

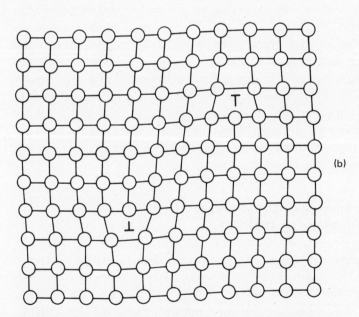

(b)

Fig. V.17 Dislocation dipole, consisting of two edge dislocations on different planes which have their extra half-planes of atoms extending in opposite directions. (a) When the extra half-planes fail to meet, the dipole is of the vacancy-producing type. (b) When the extra half-planes overlap, the dipole is of the interstitial producing type.

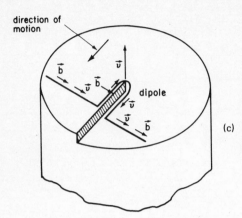

direction of
motion

dipole

(c)

Fig. V.17 (*Continued*) (c) When two screw dislocations cut through each other, the segments of the moving dislocation which pass on opposite sides of the fixed dislocation can form a dipole. In such a dipole, the two dislocations of the dipole are separated by only one plane.

Since the forces produced by the applied stress on the two dislocations of the dipole balance, the applied stress produces no net force on the dipole. The dipole does not move as the deformation proceeds. As more dislocations move through the crystal the possibility that some will pass close to the dipole and be trapped in its stress field increases. Therefore, the initial dipole gradually accumulates more and more trapped dislocations. As the process continues the dislocations in the group tend to become tangled so that the stability of the group and its effectiveness as an obstacle increase. Eventually, these trapped dislocations tend to collect into walls which partition the crystal into small cells. The cell walls are effective barriers to dislocation motion. It is the creation of these groups of trapped dislocations and the development of the cell structure which causes work hardening.

V-6 STRENGTHENING MECHANISMS BY METALLURGICAL MEANS

The flow stress of metals can be increased by such metallurgical means as alloying and inclusion of secondary phase particles. The strengthening mechanisms involve either introduction of physical obstacles to dislocation motion or creation of internal stress fields. Alloying, which is called solution hardening, creates internal stress fields by replacing the original atoms with other atoms of different sizes and by inserting small nonmetallic atoms in interstitial sites. Secondary-phase particles are incorporated by either precipitation of secondary phases during cooling and aging, or by inclusion of insoluble secondary-phase particles during the melting process.

V-6-a Solution Hardening

In substitutional alloying, foreign atoms whose atomic radii are within 15% of the host atoms are mixed with the latter. These atoms occupy regular lattice sites and create internal stress fields due to the difference in the atomic radii. When such metals are plastically deformed, the dislocation motion is impeded by the internal stress field. Therefore, a greater external stress must be applied to deform the metal plastically than when it is free of foreign atoms.

The distortions of the crystal lattice near a dislocation make it easier for impurity atoms which occupy interstitial locations to fit into the crystal lattice. Therefore, solute atoms tend to diffuse to dislocations and collect along them so as to lower the strain energy. When the dislocation moves away from these impurity atoms, the misfit energy of the impurity atoms rises. The applied stress must then do additional work to provide this increase in energy. The impurities which have collected about the dislocation in this way tend to pin the dislocations to their positions. This type of hardening is called Cottrell locking, and the impurities which have collected at the dislocation are called the impurity atmosphere or the Cottrell atmosphere of the dislocation.

An important example of the hardening effect of impurities is the case of carbon in iron. Carbon exists as an interstitial impurity in the iron crystal. Because of this, it can diffuse rapidly through the iron crystal and is strongly attracted to the dislocations.

Atmosphere formation is a static form of hardening which only resists the initial motion of the dislocations. Once the dislocations are pulled from their atmospheres they can move freely, unless the temperature is raised to increase the diffusivity of impurity atoms. A similar type of effect, though much weaker, occurs dynamically. As a dislocation approaches an atom of impurity, the stress fields of the dislocation and the impurity interact. In some cases this interaction is an attraction and in others a repulsion. In either case the dislocation must overcome these interaction forces in order to continue moving. This effect will create a minimum stress necessary to keep a dislocation moving even after it is pulled loose from a static atmosphere. This form of hardening is generally thought to be quite weak, but it has been suggested that it may be significant in cases

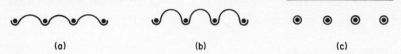

(a)   (b)   (c)

Fig. V.18 Motion of a dislocation through a row of obstacles by the Orowan process.

such as carbon in iron where the impurity has a large distortional stress field.

V-6-b Precipitation Hardening

In any two-phase material where the second phase exists as a dispersion of fine particles in a matrix of the major phase, dislocation motion is impeded. The origin of the resisting force depends on the materials in question and may arise from any of a number of reasons.

In some cases, the particles are a hard, brittle, nonmetallic phase which cannot be sheared. Therefore, the dislocations cannot pass through the particles but must surround the particles by what is known as the Orowan process. The mechanism of surrounding the particles is very similar to the first stage of dislocation generation by a Frank-Read source. A dislocation trapped against a line of particles and acted on by an applied stress will bow out between the particles, as shown in Fig. V.18. If sufficient stress is applied to make the bow-out semicircular, the segments of dislocation which have passed on opposite sides of the particle can combine to annihilate each other. This produces a free dislocation beyond the particles and leaves loops of dislocation around each of the particles. The stress required for this process is the same as that given in Sec. V-3-i for generation from a source.

When the particles are incoherent* but shearable, the process is somewhat different. Since the crystal structure of the incoherent particles is different from the crystal structure of the matrix, the Burgers vectors of dislocations in the matrix are different from the Burgers vectors of dislocations in the particle. Therefore, a matrix dislocation cannot pass through the particles until it is converted to a particle dislocation. This can be accomplished if a dislocation with Burgers vector equal to the difference between the particle and matrix Burgers vectors is left in the particle interface. The process of

---

*A coherent particle is one whose crystallographic structure is compatible with that of the matrix.

creating this interface dislocation resists the motion of the dislocation.

Both of the above processes are short range in nature, since the dislocation must make actual contact with the particles. A longer-range interaction occurs when the particles have self-stress fields or when they distort the applied stress field. In the first instance, the interaction is simply the interaction of the dislocation with an internal stress field. The second instance occurs when there is a modulus difference between particle and matrix. The interaction is a boundary condition interaction similar to the interaction which occurs between a dislocation and a free surface.

When the precipitation hardening metals are held at high temperature, the number of the secondary phase particles becomes smaller. As a consequence, the distance between these secondary-phase particles becomes larger, making it easier for dislocations to extrude through the particles by the Orowan mechanism. The metal then deforms plastically at low flow stresses. This is called *overaging*.

EXAMPLE V.4—Calculation of the Strength of a Precipitation Hardened Material

A material contains a constant volume fraction of precipitate, but by heat treating the size of the precipitate particles and their spacing can be changed. That is

$$nv_p = \frac{n4\pi r_p^3}{3} = C \tag{a}$$

where C is a constant, n is the number of particles per unit volume, v_p is the particle volume, and r_p is the radius of the particle. If the particles are too small, they can be sheared by the dislocation, and if they are too large and far apart, they can be easily bypassed by the Orowan process. If the force which a dislocation must exert on a particle to shear it is proportional to its cross-sectional area, πr_p^2, find an expression for the particle size which gives maximum shear stress to move dislocations through the particles.

Solution:
The mean distance between particles is

$$l = \left(\frac{1}{n}\right)^{1/3} = \left(\frac{4\pi r_p^3}{3C}\right)^{1/3} = Ar_p \tag{b}$$

where $A = (4\pi/3C)^{1/3}$.

The Orowan stress is the stress necessary to bow the dislocations segment out to a semicircle. From Eq. (V.25), this is

$$\tau = \frac{2Gb}{l} = \frac{2Gb}{Ar_p} \tag{c}$$

The force which a dislocation exerts on the particle is found from the line tension T, as shown in Fig. V.19, using

$$f = 2T \cos \frac{\alpha}{2} \leq B\pi r_p^2 \tag{d}$$

where $B\pi r_p^2$ is the strength of the particle. For a regular array of particles, such as that shown in Fig. V.19, the function $\cos \alpha/2$ can be related to the resolved shear stress τ and the particle spacing l. Since the dislocation segments form a segment of a circle of radius $R = Gb/\tau$, the angle $\alpha/2$ is determined as shown in Fig. V.19.

$$\cos \frac{\alpha}{2} = \frac{l/2}{R} = \frac{l\tau}{2Gb} \quad \text{for} \quad \tau \leq \frac{2Gb}{l} \tag{e}$$

For $\tau > 2Gb/l$, a stable configuration does not exist because the dislocation will bypass the particle by the Orowan process. By applying Eq. (e) to Eq. (d) and solving for τ, where the force f is equal to the strength of the particle, one obtains the stress necessary to shear the particle, τ_s.

$$\tau_s = \frac{Gb}{Tl} B\pi r_p^2 \tag{f}$$

For maximum strength the stress for shearing the particle, Eq. (f), should just equal the Orowan stress, Eq. (c), which gives, when l is evaluated using Eq. (b),

$$\frac{2Gb}{Ar_p} = \frac{Gb}{TA} B\pi r_p \tag{g}$$

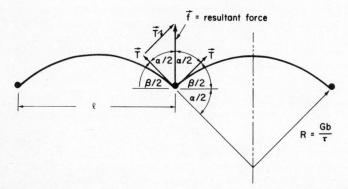

Fig. V.19 Line tension of the adjacent segments of a bowed out dislocation, producing a force f on a particle.

which can be solved for the optimum particle size r_p

$$r_p{}^2 = \frac{2T}{B\pi} \simeq \frac{2Gb^2}{B\pi} \qquad\qquad (h)$$

V-7   SECONDARY EFFECTS PRODUCED BY DISLOCATIONS

V-7-a   Microstrain

The discussion of Sec. V-4-a indicated that there is always a small plastic deformation even in the elastic regime of deformation. These small strains have been measured experimentally. The smaller the detectable plastic strain component, the lower was the stress necessary to produce it. Even materials, such as mild steel, which exhibit a sharp yield point have been found to undergo small amounts of permanent deformation (small compared to the elastic strain at yield) at stresses well below the upper yield point.

This small amount of plastic strain, which precedes the general yielding of the material, is never greater than the elastic strain at yield and is usually much smaller. Because of its small magnitude, this strain is called preyield microstrain or, simply, microstrain. The microstrain has very little engineering importance in cases of monotonic loading because it is so small. In the elastic region it contributes only a small amount to the elastic deformation, and as soon as macroscopic yielding occurs the microstrain becomes an insignificant proportion of the total plastic strain. However, microstrain and the processes which produce it play an important role in cases of oscillatory loading. Microstrain contributes to the damping of vibrations, the accumulation of fatigue damage, and is closely related to transient creep which will be discussed in Chapter VII.

The mechanisms which produce microstrain are associated with the fact that both the stress distribution and the distribution of dislocations are not homogeneous when considered on a size scale of the order of the spacing between the dislocations. When a shear stress is applied to a crystal, not all of the dislocations in the material are subjected to the same glide force and not all of the dislocations are held by obstacles of the same strength. Because of favorable interactions' with neighboring dislocations, some dislocations may experience a local stress nearly equal to the critical shear stress when the average stress is still quite low. Other dislocations may be in a configuration where they can overcome local obstacles at stresses far below the average critical shear stress. Dislocations in either of these situations will break loose from their retarding obstacles at a stress below the yield strength and their motion will cause a small amount of plastic strain. In most cases these dislocations, which were initially in favorable configurations, will only move a small distance. After they move a few interdislocation or interobstacle spacings, they reach positions unfavorable for continued motion. As the stress is raised, additional dislocations in situations favorable for motion will move through small distances contributing to the microstrain.* However, these dislocations will always stop moving when they reach unfavorable locations. General plastic flow, which requires dislocation motion over long distances, will proceed only when the stress is sufficiently large to allow the dislocations to move through regions of low local stress or high obstacle strength. This stress will be the macroscopic yield strength.

As a specific example of a microstrain mechanism, consider the motion of a dislocation through a random array of small particles, as shown in Fig. V.20. The initial configuration of the dislocation is shown in (a). When a stress is applied to the material, a force is exerted on the dislocation tending to move it to the right. After a small motion the dislocation encounters a number of obstacles in (b). As the stress increases the dislocation bows out between the obstacles, as shown in (c). Note that the radii of curvature of all segments are the same, independent of the length of the segment, as shown in Sec. V-3-i. Therefore, the dislocation will first break through the obstacles adjacent to segments with the longest span where the included angle α is smallest. In Fig. V.20, the obstacle A will be overcome first. After this obstacle is passed, the dislocation moves to a new configuration (d) where no further obstacles can be broken. Therefore, the stress must be increased to cause additional dislocation motion. As additional obstacles are overcome, the dislocation assumes a configuration where additional motion in the

---

*The magnitude of the microstrain can be estimated from the equation $\epsilon = \rho b \tilde{d}$, where ρ is taken to be the density of dislocations which move before yield, and $\tilde{d}$ is the average distance moved. Assume that somewhere between 10% and 50% of the total number of dislocations move and $\tilde{d}$ is less than ten times the average dislocation spacing. Reasonable values are:

$$\rho \simeq 10^6 \, \text{cm}^{-2}, b \simeq 10^{-8} \, \text{cm}, \tilde{d} < 10^{-2} \text{cm}. \quad \text{Then,} \quad \epsilon < 10^{-4}$$

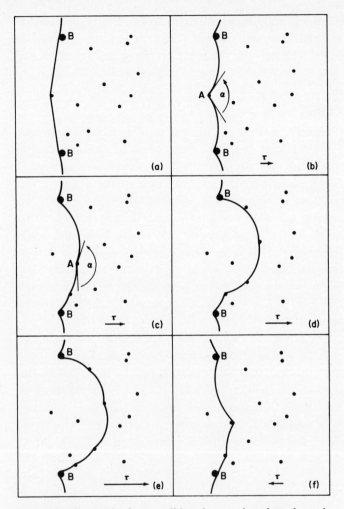

Fig. V.20 Microstrain from a dislocation cutting through weak obstacles between strongly pinned points. A dislocation which is strongly pinned at the points marked B can produce a small amount of strain by cutting through the dispersion of weak obstacles at stresses which are too small to release it from the strong obstacles. As the stress is increased from zero in (a) to its maximum in (e), the dislocation cuts through more and more of the weak obstacles. At (e) the microstrain is exhausted, because the dislocation cannot expand further without breaking loose from the strong pinning points. When the sign of the stress is reversed, the dislocation can move back through the weak obstacles at a lower magnitude of the shear stress (f).

same direction is increasingly more difficult, as in Fig. V.20. In this example general yielding will occur only when the stress is sufficiently high to release the dislocation from the two strong obstacles B.

V-7-b   Bauschinger Effect

A phenomenon closely related to microstrain is the Bauschinger effect. In Chapter IV it was assumed that the yield strength of the material at any point on the stress-strain curve depended only on $\bar{\sigma}$ and $\bar{\epsilon}^p$ but was independent of the exact details of the deformation history. When a sample is deformed in uniaxial tension to a point on the stress-strain curve where $\bar{\sigma} = \sigma_{11} = C_1$ and then is subjected to compression, the assumptions of Chapter IV imply that additional plastic deformation will occur only when $\sigma_{11} = -C_1$. This is shown by the dashed line in Fig. V.21. Actual experiments show that the yield in compression occurs at a smaller compressive stress, as shown by the solid line in Fig. V.21. Subsequent reversals of the sign of the stress show the same effect.

Consider first the transient response associated with the apparent reduction in the yield strength on reversed stressing. This effect is partially caused by the existence of microstrain. In many cases the dislocation configurations which are least favorable for motion in one direction are most favorable for motion in the opposite direction. Therefore additional microstrain can occur when the stress is reversed. Consider, for example, the case illustrated in Fig. V.20. If

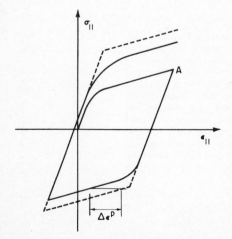

Fig. V.21 Schematic representation of the Bauschinger effect. The dashed line represents the theory presented in Chapter IV. The solid line is a typical experimental measurement. The strain lag $\Delta\epsilon^p$ in the work hardening is typically less than 0.01.

the stress is reversed after it reaches the configuration shown in (e), the configuration becomes that shown in (f). Note that the average curvature of the dislocation which was unfavorable for motion to the right is now favorable for motion to the left. Microstrain of the reverse sign is now favored at stresses well below the macroscopic yield. Other microstrain mechanisms are affected in a similar fashion by reversing the sign of the stress. All of these microstrain processes must be exhausted before the material reaches a state equivalent to what it was at point A in Fig. V.21. An additional reason for the strain lag is that the plastic deformation does not create dislocation configurations which resist dislocation motion in all directions equally. Thus, while the general increase in dislocation density which accompanies deformation makes all plastic strain more difficult, work hardening is somewhat less effective against dislocation motion opposite to the original direction.

Since $\Delta \epsilon^p$ is of the order of 1%, the Bauschinger effect will play a dominant role in any cyclic process where the plastic strain per cycle is less than a few percent. An important example where this is the case is low-cycle fatigue, which will be discussed in Chapter IX. Since the Bauschinger effect was ignored in the treatment given in Chapter IV, the methods of that chapter are not applicable to these small plastic-strain-amplitude cyclic processes.

The Bauschinger effect also occurs for changes in loading other than for complete reversal of loading from tension to compression. That is, there will be a Bauschinger effect on the yield stress when the tensile axis is changed, or when torsion follows plastic deformation in tension. The Bauschinger effect for these more general cases can be represented by a distortion and shifting of the yield locus away from the usual Mises cylinder by deformation in one direction. Just as the Bauschinger effect in tension-compression only applies to small plastic strains and disappears quickly as the plastic strain is increased, the Bauschinger effect on the yield locus applies only to the microstrain yield locus.

REFERENCES

1. Cottrell, A. H.: "Dislocations and Plastic Flow in Crystals," Clarendon Press, Oxford, 1953.
2. Cottrell, A. H.: "Theory of Crystal Dislocations," Gordon and Breach, New York, 1964.
3. Friedel, J: "Dislocations," Addison-Wesley, Reading, Mass., 1964.
4. Hirth, J. P., and J. Lothe: "Theory of Dislocations," McGraw-Hill, New York, 1967.
5. Hull, D.: "Introduction to Dislocations," Pergamon Press, New York, 1965.

6. McLean, D.: "Mechanical Properties of Metals," Wiley, New York, 1962.
7. Nabarro, F. R. N.: "Theory of Crystal Dislocations," Clarendon Press, Oxford, 1967.
8. Read, W. T.: "Dislocations," McGraw-Hill, New York, 1953.
9. Weertman, J. and J. R. Weertman: "Elementary Dislocation Theory," Macmillan, New York, 1964.

PROBLEMS

V.1   A tensile specimen is made of single-crystal copper. The axis of the specimen is along the [001]-direction. A glissile (mobile) dislocation with dislocation vector $t = [\bar{1}01]$ and Burgers vector $b = a/2[\bar{1}01]$ lies on the (111)-plane. Find the force acting on this dislocation if the axial stress is 100 psi. The symbol a is the lattice constant.

V.2   Determine whether the reaction given below is favorable in an f.c.c. crystal. On what plane or planes must such a reaction take place if the $a/2 \langle 110 \rangle$ dislocations are to move by conservative motion?

$$\frac{a}{2}[110] + \frac{a}{2}[1\bar{1}0] \rightarrow a[100]$$

V.3   In low-carbon steel the hard, unshearable, secondary-phase particles are uniformly distributed. The distance between the particles is 500 Å. Estimate the yield strength of this steel.

V.4   The creep rate of a medium-carbon forged steel is given in Fig. VII.3. Estimate approximately the average velocity of dislocations in motion at $800°F$ and at 10,000 psi, if the mobile dislocation density is 10^8.

V.5   The fact that the yield stress of annealed low-carbon steel decreases inversely as the square root of the grain size may be explained in terms of the dislocation pile-up at grain boundaries. The stress at the head of a dislocation pile-up is given by $n\sigma$, where n is the number of dislocations piled up, and σ is the applied shear stress. The distance of the n-th dislocation from the grain boundary is approximately given by

$$L = \frac{nGb}{\pi\sigma}$$

Determine the number of dislocations and the stress required to penetrate through the grain boundary if the shear strength of the grain boundary is $G/10$. Assume that the dislocations are generated at the center of a grain. The grain size is 50 μm.

V.6   If the dislocation friction stress, which is defined as the minimum stress a dislocation has to overcome in order to glide, is 4 kg/cm^2 in very pure copper, determine the thickness of the low dislocation zone near the surface caused by the attraction of dislocations to the free surface. In 3% silicon-iron the stress is 1500 kg/cm^2. What is the thickness of the low dislocation zone in this case? The shear moduli of copper and 3% silicon-iron are 2.2 $\times$ 10^6 and 3.9 $\times$ 10^6 kg/cm^2, respectively.

V.7   Why doesn't the elastic modulus change much when a small amount of solute atoms are added to a metal to make an alloy, even though the flow stress is changed substantially?

V.8   When metals are subjected to cyclic plastic deformation through a strain of $\pm 10^{-4}$ to $\pm 10^{-3}$, the material develops a dislocation structure called a cell structure. It is characterized by the formation of dense dislocation tangles in walls which surround relatively dislocation-free regions (cells). Between 30% and 50% of the total volume of the material is in the cell walls, with the remainder being the dislocation-free cell interior. After a cell structure forms, it is believed that cyclic strain is produced by dislocations being pulled out of one cell wall, crossing the cell, and becoming trapped in the opposite cell wall. When the strain is reversed, many of the same dislocations traverse back across the cell.

In an experiment with iron, a strain amplitude of 10^{-3} caused the formation of cells 1 μm across. How many dislocations must cross each cell to account for this magnitude of strain? The Burgers vector is iron is 2.48 Å long.

V.9   In a crystal containing a network of pinned dislocations, the application of a stress to the crystal can cause recoverable plastic strain to occur due to the dislocations bowing out between the pinning points. Because this strain is indistinguishable from elastic strain, this effect apparently shifts the elastic modulus from what it would be for a perfect crystal. For the special case of an array of dislocations parallel to the x_3-axis, with Burgers vector $\mathbf{b}$ parallel to the x_1-axis and pinned at points uniformly spaced a distance l apart along the dislocation, what plastic strain is produced as a function of the shear stress σ_{21}? What is the apparent shear modulus? What is the maximum stress which can be applied so that all deformation is reversible? The dislocation density is ρ, and the shear modulus of the perfect crystal is G (see figure below).

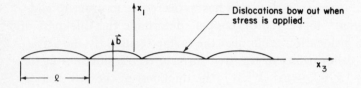

V.10 Two dislocations on the (111)-plane of copper meet to form a junction. The two intersecting dislocations have Burgers vectors $a/2[1\bar{1}0]$ and $a/2[101]$. The junction dislocation has Burgers vector $a/2[0\bar{1}1]$. When a suitable shear stress is applied to the crystal, the junction can be pulled apart to form the original two dislocations, as shown. The force on the junction can be calculated from the line tension and the angle between the intersecting dislocations, α. What is the critical α at which the junction is broken?

V.11 The rate of plastic deformation is sometimes approximated by

$$\dot{\gamma} = \nu A N b \exp\left(-\frac{\Delta E}{kT}\right)$$

where A is the area swept out by a dislocation after it has been freed by thermal activation, and N is the number of sites per unit volume which are susceptible to thermal activation. Consider the dislocation structure as a network which is primarily held up by its own stress fields, without effects due to precipitate particles (see sketch). ν is the frequency of vibration of a dislocation segment, and ΔE is the activation energy for the intersection process. Give an order-of-magnitude approximation to A and N in terms of the dislocation density Λ.

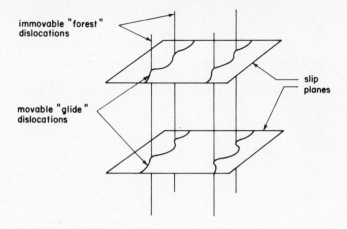

immovable "forest" dislocations

movable "glide" dislocations

slip planes

APPENDIX V-A

Miller and Miller-Bravais Indices

When discussing planes, such as slip planes, and directions in a crystal lattice, it is convenient to use the system of designation called Miller indices (or in the special case of the hexagonal lattice the Miller-Bravais indices). Miller indices are defined in a general way and can be used for any type of crystal structure.

Miller Indices for Planes

Consider a coordinate system in a crystal where the coordinate axes are parallel to the edges of the unit cell of the crystal and where a unit length along the axis is equal to the length of the edge of the unit cell. The case of a cubic crystal structure is shown in Fig. V.A1. Any plane which passes through the unit cell must intersect at least one of these axes. For example, consider the face of the cube which is parallel to the x_2- and x_3-axes and intersects the x_1-axis at 1. This plane intersects the x_1-, x_2-, and x_3-coordinate axes at 1, ∞, and ∞, respectively. The Miller indices of this plane are then defined as the reciprocals of the coordinate intercepts, so that this plane is designated the (100)-plane. The parentheses, ( ), indicate that this is a specific plane which is perpendicular to the x_1-axis. In the cubic system all of the planes which are faces of the cube are equivalent, so that one might wish to refer to this entire class of planes. Thus, $\{100\}$ signifies any plane which is a face of the cube, i.e., (100), (010), or (001). As another example, consider the shaded

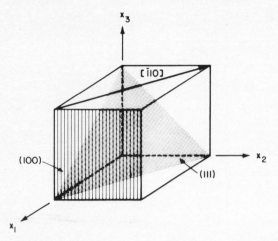

Fig. V.A1 Cubic crystal showing the (100)- and (111)-plane and the [$\bar{1}$10]-direction.

plane in Fig. V.A1. This is the body diagonal plane which intersects each coordinate axis at 1. It is therefore designated the (111)-plane. There are three other equivalent planes in the cubic system so that the set of planes {111} includes (111), (11$\bar{1}$), (1$\bar{1}$1), ($\bar{1}$11), where the bar above the number indicates that the intersection is on the negative portion of the axis. In most cases where Miller indices are used, it is only the direction of the plane relative to the crystal lattice which matters, not its distance from the origin. It is therefore common to divide out any common factor from the indices. Thus, the (200)-plane is usually considered to be the same as the (100)-plane. However in specifying the planes which give rise to x-ray diffraction it is often convenient to retain as specific planes the multiple values of the indices, so that designations (200), (220), etc. are often encountered in this context.

Miller Indices for Directions

Miller indices for directions in a crystal lattice are the components in the crystal coordinate system of a vector drawn from the origin parallel to the specified direction. Thus, the direction which is indicated as the diagonal of the face perpendicular to the x_3-axis in Fig. V.A1 has components -1, 1, and 0 along the x_1-, x_2-, and x_3-axes, respectively, when shifted so that it begins at the origin. This direction is denoted as [$\bar{1}$10], where the square bracket, [ ], indicates that a specific direction is meant. The symbolism ⟨110⟩

means any direction which is crystallographically equivalent to the direction $[110]$. There are six such directions $[110]$, $[101]$, $[011]$, $[\bar{1}10]$, $[\bar{1}01]$, and $[0\bar{1}1]$, if both the negative and positive directions are considered to be the same; i.e., $[10\bar{1}]$ is the same as $[\bar{1}01]$. *In the cubic system*, the direction $[hkl]$ is perpendicular to the plane (hkl), but this is only true for the cubic system.

Miller-Bravais Indices

The case of the hexagonal crystal system presents a special problem. In this system there are three equivalent directions in the basal plane of the unit cell as shown in Fig. V.A2. Any two of these directions could be taken as two of the coordinate axes for the designation of Miller indices, with the third axis being perpendicular to the basal plane. This would give a unique system for specifying planes. However, if one uses this system, crystallographically equivalent planes do not always have similar forms for their Miller indices. Thus, the three faces of the hexagon parallel to the axis designated x_4 in Fig. V.A2 are equivalent, but in a three-index system using axes x_1, x_2 and x_4 as the coordinate axes, the three faces would be

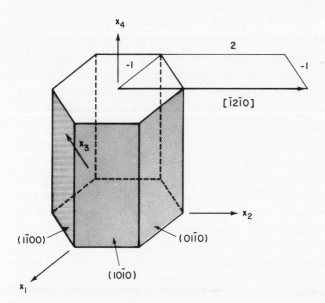

Fig. V.A2 Hexagonal crystal showing the four-index Miller-Bravais system. The $(1\bar{1}00)$-, $(10\bar{1}0)$-, and $(01\bar{1}0)$-planes are indicated. The $[\bar{1}2\bar{1}0]$-direction, defined to follow the rule $i = -(h + k)$, is also shown.

designated $(1\bar{1}0)$, (100), and (010). In order to make a system which has equivalent form for equivalent planes, one includes as an additional index the reciprocal of the intercept with the x_3-axis after the first two Miller indices. Thus, in the four-index Miller-Bravais index system, the three hexagon faces are denoted by $(1\bar{1}00)$, $(10\bar{1}0)$, and $(01\bar{1}0)$. It can be easily shown that the third index in this system is always given by

$$i = -(h + k) \tag{A1}$$

where the notation is $(h \ k \ i \ l)$.

A similar system is used for directions by following the rule given in Eq. A1. Thus, the direction which is parallel to the x_2-axis is designated $[\bar{1} \ 2 \ \bar{1} \ 0]$ rather than $[0100]$, since the latter does not obey the rule.

APPENDIX V-B

Crystal Structures and Slip Systems of the Common Metals

Metals are crystalline solids, so that their atoms are arranged in a regular repetitive fashion within each crystal grain of the poly-crystalline aggregate. Many of the mechanical properties of metals which have the same crystal structure are similar. For example copper, aluminum, and gold, which all have the same crystal structure, are relatively soft metals.

The basic structural unit of a crystal is the unit cell. The entire crystal is made up by repeating the structure of the unit cell over and over again in all three directions. Different types of crystal structures are classified on the basis of the symmetry possessed by the unit cell. There are fourteen basic types of unit cell which can exist called the Bravais lattices. Fortunately, most metals have relatively simple, highly symmetric structures, so that only three of the fourteen lattices need to be considered here.

a) Face-Centered Cubic Structure (f.c.c.)

Copper, aluminum, gold, and austenitic stainless steel have the face-centered cubic structure. (Austenite is the f.c.c. form of iron. In normal iron or plain carbon steel, iron has this form only at elevated

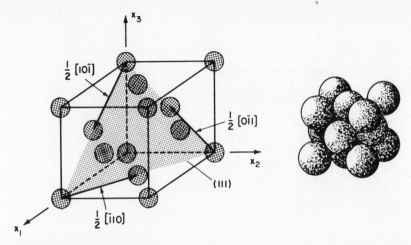

Fig. V.B1   Face-centered cubic (f.c.c.) crystal structure showing the 1/2 ⟨110⟩-type Burgers vectors on the {111}-type glide plane.

temperature. The addition of Ni and Cr to make stainless steel causes the f.c.c. structure to exist at room temperature.) Figure V.B1 shows the unit cell of the f.c.c. structure. The atoms are located at the corners of the cube and at the centers of the cube faces. In this structure, the dislocation Burgers vectors are the 1/2 ⟨110⟩ type, and the slip planes are of {111} type. Each of the four slip planes, (111), (11$\bar{1}$), (1$\bar{1}$1), and ($\bar{1}$11), contain three of the Burgers vectors, i.e., 1/2[10$\bar{1}$], 1/2[1$\bar{1}$0], and 1/2[01$\bar{1}$] lie in the (111)-plane. There are therefore twelve slip systems, if each possible combination of a plane and a Burgers vector is called a slip system. As mentioned above, the f.c.c. metals tend to be soft and ductile. This property is attributed to the fact that the {111}-planes are close-packed planes, making it easier for dislocations to slip.

b) Body-Centered Cubic Structure (b.c.c.)

Iron (at room temperature), molybdenum, tantalum, niobium, and tungsten have body-centered cubic structure. Figure V.B2 shows the unit cell of the b.c.c. structure. The atoms are located at the corners and the center of the cube. In this structure, the Burgers vectors are the 1/2 ⟨111⟩-directions. The slip planes vary from material to material. In iron, the most common slip planes are {110}-planes, but slip on {112} and other planes is also observed sometimes. There is some evidence that screw dislocations

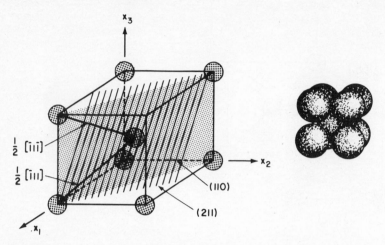

Fig. V.B2 Body-centered cubic (b.c.c.) crystal structure, showing two of the $1/2 \langle 111 \rangle$-type Burgers vectors, and the $\{110\}$- and $\{211\}$-type glide planes.

can move on any crystallographic plane which contains the Burgers vector. This is referred to as pencil glide. In some of the b.c.c. metals, slip on $\{112\}$-planes predominates. It should be noted that there are no truly close-packed planes in b.c.c. metals. These metals are harder and more brittle than f.c.c. metals.

c) Hexagonal Structure

This structure is often called hexagonal close packed (h.c.p.). Most of the metals which have the structure do not have the perfect ratio, $c/a = 1.633$, which is derived for close packing of spheres. Zinc, cadmium, magnesium, beryllium, and titanium have hexagonal structures. The atoms are situated at the corners of the unit cell, in the center of the top and bottom faces, and halfway between the top and bottom surfaces below the centers of three of the six triangles formed by the diagonals of the top surface. The unit cell is shown in Fig. V.B3.

The easiest slip plane in this structure is the basal plane, (0001), with the Burgers vectors being the a directions, $1/3 \langle \bar{1}2\bar{1}0 \rangle$. Since this is not a sufficient number of slip systems to accommodate an arbitrary shape change, other modes of deformation must occur if these materials are to deform. Secondary slip systems which are observed are $1/3 \langle \bar{1}2\bar{1}3 \rangle$-type Burgers

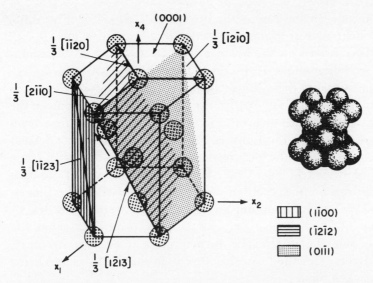

Fig. V.B3 Hexagonal close packed (h.c.p.) crystal structure, showing the $1/3 \langle \bar{1}\bar{1}20 \rangle$-Burgers vectors of the basal slip systems and the $1/3\,[1\bar{2}13]$-Burgers vector, which can move on the $(1\bar{1}00)$-, $(\bar{1}2\bar{1}2)$-, or $(01\bar{1}1)$-planes.

vectors moving on $\{10\bar{1}0\}$-, $\{01\bar{1}1\}$-, or $\{\bar{1}2\bar{1}1\}$-planes. Deformation can also occur by the formation of lattice twins.*

The c/a ratios for zinc and cadmium are greater than the ideal value of 1.633, being 1.866 and 1.856, respectively. Magnesium is close to the ideal ratio at 1.623, while the ratios for titanium and beryllium are less than the ideal at 1.591 and 1.568, respectively. Since slip is usually easier the larger the ratio of the slip plane spacing to the Burgers vector magnitude, materials such as zinc and cadmium exhibit easy basal slip, while materials such as titanium and beryllium are harder since basal slip is more difficult.

Both the f.c.c. and h.c.p. structures are close packed, meaning that if hard spheres are packed together, these two structures give the maximum number of spheres in a given volume. The difference between the two structures is the stacking sequence. If one imagines an arrangement of spheres in a plane so that they are packed as closely as possible, the characteristic hexagonal pattern is obtained,

---

*A lattice twin is a plate-shaped region inside which the structure is the mirror image of the rest of the crystal outside the twin. They can be formed by deformation when the lattice shears by less than a complete Burgers vector.

such as the one on the top of the hexagonal unit cell in Fig. V.B3. When the next row of atoms is added, the atoms fit in the "holes" between three atoms on the bottom row. However, once an atom is placed in a "hole," none of the three immediately adjacent "holes" can be filled because there is insufficient space. Therefore, the atoms in the second layer fill only half of the "holes" between atoms in the first layer. Call the atom positions in the plane of the first layer A-type positions, the positions of the atoms in the second layer B-type positions, and the positions of the unfilled "holes" C-type positions, as shown in Fig. V.B3. When the third layer of atoms is added, they can be placed either above the atoms of the first layer, i.e., in A-type positions again, giving ABAB . . . stacking order, or above the unfilled holes, i.e., in C-type positions, giving an ABCAB . . . stacking order. The former gives the h.c.p. structure, while the latter gives the f.c.c. structure with the $\{111\}$-planes being the close packed planes.

The stacking sequence in a given metal can be violated when a dislocation does not have a complete Burgers vector. For example, in an f.c.c. metal a complete Burgers vector extending from position A to the nearest position A may be split into two partial dislocations, one going from an A to a B position and the other from a B to an A position. When two such partial dislocations are created by dissociation of a dislocation with a complete Burgers vector, they cannot stay next to each other because of the repulsive forces acting between them. As a consequence, these two partial dislocations separate from each other. Then, the region between these two partial dislocations has a different stacking sequence from the rest of the crystal. This is called a stacking fault, and the energy associated with the stacking fault as a result of the disruption of the atomic order is called the stacking fault energy. The size of the stacking fault region is determined by the stacking fault energy and the repulsive force acting between the two partial dislocations. The two partial dislocations will move apart until the repulsive force is insufficient to do the work necessary to create the additional stacking fault energy.

CHAPTER SIX

Visco-Elastic-Plastic Deformation of Polymers

VI-1 INTRODUCTION

In Chapters III and IV, the time-independent elastic and plastic deformation of metals was discussed from the continuum point of view. In discussing the elastic deformation, it was pointed out that elastic deformation is reversible and that it is induced by all stress components. Plastic deformation was shown to be time-independent permanent deformation which is caused only by the distortional (shear) stress components. The term time-independent permanent deformation was used to signify the fact that a given shear stress increment in the plastic regime produced a corresponding plastic shear strain increment, regardless of the duration of loading. Implicit in this statement is the assumption that the plastic deformation occurs within a time period shorter than the usual loading period.

There is yet a third type of deformation which is *irreversible*, *permanent*, and *time dependent*, called *viscous deformation*. Those materials which exhibit viscous behavior deform further and further as the duration of loading increases. This type of viscous behavior is exhibited by metals at high temperature and by plastics even at room temperature. This chapter examines the visco-elastic-plastic behavior of polymers for the purpose of applying this knowledge to the design and processing of plastics.

In a somewhat oversimplified manner, it may be stated that the genesis of viscous deformation lies in the thermal energy possessed by atoms and molecules in solids. The thermal energy of atoms is manifested in the vibration of atoms about their equilibrium sites. Some atoms have sufficient energy to overcome the bonding energy that holds them in their equilibrium sites. If the bonding energy is denoted by E, the probability of overcoming this energy barrier is given by $\exp(-E/kT)$,* where kT represents the average thermal energy possessed by each atom. In the absence of any external load this probability is the same in *all* directions and, therefore, there is no net deformation of the solid in any particular direction. However, if an external shear load is applied along a given direction, the permanent switching of atomic and molecular positions along the stressed direction is favored, resulting in a gradual deformation along the loaded direction. If a tensile specimen made of polystyrene is loaded axially at room temperature, it will stretch or *creep* continuously at a rate determined by the probability that the energy barrier can be overcome by the thermal energy of atoms and molecules.

As in the plastic deformation of metals, viscous deformation is only caused by shear (distortional) stresses, and the hydrostatic component of stress primarily induces elastic volume change. The mechanical work done during viscous deformation is dissipated in the form of thermal energy. Since it was shown that the creep rate is an exponential function of temperature, the internal dissipation of mechanical work drastically affects the overall response of a viscous material by influencing the temperature of the body.

From the foregoing discussion, it is obvious that the creep of solids depends a great deal on temperature of the solid relative to its

---

*The rate of change of any thermally activated process such as diffusion, evaporation, or decomposition of a solid, or thermally assisted deformation can be expressed in terms of the Arrhenius relationship $N_0 \nu \exp(-E/kT)$, where N_0 is the number of active sites and ν is the characteristic frequency of oscillation. The term $N_0 \nu$ is often called the frequency factor of the process. Thermal activation is discussed further in Sec. VI-7.

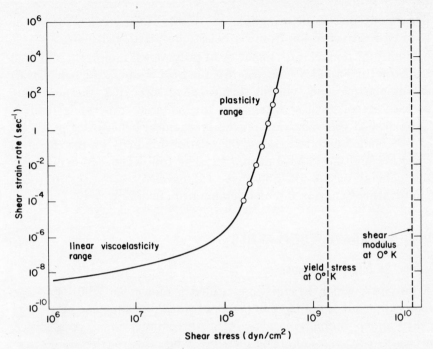

Fig. VI.1 Strain rate of glassy polyethylene terephthalate (PET) as a function of the applied shear stress, showing the transition from linear viscoelastic to plastic behavior. (*After A. S. Argon.*)

melting point (or, in the case of glassy amorphous polymers, on the glass-transition temperature*). This dependence is due to the fact that the bonding energy of atoms in a solid is directly proportional to the melting temperature of the solid. Therefore, solids with low melting points, such as polymers and certain metals like lead, cadmium, etc., creep even at room temperature, whereas metals with high melting points, such as steel and tungsten, do not creep at noticeable rates at room temperature.

When a shear load is applied to a polymer, the immediate mechanism of deformation is elastic. This is followed by thermally activated creep or viscous deformation. In many structural applications, the viscous strain eventually greatly exceeds the elastic strain, so that designs can be based on the viscous component of strain exclusively. When very high loads are applied to a polymer, immediate

---

*The glass transition temperature is the temperature where polymer behavior changes from brittle to rubbery or viscous.

permanent plastic deformation can result. Figure VI.1 shows the nature of the transition from viscoelastic to plastic behavior.

The room temperature behavior of polymers is important in design and structural applications, whereas the high temperature behavior of thermoplastics and viscous behavior of precured thermosetting plastics are of greatest interest in the materials processing field. A great deal of generalization on the mechanical behavior of polymers can be made by understanding the relationship between the chemical structure and mechanical properties. The following section discusses the general nature of polymers, followed by sections on room temperature behavior and high temperature behavior.

VI-2   AN OVERVIEW OF PLASTICS

VI-2-a   Structure of Polymers

A polymer, which is sometimes called a molecular solid, is a large molecule made up of large numbers of small repeating chemical units called *mers*. Some polymers form three-dimensional cross-linked networks (i.e., thermosetting plastics), while others (thermoplastics) form "two-dimensional" linear networks. All of the mers in the network are joined together by covalent bonding. Although the polymers (plastics) of most common knowledge are synthetic polymers, all living systems are made of natural polymers, e.g., protein in animals and cellulose in wood. The molecular weight of a typical "two-dimensional" polymer is in the range of 10^4 to 10^7 gm/gm-mole. Nylon has a molecular weight of about 15,000, while cellulose and polymethylmethacrylate (Lucite) have molecular weights of 2×10^5 and 5×10^5, respectively. Ultrahigh molecular weight polyethylene has a molecular weight as high as 10^7. Polymers with extremely low molecular weights lack the mechanical strength required in engineering applications, while polymers with molecular weights that are too high are difficult to process because of their high viscosity. If a metal is analogous to hard spheres stacked together in a systematic way with cohesive forces between them, a polymer may be likened to a dish of spaghetti where each noodle is a molecule and the sauce represents the plasticizer.*

---

*Plasticizers are high boiling point liquids or low molecular weight polymers, used as solvents for plastics, which weaken the attractive forces between the molecules, making thermoplastics more pliable.

VI-2-b Classification of Polymers

Polymers may either be classified in terms of the nature and the arrangement of the chemical bonds or in terms of their crystallinity and morphology. In terms of the nature and the arrangement of chemical bonds (or in terms of thermomechanical properties), polymers may be classified into the following four groups: *thermoplastics*; *thermosetting plastics*; *elastomers*; and *ionomers*.* Thermoplastics are sensitive to temperature changes; they soften with increased temperature and, if crystalline, have melting points. Thermosetting plastics, on the other hand, do not soften at high temperatures. They decompose at high temperatures before appreciable softening. The difference stems from the fact that separate linear chains of molecules are held together only by weak secondary forces, in the case of thermoplastics, enabling the chains to move individually. Strong primary covalent bonds between chains, in the case of thermosets, prevent individual movement of each chain. When the temperature of a thermosetting plastic exceeds a critical value, it decomposes. An elastomer is similar to a thermosetting plastic, except that it can undergo large nonlinear elastic deformations. Elastic nature is introduced into thermosetting-type polymers by limiting the number of cross links between the molecular chains. The neighboring molecules are free to move relative to each other, since the chains are not everywhere tied down. Yet the limited number of cross links that are present do prevent permanent changes on a larger scale. Ionomers, which have been introduced very recently, are similar to thermoplastics in that they can deform at high temperatures. They are also similar to elastomers in that the neighboring chains are held together by occasional primary bonds, except that, in this case, the cross links are ionic bonds rather than covalent bonds.†

Polymers may also be classified in terms of the following states of aggregation: amorphous (or glassy type) and crystalline type. Certain thermoplastics, such as polyethylene and polypropylene, are partially crystalline. Therefore, they have melting points, are ductile, and are

---

*This type of classification is somewhat arbitrary. Sometimes ionomers and rubbers are treated as thermoplastics and thermosetting plastics, respectively.

†There is also so-called thermoplastic rubber, which softens like a thermoplastic, but is rubbery at room temperature. This type of thermoplastic rubber is easy to process. Cross linking in such a rubber can be reversibly broken and reformed.

translucent. On the other hand, certain other thermoplastics, such as polystyrene, polymethylmethacrylate, and polycarbonate and all thermosetting plastics, are amorphous or glassy. Consequently, these polymers tend to be brittle (except polycarbonate), are usually transparent, and have no melting point.

VI-2-c Polymers as Engineering Materials

Polymers are likely to find ever-increasing application in the future, since their physical properties have been improving, their price has been decreasing, and they are easily processed. If one can look into a crystal ball and predict the future, it may not take too much searching to see that the future of polymers is indeed quite bright, since their availability is not likely to be limited in the same sense that the supply of metals is limited. Although the natural resources for polymers are not as abundant as those for some of the ceramics, they can be replenished, at least in theory, by growing organic materials such as trees and algae. If the population growth can be checked, the future of polymeric materials, other than those based on carbon, also looks promising.

The major disadvantages of polymers as engineering materials are that they are thermally unstable, have low moduli, have low strength, and are difficult to recycle. Also, resistance to prolonged exposure to ultraviolet light is, in general, poor in comparison to other engineering materials. Because of the covalent nature of their bonding, they cannot conduct electricity or heat.

The great challenges for mechanical engineers whose interest is in polymers seem to lie in being able to devise really ingenious new processing techniques which can substantially lower the manufacturing cost, and also in understanding and controlling the processes of polymeric deformation and fracture in order to improve the mechanical properties of the resulting plastics.

VI-2-d What Controls the Physical Properties of Polymers?

The physical properties of a polymer are primarily controlled by such basic chemical properties as:

1. size of molecule
2. molecular structure and shape
3. primary and secondary bond strengths

through their influence on
1. crystallinity
2. viscosity, surface tension, volatility, etc.
3. flow stress
4. melting and glass transition temperatures
5. brittleness
6. decomposition temperature
7. optical properties
8. thermal conductivity.

The relationship between the chemical bonds, chemical structures, and physical properties may be summarized as follows:

1. The physical properties of the thermoplastics are governed by secondary bond forces between the molecules, rather than by the primary covalent bond forces which hold all the atoms in a given molecule to each other. Polymers with low secondary bond forces generally have low melting points, coefficient of friction, and cohesive energy. Since the coefficient of friction is related to the bulk shear stress, the bulk shear stress also decreases with decrease in the secondary bond forces. This can be accomplished by increasing the distances between the molecules through the incorporation of plasticizers or by increasing the thermal energy of the molecules through heating.

2. The physical properties of thermosets which derive all of their strength from primary covalent bonds do not change a great deal with temperature until it is high enough to break these primary bonds. At such a high temperature, the plastic decomposes, thereby losing its original physical properties.

3. Deformation of polymers by creep is a sensitive function of the degree of cross linking. Thermoplastics creep extensively, whereas thermosetting plastics and elastomers do not.

4. Thermoplastics are either completely amorphous, glassy, or partially crystalline. Molecules with symmetric, short repeating units exhibit local crystallinity, e.g., polyethylene and polypropylene. Crystalline thermoplastics are tougher and translucent. Amorphous thermoplastics, e.g., polystyrene and polymethylmethacrylate, are in general brittle and transparent. However, polycarbonate, a glassy polymer, is relatively tough. Ionomers, which are similar to polyethylene, are thermoplastics with greater tensile strength, stiffness, and toughness than conventional thermoplastics. Ionomers are less crystalline than

linear polyethylene and, thus, more transparent. Desired physical properties are often obtained by copolymerization, e.g., brittle polystyrene is toughened by copolymerizing it with rubbery material. Impact-grade polystyrene and ABS are examples of such a copolymer. Thermosetting plastics are amorphous.

5. Since the intermolecular distances are affected by the hydrostatic pressure, the flow stress of polymers depends much more on the hydrostatic stress than does the flow stress of metals.

6. The frictional behavior of thermoplastics differs sharply, depending on whether the interface is above or below the melting temperature. At temperatures higher than the melting point, hydrodynamic mechanisms govern the shear force, whereas at lower temperatures, the shearing of interface junctions governs the friction mechanism. The hydrostatic pressure dependence of the flow stress causes the coefficient of friction to depend on the normal load, i.e., the coefficient of friction decreases with the normal load.

VI-3 RELATIONSHIP BETWEEN PHYSICAL PROPERTIES AND CHEMICAL BONDS AND STRUCTURES

VI-3-a Chemical Bonding and Cohesive Strength of Polymers

In order to appreciate the relationship between atomic bonds and physical properties, consider the molecular structures of polyethylene and phenol-formaldehyde shown in Fig. VI.2. It should be recalled that polyethylene is a thermoplastic, while phenol-formaldehyde (Bakelite) is a thermosetting plastic. The bars between the atoms indicate covalent bonds.* The free bar, which does not end at an atom, is used to indicate that it is a reaction site which is available for reaction with other molecules. Those reaction sites may be generated either by breaking a double bond (addition polymerization) or by letting a part of the molecule react† with other molecules

---

*Covalent bonds are formed when one or more pairs of valence electrons are shared between the atoms, resulting in stable electronic shells. It is a *primary* bond. The other two types of primary bonds are the ionic bond and the metallic bond.

†The part of a molecule which readily reacts with other molecules (e.g., OH, the hydroxyl group in ethyl alcohol) is called the functional group, and the group that survives a reaction without undergoing any change is called the radical group (i.e., the C_2H_5 ethyl group in ethyl alcohol).

(condensation polymerization), or by converting cyclic compounds to open chains for addition polymerization (cyclic polymerization), as shown in Fig. VI.3. A molecule with two reaction sites is called bifunctional, and one with three reaction sites is called trifunctional, etc. In polyethylene there are two reaction sites, created by the rupture of the double bond of ethylene monomer, and, therefore, ethylene is bifunctional. When a monomer reacts with other similar monomers, the molecule increases in length until it terminates by

(a) Thermoplastics

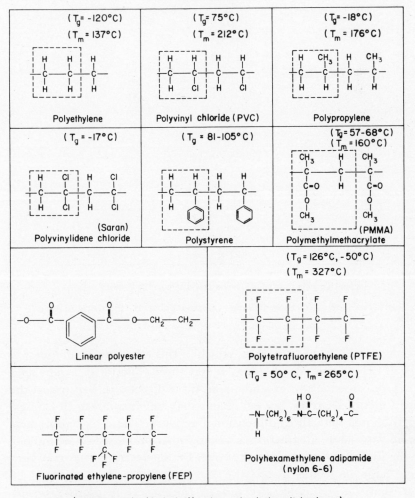

(repeating units blocked off, unless only single unit is shown)

Fig. VI.2 Structure of some common polymers. (a) Thermoplastics.

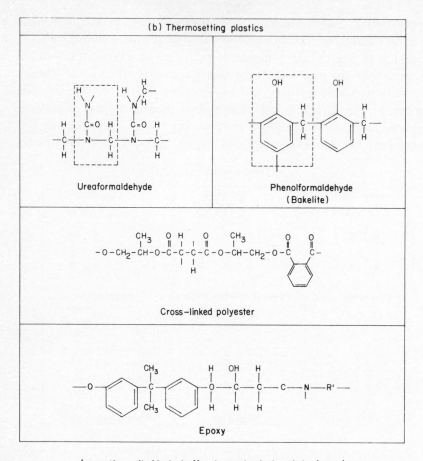

(repeating units blocked off, unless only single unit is shown)

Fig. VI.2 (*Continued*) Structure of some common polymers. (b) thermosetting plastics.

reacting with a terminal group, which in this case may be the methyl radical, CH_3. Phenol-formaldehyde has three functional groups and, therefore, the reaction extends in all three directions. Because of the third functional group, several chains may be *cross linked*, forming a large polymer with three-dimensional network. These space polymers can be very large, all the bonds being primary bonds. For example, a molded Bakelite part such as the cover for an electric outlet may be a single large molecule. From the foregoing discussion it should be clear that *the cohesive strength of thermosetting plastics results principally from the primary covalent bonding*.

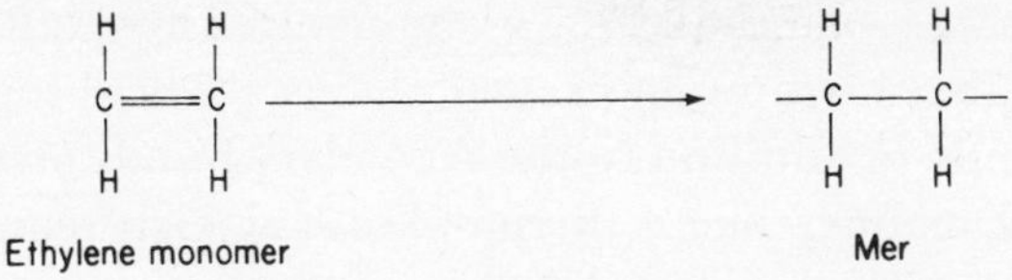

Ethylene monomer   Mer

(a)

[Generation of two reaction sites (bi-function)
by breaking double bonds in addition polymerization]

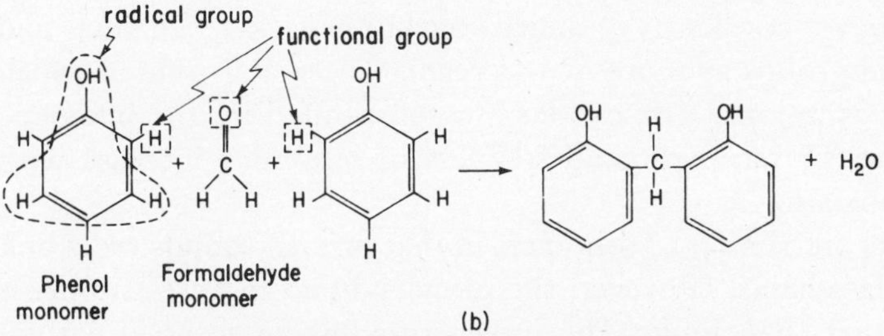

radical group

functional group

Phenol monomer   Formaldehyde monomer $+$ H_2O

(b)

(Reaction sites provided by the functional group
in condensation polymerization)

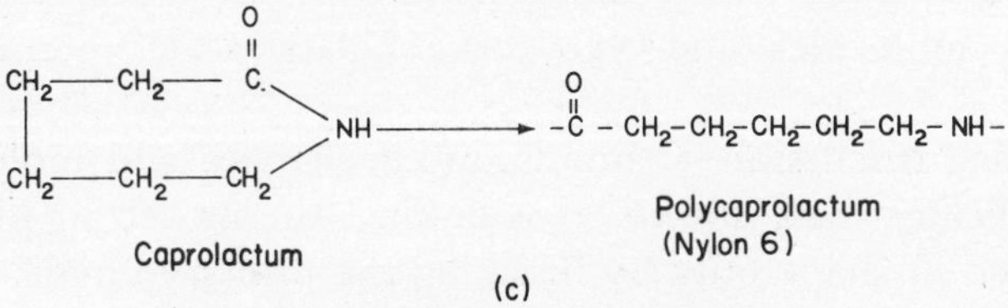

Caprolactum   Polycaprolactum
(Nylon 6)

(c)

(Reaction sites provided by cyclic
compounds in cyclic polymerizaton)

Fig. VI.3 Schematic representation of types of polymerization reactions.

A linear molecule such as polyethylene does not have such a large molecular weight. Therefore, an item made of polyethylene has a large number of polyethylene molecules which are not joined together by primary bonds. *The cohesive force (the intermolecular force) between the molecular chains in this case results from the secondary bonds.** Many of the physical properties, such as viscosity, miscibility, volatility, surface tension, frictional properties, and solubility, are closely related to the intermolecular forces and, thus, to the secondary bond energy.

*Secondary bonds arise from short-range electrostatic attractions between molecular dipoles. Hydrogen bonds and van der Waals forces are examples of secondary bonds.

An ionomer (trade name Surlyn, manufactured by duPont) is a modified thermoplastic whose molecular chains are held together by primary ionic bonds in addition to the secondary bond forces. These flow in a viscous manner like a thermoplastic at high temperatures but have higher strengths and moduli. Ionic bonding is accomplished by adding an acid group such as the carboxylic acid group (–COOH) to the chain and by forming the carboxylic ion through neutralization of the acid group with an inorganic base, as shown in Fig. VI.4. Ionomers have covalently bonded backbones (i.e., chains) and pendant acid groups at more or less regular intervals, some of which are present as anions. The cations are contributed by the inorganic base. The tensile strength and stiffness of ionomers increase with degree of ionization.

Elastomers are similar to ionomers in that primary bonds cross link the molecular chains. However, the primary bond in an elastomer is covalent rather than ionic. The usual cross linking agent in natural and synthetic rubber is sulfur. The chemical structure of natural rubber is shown in Fig. VI.5. The elastic nature is due to the limited number of cross links which do not hinder local chain mobility between the molecules and yet prevent large-scale, permanent distortion.

The foregoing discussion of bonds, bond energies and molecular structures indicates that, due to cross linking, thermosetting plastics can be changed in shape only by breaking the primary bonds. When the primary bonds are broken, the molecules are no longer the same. This process of breaking primary bonds at high temperatures is called thermal *decomposition*. All plastics decompose when heated to high temperatures. Partially decomposed plastics are in general brittle and yellowish, indicating a loss of optical clarity. In the case of thermoplastics, however, the force that needs to be overcome for deformation is the secondary bond force. Since the secondary bond energies are much smaller than the primary bond energies, as shown

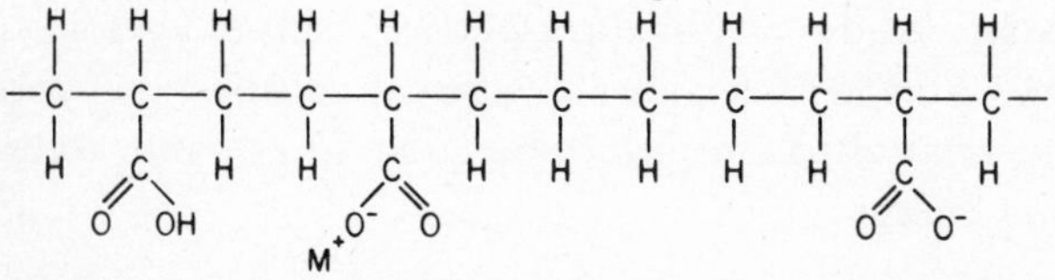

Fig. VI.4 Backbone of an ionomer molecule (tradename Surlyn).

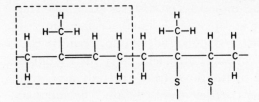

Fig. VI.5 Structure of natural rubber. Sulfur reacts with the rubber by breaking the carbon-carbon double bond.

in Table VI.1, thermoplastics can be deformed at temperatures below the decomposition temperature. In Table VI.1, it should also be noted that the bond energies of double and triple bonds are much greater than those of single bonds.

VI-3-b Cohesive Strength of Polymers and Other Physical Properties

Because the secondary bond forces stem from the electrostatic force, the cohesive force of thermoplastics depends on the distance between these molecules and on the effectiveness of shielding accorded to "charged" ions by other atoms of the molecule. In the case of polytetrafluoroethylene, Teflon, the large fluoride ion shields the positive charge on the carbon atom, resulting in small intermolecular cohesion. The cohesive force in polyethylene is larger than that in Teflon, since the small hydrogen ions cannot shield as

TABLE VI.1 Energies of Various Bonds

Primary bonds		Secondary bonds				
		Hydrogen bonds		Dispersion		Dipole	Induction
Bonded atoms	Dissociation energy kcal/mole	Bonded atoms	Dissociation energy kcal/mole	Molecules	Inter-molecular energy kcal/mole	Inter-molecular energy kcal/mole	Inter-molecular energy kcal/mole
C-C	59–70	F-H-F	6.2	A	2.03	0	0
C=C	100–125	O-H-O	5.9–10.2	CO	2.09	0.0001	0.0002
C-H	87–94	N-H-O	6.0– 6.8	HI	6.18	0.006	0.002
C-N	49–60			HBr	5.24	0.16	0.12
C=N	150–185			HCl	4.02	0.79	0.24
C-O	70–75			NH_3	3.52	3.18	0.37
C=O	142–166			H_2O	2.15	8.69	0.46
N-H	84–97						
O-H	101–110						

well as fluorine ions, and the molecules are closer together. The bonding forces of polystyrene and polymethylmethacrylate are still higher due to the benzene ring and polar side group, respectively. The flow stress increases with cohesive force, and, since the frictional properties are directly related to the flow stress, the polymers with higher flow strength have higher coefficients of friction. The cohesive energy densities of polytetrafluoroethylene, polyethylene, polystyrene, and polymethylmethacrylate are 38, 62, 74, and 83 cal/cm^3, respectively. The coefficient of friction increases in a similar manner.

Another important physical property of thermoplastics resulting from the nature of the cohesive force is the dependence of flow stress on hydrostatic pressure. The flow stress in polymers depends much more strongly on the hydrostatic pressure than in metals, since the intermolecular distances are more easily affected by the hydrostatic pressure. This is shown by the relatively low bulk moduli of plastics compared to those of metals. The flow stress is a nearly linear function of the applied pressure.

Secondary bond forces, and thus the cohesive strength, also influence the degree of crystallinity and the *melting point**** of thermoplastic polymers. Crystallinity is important since it affects the mechanical properties. In polymers long-range ordering is completely absent, although a crystalline structure is the lowest energy structure of a polymer. This is a result of the size of molecules, steric hinderance of the side groups, and the stiffness of chains and entanglements. There is, however, short-range ordering in small regions of polymers (of the order of a few hundred angstroms). Depending on the conditions of crystallization, polymers can be crystallized in different morphologies. Figure VI.6† shows the spherulite structure of a crystalline polymer as nucleated from melt and the folding of molecules in single polymer crystals. Unlike metals, polymers do not have a sharp melting temperature, but there is a range over which crystallinity gradually disappears. At the melting point some of the physical properties, such as density, index of refraction, heat capacity, and optical properties, undergo discontinuous changes, as shown in Fig. VI.7 in terms of the volume

---

*Melting temperature (a melting point) is the temperature at which crystallinity disappears completely. The melting temperature is sometimes referred to as the first-order transition temperature.

†The crystalline region is where the molecule takes the form of a folded chain.

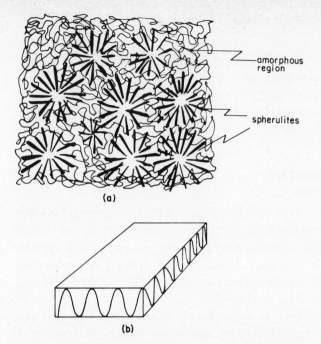

(a)

(b)

Fig. VI.6 Schematic representation of a crystalline polymer. (a) Small crystallites surrounded by amorphous polymer. (b) A folded chain single crystal.

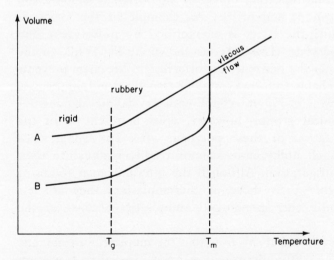

Fig. VI.7 Schematic representation of the effect of temperature on the volume of amorphous and crystalline phases of a polymer. The amorphous phase can be formed by rapid cooling from above the melting temperature even in the case of crystalline polymers. (Curve A) amorphous material; (curve B) crystalline material; (T_g) glass transition temperature; (T_m) melting temperature of the crystalline material.

change. There are certain molecular solids, such as most of the commercial polystyrenes, which, because of their molecular structure, do not have any crystalline region. These polymers do not have a melting point, but undergo a secondary transition from the glassy to the rubbery or viscous state.

VI-3-c   Molecular Structure and Physical Properties

So far the discussions have mainly concerned the relationship between the physical properties and the secondary bond forces, in the case of thermoplastics, and the relationship between the physical properties and cross linking by primary bonds, in the case of thermosetting plastics. Now the relationship between the molecular structure and physical properties will be discussed.

The earlier discussion on crystallinity dealt primarily with the effect of secondary bond forces on short-range ordering. However, a polymer's crystallinity also depends on its molecular structure. Only the thermoplastics with reasonably symmetric, short, repeating units exhibit any extended crystallinity. Thermoplastics with low symmetry, or with long repeating units, are not often crystalline and tend to be amorphous and brittle. An example of the former is polyethylene, while the latter is exemplified by polystyrene and polymethylmethacrylate (PMMA*), as shown in Fig. VI.2. In the case of polystyrene, the benzene ring attached to the chain prevents close packing of chains to form crystalline structure; and, in the case of PMMA, the ester group interferes with crystalline alignment.[†] These large chemical groups, however, assist in retention of the mechanical orientation of these polymers after stretching, since molecules with such bulky units cannot readily readjust to their original chain configuration. Although the intermolecular forces of ionomers are higher than those of thermoplastics, they are less crystalline than the corresponding thermoplastics because of the large acid group.

It should be mentioned at this time that thermoplastics exhibit the *memory effect*, i.e., after deformation below the glass transition

---

*Tradenames are Lucite (duPont) and Plexiglas (Rohm and Haas).

†Even polystyrene can be crystalline if the benzene rings are all attached to one side of the backbone chain in a regular manner. However, in most commercially available polystyrene, the arrangement of the benzene rings is not this regular, and crystallization does not occur.

TABLE VI.2   Chemical Structure and Melting Point of Polymers (Ref. 11)

Structure	Melting point °C
$-CH_2-CH_2-CH_2-$	115
$-\langle\bigcirc\rangle-CH_2-CH_2$	380
$-O(CH_2)_2OCO(CH_2)_6CO-$	45
$-OC(CH_2)_3OCO-\langle\bigcirc\rangle-CO-$	264
$-O(CH_2)_2OCO(CH_2)_{12}CO-$	80
$-O(CH_2)_2OCO-\langle\bigcirc\rangle\langle\bigcirc\rangle-CO-$	330

temperature, the polymers recover their original shape upon reheating above that temperature. This is caused by the fact that the molecules strained during deformation tend to return to their minimum energy configuration when the thermal energy is sufficient to overcome constraining forces.

The melting point and the glass transition temperature of a linear polymer are also controlled by the molecular structure of its chain, as shown in Table VI.2. In general, factors affecting chain flexibility or mobility (whether of the chain itself or of the secondary interactions between chains) will affect both the glass transition temperature and the melting temperature. Inclusion of divalent bond, such as oxygen, in the backbone of the chain decreases the melting point, due to the increased chain flexibility resulting from free rotation around the bond. On the other hand, a ringlike structure, such as a benzene ring, in the chain hinders the bond rotation and thereby increases chain stiffness, thus raising the melting point. Higher secondary bondings, such as hydrogen bonding, will decrease the chain mobility, thereby increasing the melting temperature. Bulky side groups as well as plasticizers prevent chains from packing close together and thus reduce the secondary bonding forces. Therefore, both melting point and glass transition temperatures are decreased by bulky side groups. By a similar reasoning, one can readily conclude that double and triple bonds should increase the melting points of polymers. Recent advances in polymer chemistry for high-temperature applications have been made by incorporating into the chain a series of interconnected rings called ladder polymers.

Crystallinity was shown earlier to depend on both the secondary bond energies and the molecular structure. Crystallinity, in turn, determines the optical properties of polymers. Most polymers are clear and transparent in the visible spectrum of electromagnetic radiation, because polymers are nonconductors without any free electrons. In nonconductors, the electrons are tightly bound to the nucleus and cannot easily be excited by the low-energy photons. Therefore, little absorption or reflection of incident electromagnetic energy takes place. Since amorphous polymers are also free of internal light-scattering centers, they are transparent. However, when a polymer is partially crystalline, the forward scattering of light occurs at the interfaces between the amorphous and crystalline regions, and the material becomes translucent. This effect can be clearly demonstrated in polyethylene or polypropylene. Polyethylene is translucent at room temperature, but becomes transparent at its melting point.

VI-3-d Crystallinity, Chain Orientation, and Mechanical Properties

Crystalline polymers are generally stronger and tougher than amorphous polymers. This is because cracks cannot readily propagate through the crystalline regions when the direction of crack propagation is normal to the molecular axis. In many polymers, it is desirable to have high degree of crystallinity but with smaller crystallite sizes, because the impact strength increases as the size of crystallites decreases.

The orientation of molecular chains also increases the strength and toughness of polymers. Textile fibers are oriented mechanically by a stretching along the fiber axis to improve strength and toughness. This is commonly called cold (or hot) drawing. Brittle polystyrene sheets can be strengthened by stretching them biaxially in the plane of the sheet. However, merely stretching a polymer does not guarantee orientation of the molecules unless the stretching temperature is in a range where the thermal energy is too low to permit free movement of the polymer chains and yet is high enough to allow viscoplastic deformation to occur during stretching. The mechanical orientation of a polymer is usually done near the *glass transition temperature* (T_g),* sometimes called the *second-order transition temperature*. It is

---

*As a rough rule of thumb, the glass transition temperature T_g is approximately $1/2$–$2/3$ T_m for crystalline polymers.

the temperature at which the polymer makes the transition from its glassy state to the rubbery or viscous state.

Below the glass transition temperature, the molecules are in a frozen state where molecular movement other than localized bond rotation and vibration is limited. At the glass transition temperature, it is believed that local movement of molecular segments and rotation of side groups takes place. Above T_g, the molecules have more energy, and the motion of large segments of a molecule becomes possible. However, movement of the complete molecule as a unit does not take place. Flow is believed to occur only when there is a general correlated movement of many molecular segments.

VI-4   GENERAL DEFORMATION CHARACTERISTICS OF POLYMERS

The phenomenological constitutive relationships of real polymers are quite complex. Except under some special circumstances, it is difficult to generalize the behavior of polymers over a long time period, because load, temperature, and loading time affect the constitutive relationship. However, it is clear that polymers behave in a viscoelastic-plastic manner. At low temperatures and high loads, the elastic and plastic deformations dominate, whereas at high temperatures and high loads, the viscous behavior is more important. In this section the viscoelastic nature of polymers will necessarily be oversimplified in order to determine the important physical parameters. The reader should gain an appreciation of this viscoelasticity from the following presentation but should be careful not to draw conclusions that are too sweeping concerning the behavior of real materials.

VI-4-a   Models and Constitutive Relations

One of the distinguishing characteristics of the mechanical behavior of polymers at room temperature under a moderate load is *viscoelasticity*. The viscoelastic nature of polymers may be approximately modeled by spring-dashpot systems. Two of the well-known models are the Maxwell model and the Voigt-Kelvin model (Fig. VI.8). These models describe *linear* viscoelasticity, which is an approximation to real viscoelastic behavior of polymers. Certain generalizations can be made by using these models, although the past attempts to develop a constitutive relationship for real solids using a

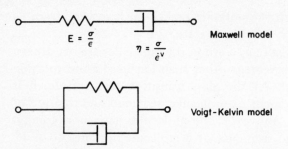

Fig. VI.8 Spring and dashpot models for a viscoelastic solid.

combination of many springs and dashpots have been largely unsuccessful. Qualitatively, it is shown by the Maxwell model that at the instant of loading, the solid behaves elastically, and then viscous deformation occurs, since the spring deforms *instantaneously* and the deformation *rate* of the dashpot is proportional to the applied stress. In the case of the Voigt-Kelvin solid there is no instantaneous deformation, and the maximum strain is limited by the spring.

When the magnitude of the applied load is very high, plastic deformation may also take place immediately upon loading. Such a solid may be represented by a model which also includes a friction element, as shown in Fig. VI.9.

Although the mechanical behavior of real polymers cannot be modeled perfectly by combinations of springs and dashpots, many characteristics of real polymers can be understood from the characteristics of the simple Maxwell and Voigt-Kelvin models. Consider the constitutive relationship for the Maxwell model. The total strain ϵ_{ij} is the sum of elastic and viscous deformation, i.e.,

$$\epsilon_{ij} = \epsilon_{ij}^{e} + \epsilon_{ij}^{v} \qquad (VI.1)$$

where ϵ_{ij}^{e} is the elastic strain, and ϵ_{ij}^{v} is the strain due to the viscous part. These may be related to the stresses by

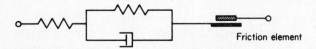

Fig. VI.9 Model for a visco-elastic-plastic solid.

$$\epsilon_{11}{}^{e} = \frac{\sigma_{11}}{E} \qquad \text{(VI.2)}$$

$$\frac{d\epsilon_{11}{}^{v}}{dt} = \frac{\sigma_{11}}{\eta} \qquad \text{(VI.3)}$$

Differentiating Eqs. (VI.1) and (VI.2), and combining them with Eq. (VI.3), the constitutive relation for the Maxwell solid may be written as

$$\frac{d\epsilon_{11}}{dt} = \frac{1}{E}\frac{d\sigma_{11}}{dt} + \frac{\sigma_{11}}{\eta} \qquad \text{(VI.4)}$$

The constitutive relationship for the Voigt-Kelvin model may similarly be derived. From Eq. (VI.4) it is seen that the usual stress-strain relationship obtained by tensile testing will be very sensitively dependent on the strain rate. For a material with very large viscosity or under high loading rates, the Maxwell solid behaves as a Hookeian solid and the Voigt-Kelvin solid behaves as a rigid solid. In this respect, the behavior of polymers is more closely represented by the Maxwell model than by the Voigt-Kelvin model.

Of course, real polymers cannot adequately be described in terms of these simple models. By adding more springs and dashpots, one can approximate the real behavior, at least in theory. The constitutive relation for any arbitrary number of these elements may be represented as

$$\sum_{m=0}^{a} P_m \frac{d^m \sigma'_{ij}}{dt^m} = \sum_{n=0}^{b} q_n \frac{d^n \epsilon'_{ij}}{dt^n} \qquad \text{(VI.5)}$$

where the terms P_m and q_n are constants. Equation (VI.5) only includes the deviator components of stress and strain. The hydrostatic compression of viscoelastic materials is elastic, and, therefore, the hydrostatic components of stress-strain can be related in terms of the bulk modulus of the material.

Equation (VI.5) is a linear ordinary differential equation. A solid whose constitutive relationship can be represented by Eq. (VI.5) is called linear viscoelastic. Note the proportionality constants P_m and q_n are independent of the magnitude of the applied stress. Therefore, in a linear viscoelastic solid under a constant stress distribution, strain is always linearly proportional to stress at any given time. Real polymers behave in a linear viscoelastic manner over a limited range of load; over a wider range of load, the behavior is nonlinear.

VI-4-b Deformation of the Maxwell Solid under Constant Stress (Creep)

The creep of the Maxwell solid will be examined now using Eq. (VI.4). In a creep test, the engineering stress is held constant, i.e., σ_0 = constant, and the strain is measured as a function of time. Solving Eq. (VI.4) and noting that at $t = 0$ we have $\epsilon = \sigma_0/E$, the strain is given by

$$\epsilon = \left(\frac{1}{E} + \frac{t}{\eta}\right)\sigma_0 = \frac{\sigma_0}{E}\left(1 + \frac{Et}{\eta}\right) = J(t)\sigma_0 \qquad \text{(VI.6)}$$

The term $J(t)$ is sometimes known as the *time-dependent compliance*. Note that the creep rate is dictated by the ratio η/E. In real polymers, the rate depends on stress, loading time, and temperature.

VI-4-c Stress Relaxation of the Maxwell Solid

An alternative to the creep test is the stress relaxation test, particularly when the deformation of the specimen is so large that the true applied stress varies substantially with deformation. In the stress relaxation test, a specimen is stretched to a constant strain, and the change in the applied stress is measured. The stress relaxation of the Maxwell solid, when the solid is suddenly stretched to a given ϵ_0, can be expressed by solving Eq. (VI.4) as

$$\sigma = E\epsilon_0 \exp\left(-\frac{Et}{\eta}\right) = E_r(t)\epsilon_0 \qquad \text{(VI.7)}$$

The relaxation rate is governed by the *relaxation constant* τ, which is defined as

$$\tau = \frac{\eta}{E} \qquad \text{(VI.8)}$$

According to Eq. (VI.7), the stress decays exponentially. The expression $E_r(t)$ is defined as the *time dependent relaxation modulus*.

VI-4-d Maxwell Model vs. Reality

In real polymers, both the time-dependent creep compliance $J(t)$ and the time-dependent relaxation modulus $E_r(t)$ vary in complicated fashions, and must be determined experimentally. However, the simple model does indicate that these time-dependent properties are a consequence of coupling between the elastic and the viscous behavior. Thus, the creep rate is not simply dictated by the viscous part of the behavior, but rather by the ratio η/E. Note that $E_r(t)$ is approximately equal to $1/J(t)$, except at large strains. It has been shown that for real polymers both η and E are functions of the applied load and the duration of loading. The physical reason for this dependence is that a large number of mechanisms—including sliding, entanglement, and stretching of molecules, as well as changes in crystallinity—affect the deformation of polymers. Since these processes are strongly controlled by thermal activation, the overall material behavior is qualitatively similar to the spring dashpot models.

VI-5 STRESS–STRAIN RELATIONSHIPS FOR VISCOELASTIC SOLIDS

VI-5-a Maxwell Solid

The stress-strain relationship for a linear viscoelastic Maxwell solid may be written for the three-dimensional case as

$$\frac{d\epsilon_{11}}{dt} = \frac{1}{E}\left[\frac{d\sigma_{11}}{dt} - \nu\left(\frac{d\sigma_{22}}{dt} + \frac{d\sigma_{33}}{dt}\right)\right] + \frac{1}{\eta}\left[\sigma_{11} - \tfrac{1}{2}(\sigma_{22} + \sigma_{33})\right]$$

$$\vdots$$

$$\text{(VI.9)}$$

$$\frac{d\epsilon_{12}}{dt} = \frac{1}{2G}\frac{d\sigma_{12}}{dt} + \frac{3}{2\eta}\sigma_{12}$$

$$\vdots$$

Note that these relationships can be derived from the assumption that the deviator viscous strain rate components are proportional to the corresponding deviator viscous stress components. Then, writing the above equations in terms of the deviator and hydrostatic components of stress and strain, using subscript notation, we have

$$\frac{d\epsilon'_{ij}}{dt} = \frac{1}{2G}\frac{d\sigma'_{ij}}{dt} + \frac{3\sigma'_{ij}}{2\eta}$$

$$\epsilon = \frac{3(1 - 2\nu)\sigma}{E}$$

(VI.10)

VI-5-b   General Viscoelastic Solids

Since, just as in the case of the plastic deformation of metals, the viscous deformation of viscous solids is also caused by shear stresses, the viscous deformation of a linear viscoelastic material can be written more generally, using the Prandtl-Reuss equation, as

$$d\epsilon_{11}{}^v = \frac{d\bar{\epsilon}^v}{\bar{\sigma}}\left[\sigma_{11} - \tfrac{1}{2}(\sigma_{22} + \sigma_{33})\right]$$

$$\vdots$$

(VI.11)

$$d\epsilon_{12}{}^v = \frac{3}{2}\frac{d\bar{\epsilon}^v}{\bar{\sigma}}\,\sigma_{12}$$

$$\vdots$$

The equivalent stress and equivalent strain for multiaxial states of stress can be related through approximate expressions involving the creep compliance and the relaxation modulus. Under creep conditions, i.e., when all stresses are constant.

$$\bar{\epsilon} = J(t)\bar{\sigma} = \bar{\epsilon}^e + \bar{\epsilon}^v = J_o\bar{\sigma} + \bar{\epsilon}^v \qquad \text{(VI.12a)}$$

where J_0 is the compliance at $t = 0$, when the load is applied. Under stress relaxation conditions, when the strains are held constant,

$$\bar{\sigma} = E_r(t)\bar{\epsilon} = E_r(t)(\bar{\epsilon}^e + \bar{\epsilon}^v) = E_r(t)\left(\frac{\bar{\sigma}}{E_o} + \bar{\epsilon}^v\right) \qquad \text{(VI.12b)}$$

where $E_0 = 1/J_0$ is the modulus at $t = 0$, when the load is applied. These expressions are not exactly correct because the uniaxial strain in a creep or stress relaxation test is not exactly equal to $\bar{\epsilon}$ when the elastic strain includes a volume change. These expressions become increasingly accurate as the time increases and the viscous components of the strain become large compared to the elastic components.

For a linear viscoelastic solid the time-dependent strain (or conversely the time-dependent stress relaxation) under various time-dependent loading conditions can be superimposed using Boltzmann's superposition principle. For example, if a constant stress σ_{ij_0} is applied at time $t = 0$, σ_{ij_1} at $t = T_1$, σ_{ij_2} at $t = T_2$, etc., the strain at time t is given by

$$\epsilon_{ij}(t) = J(t)\sigma_{ij_0} + J(t - T_1)\sigma_{ij_1} + J(t - T_2)\sigma_{ij_2} + \cdots$$

$$= \sum_{N=1}^{n} J(t - T_N)\sigma_{ij_N} \qquad (VI.12c)$$

VI-6 CREEP AND STRESS RELAXATION OF REAL POLYMERS

The preceding sections indicate that any theory of polymer behavior must include the effects of time. Unlike the case of elastic-plastic behavior, where stress and strain can be related without regard to rate effects, a description of a viscoelastic material must include the effects of strain rate and duration of loading. A practical example of creep failure is the failure of a garden hose loaded by a static pressure for a period of time.

EXAMPLE VI.1—Failure of a Garden Hose

In a tensile stress relaxation experiment, the engineering stress required to maintain a fixed strain is obtained as a function of time. To measure the time-dependent yield stress a polyethylene specimen is strained to the yield strain and the yield stress is calculated as a function of time, as shown in Fig. VI.10. Long-time data were obtained by time-temperature superposition, which is discussed in a later section. Using the data given,* predict the maximum pressure a garden hose made of the same polyethylene can withstand for 1 year at room temperature without bursting. The internal diameter of the tube is 1 in. and the wall thickness is 1/16 in.

Solution:

If the ends of the polyethylene tube are not closed, the only stress component acting on the tube is the hoop stress, since the tube may be treated as a thin-walled tube. Then the tube is uniaxially loaded as in the tensile stress-relaxation experiment. Then, directly from the experimental data, it is

---

*Stress relaxation data are usually reported as engineering stress, i.e., load divided by unstretched area.

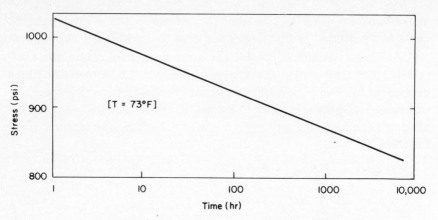

Fig. VI.10 Stress-relaxation behavior of low-density polyethylene at the yield strain. (*From E. Bäer [ed.], "Engineering Design for Plastics," p. 188, Van Nostrand Reinhold, New York, 1964.*)

seen that if the hoop stress is less than 820 psi, it will last a year without bursting. Then, the corresponding maximum internal pressure is given as

$$p = \frac{\sigma t}{r} = \frac{820 \times 1/16}{1/2} = 102 \text{ psi} \qquad (a)$$

If the ends of the tube are closed, there also exists an axial stress, which is 1/2 of the hoop stress. In this case, the Mises yield criterion (or the definition of the equivalent stress) may be used to find the internal pressure which gives the equivalent stress a value of 820 psi.

Typical creep data for polyethylene and acrylic sheet are given in Figs. VI.11 and VI.12. It should be noted that the shape of the creep curve depends significantly on the applied load level, so that the strain after a given duration of creep is not proportional to the applied stress. Therefore, the constitutive relationships given in Sec. VI-5 are not applicable. However, when the change in the applied load is small, strain may be assumed to be nearly proportional to stress at a given time. Over the load range in which this assumption is valid, *creep and stress relaxation data are often given in terms of a time-dependent creep compliance and relaxation modulus, which are defined as* $J(t) = \epsilon(t)/\sigma_0$ and $E_r(t) = \sigma(t)/\epsilon_0$, respectively. It should be emphasized that $E_r(t)$ and $J(t)$ do not depend on the stress amplitude and are thus only valid for linear viscoelastic solids.

Figure VI.13 illustrates the typical creep compliance of polyacetal, while Fig. VI.14 is a plot of the stress relaxation of polymethylmethacrylate. Figure VI.14 shows that at low temperatures (40°C), the modulus is nearly constant for short times, being 6.4×10^{10} dynes/cm^2 (9.28×10^5 psi); at high temperatures (135°C), the relaxation modulus represented by the flat plateau drops to 2.0×10^7 dynes/cm^2 (290 psi). In between these two regions (in the glass transition region) the modulus changes drastically. The low-temperature region represents the glassy region, and the high-temperature region is the rubbery region. At long times, and at high temperatures, the relaxation modulus decreases rapidly, leading to creep rupture. In this region, the polymer flows like a fluid.

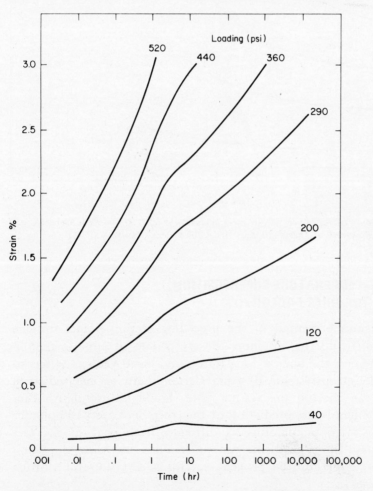

Fig. VI.11 Creep curves of polyethylene in tension at 10°C. (*From R. M. Ogorkiewicz, "Engineering Properties of Thermoplastics," Wiley-Interscience, London, 1970.*)

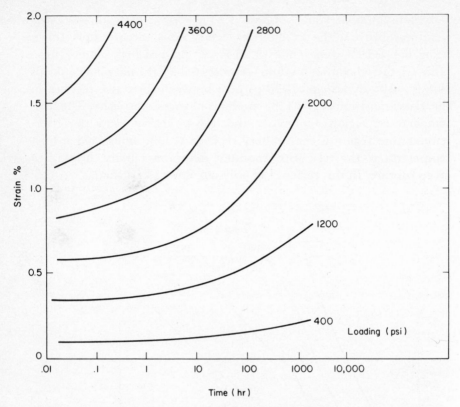

Fig. VI.12 Creep curves of acrylic cast sheet in tension at 60°C. (*From R. M. Ogorkiewicz, "Engineering Properties of Thermoplastics," Wiley-Interscience, London, 1970.*)

VI-7 TIME-TEMPERATURE SUPERPOSITION AND THE SHIFT FACTOR $a(T)$

The discussion presented in the preceding section points up one practical difficulty in obtaining the creep data. It appears that in order to obtain the data at long times, say, in 10 years, one has to perform the creep test for 10 years. Certainly any accelerated creep test that can shorten the testing time should be useful. Another related problem is the prediction of the creep behavior of a polymer when its service temperature is different from that of the test temperature. These problems can be handled by using the *time-temperature superposition* procedure. This is a very useful tool for calculations.

The basic concept of the time-temperature superposition can be illustrated by examining the complete $\log E_r(t)$ vs. $\log t$ curves at various temperatures shown in Fig. VI.15. Suppose that it is of interest to determine $E_r(t)$ at T_3, but at an accelerated rate. Many plastics behave in such a fashion that the $\log E_r(t)$ vs. $\log t$ curves are similar in shape, regardless of the test temperatures. Then, instead of determining the entire curve (called the master curve) at T_3, the test may be accelerated by determining the C-portion of the curve at T_3, the B'-portion of the curve at T_2, and the A''-portion of the curve at T_1. These latter two curves then can be translated horizontally until the curves A'', B' and C form a smooth continuous curve. The degree of the horizontal shifting, $\log a(T)$, depends solely on the relative temperatures and is independent of the time. This is the basis for the time-temperature superposition.

The theoretical basis for the time-temperature superposition may be given by considering how the temperature affects the material properties. The generalized discussion in Sec. VI-4 indicated that the creep behavior is controlled by the relaxation time. As the service temperature changes, the relaxation time τ is affected strongly,

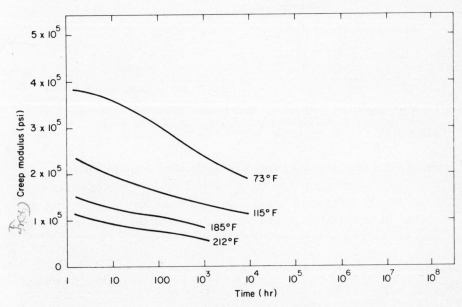

Fig. VI.13 Creep modulus of duPont Delrin 500 (polyacetal). Creep modulus is $1/J_0(t)$. (*Plotted from data taken from "Modern Plastics Encyclopedia," vol. 43, 1966.*)

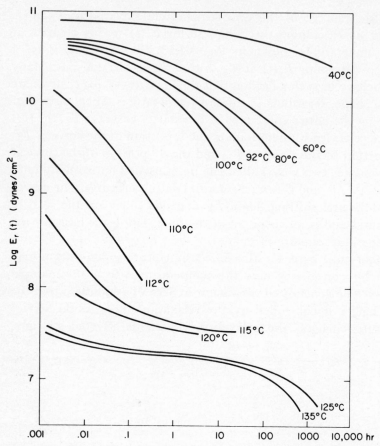

Fig. VI.14 Relaxation modulus for unfractionated polymethyl methacrylate (PMMA). (*From J. R. McLoughlin and A. V. Tobolsky*, J. Colloid. Sci., *vol. 7, p. 555, 1952.*)

mainly because the viscosity η is a sensitive function of tempera-ture.* If it is assumed that the sole effect of changing temperature is to affect the relaxation time, it is seen from Eq. (VI.13) that the change in temperature shifts the time scale of the creep test result, i.e.,

$$\epsilon = \left(1 + \frac{Et}{\eta}\right)\frac{\sigma_0}{E} = \left(1 + \frac{t}{\tau(T)}\right)\frac{\sigma_0}{E} = J_0(t, T)\sigma_0 \qquad \text{(VI.13)}$$

*The stiffness of a polymer is also affected by temperature change as a result of the changes in density and free energy of the molecules.

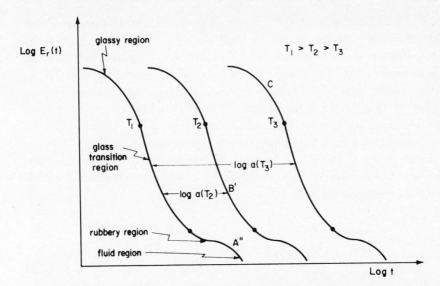

Fig. VI.15 Schematic of master curves for the relaxation modulus of a polymer at different temperatures. Note that the curves are parallel and only shifted along the log (time) axis.

Although the foregoing argument on the temperature effect is based on the Maxwell model, it can also be explained in terms of the thermal activation process, since creep is a thermally activated process. In a thermally activated process, the thermal energy kT (where k is Boltzmann's constant) must overcome the energy barrier ΔE. Since kT at room temperature is of the order of 1/40 eV, and ΔE is of the order of 1 or 2 eV, the transition from the unfavorable sites to the favorable sites cannot occur instantaneously. The probability of such a transition is given by Boltzmann's relation*

$$\dot{\epsilon} = A \exp\left(-\frac{\Delta E}{kT}\right) = \frac{\Delta \epsilon}{\Delta t} \qquad \text{(VI.14)}$$

where A is a constant sometimes known as the frequency factor. If the creep rate is governed by the same physical process, ΔE should be the same. The activation energy usually changes when the molecules assume a new state, such as occurs at the glass transition temperature and at the melting point. However, in a limited temperature regime, ΔE may be assumed to be constant. In this case,

*Equation (VI.14) is sometimes known as the Arrhenius equation.

Eq. (VI.14) states that the change in temperature affects the creep rate. Since $\dot{\epsilon} = \Delta\epsilon/\Delta t$, it can be seen that the effect of changing the temperature is to alter the time it takes to creep through a given displacement $\Delta\epsilon$. Equation (VI.14) states that a change of $10°C$ in temperature near room temperature may affect the creep rate by an order of magnitude. That means that, if it is known precisely how the temperature change affects the time scale, it is possible to get the long-time creep data by simply raising the temperature.

From Eq. (VI.14) it can be seen that the ratio of the deformation times at two different temperatures for a given creep strain depends only on temperature as

$$\frac{\Delta t}{\Delta t_0} = \frac{\exp(-\Delta E/kT_0)}{\exp(-\Delta E/kT)} = \exp\left[-\frac{\Delta E}{k}\left(\frac{1}{T_0} - \frac{1}{T}\right)\right]$$

$$= a(T) \qquad\qquad\qquad\text{(VI.15)}$$

or $\qquad\qquad \Delta t = a(T) \cdot \Delta t_0$

The expression $a(T)$ is defined as the shift factor. Equation (VI.15) states that the deformation (or the relaxation) time at temperature T is $a(T)$ times longer than that at temperature T_0. Therefore, the curves shown in Fig. VI.15 are separated by shifts of $\log a(T_2)$ and $\log a(T_3)$ from the curve obtained at T_1.

A more rigorous theoretical argument shows that the ratio of a relaxation time at temperature T to the same relaxation time at some other temperature T_0 is given by*

$$\frac{\tau(T)}{\tau(T_0)} = \frac{\eta}{\eta_0}\frac{\rho_0 T_0}{\rho T} = a(T) \qquad\qquad\text{(VI.16)}$$

where ρ denotes the density of the polymer.

The mechanics of obtaining the long-time creep data from short creep tests will now be illustrated. Figure VI.16 shows curves of the relaxation modulus $\log E_r(t)$ vs. log time data for polymethylmethacrylate (Lucite, a glassy polymer). Using the curve obtained at $115°C$ as a reference, the curves obtained at other temperatures may be shifted horizontally (parallel to the log time axis) until the curves overlap and form a continuous *master curve*. Similar shifting can also be done for other reference temperatures. For example, by shifting

---

*Ferry, J. D., *J. Chem. Soc.*, vol. 72, p. 3746 (1950).

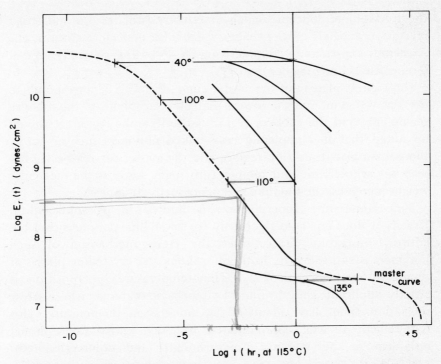

Fig. VI.16 Relaxation modulus master curve for PMMA, constructed from the data in Fig. VI.14. (*From J. R. McLoughlin and A. V. Tobolsky*, J. Colloid. Sci., *vol. 7, p. 555, 1952.*)

the curve obtained at 135°C toward the right, the relaxation modulus at 10^5 hr can be obtained, as shown in Fig. VI.16.*

Over a limited temperature range, from T_g to $T_g + 100°C$, the following universal function for the shift factor $a(T)$ for all glassy polymers is found to exist:

$$\log a(T) = \frac{-17.4(T - T_g)}{51.6 + (T - T_g)} \tag{VI.17}$$

where the temperatures are given in degrees centigrade. Equation (VI.17) is known as the WLF (Williams-Landel-Ferry) equation.† The WLF equation was originally determined empirically. The form of the WLF equation has been justified by using a theory based on

---

*When a substantial change in density occurs, such as near the melting point, it has been found that a vertical shift along the modulus axis by a factor of $\rho T / \rho_0 T_0$ must be made before the horizontal shift by $a(T)$ is done.

†See M. L. Williams, R. F. Landel, and J. D. Ferry, *J. Amer. Chem. Soc.*, vol. 77, p. 3701 (1955).

the free-volume concept, which in effect postulates that molecular relaxation depends on the space available for molecular motion. The Arrhenius expression used in deriving Eq. (VI.15) is generally valid at temperatures much higher than T_g.

The highest plateau of the master curve, Fig. VI.16, corresponds to the glassy region, and the lowest plateau corresponds to the rubbery region. Beyond the rubbery region is the fluidlike region. It should be noted that in completely cross-linked polymers such as epoxy there is no glass transition region, and the master curve is nearly flat over the whole time and temperature span. Also, as the molecular weight increases, the fluidlike region gradually disappears.

Time-temperature superposition is limited in its applicability. Usually it does not hold rigorously for crystalline polymers over the entire temperature range, since the creep mechanism of such polymers is controlled by both amorphous and crystalline phases at temperatures higher than T_g. Time-temperature superposition is strictly applicable only to linear* viscoelastic materials whose stress relaxation times have identical dependence on temperature. This requirement is well approximated by such glassy (amorphous) polymers as polystyrene, polycarbonate, and polymethylmethacrylate. However, one may use time-temperature superposition for all plastics, if approximate extrapolation is sufficient for design purposes.

EXAMPLE VI.2—Construction of the Master Curve for Polystyrene

Use the data for the creep compliance $J_p(t)$ of polystyrene, given in Fig. VI.17, to determine the shift factor $a(T)$ for the temperature range covered by the data. Construct a master curve for the compliance of the material using $109.8°C$ as the reference temperature. Note that the glass transition temperature of polystyrene is $81°C$.

Solution:

The master curve is constructed by shifting the curves adjacent to the reference curves horizontally until the slopes match reasonably well. This process is continued until the whole temperature range is covered as shown in Fig. VI.18. The plot of log $a(T)$ vs. temperature is shown in Fig. VI.19.

It should be noted at this time that in plotting the creep data it is convenient to use the engineering stress rather than the true stress, since the continuous change in the dimension of the specimen does not need to be taken into account.

---

*Some investigators claim that time-temperature superposition is also applicable to nonlinear viscoelastic materials.

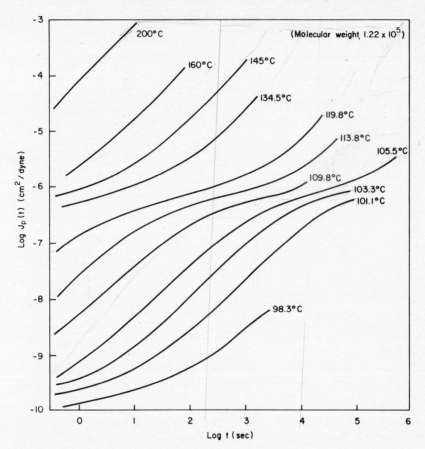

Fig. VI.17 Creep compliance $J_0(t)$ for polystyrene. (*From D. J. Plazek*, J. Polym. Sci. A-2, *vol. 6, pp. 621–638, 1968.*)

EXAMPLE VI.3—Serviceability of Polycarbonate Pipes

A reasonable design life for the plumbing in a building would be on the order of 100 years. Using the data shown in Fig. VI.20 for the relaxation modulus of polycarbonate, determine whether pipes of this material would be suitable for cold water $(T = 23°C)$ and hot water $(T = 60°C)$. A typical value of water pressure is 50 psi. The dimensions of the pipe are not fixed, but r/t should be greater than 5. Find the maximum possible value of r/t. The maximum strain should be less than 0.01. For simplicity, assume the pipe to be in plane stress (i.e., $\sigma_{zz} = 0$) and assume that the creep compliance is the reciprocal of the relaxation modulus. Are these assumptions conservative or nonconservative?

Solution:

One hundred years is equal to 3.16×10^9 sec. On the logarithmic scale, log $(3.16 \times 10^9) = 9.5$. A temperature of $100°C$ will be used as the reference

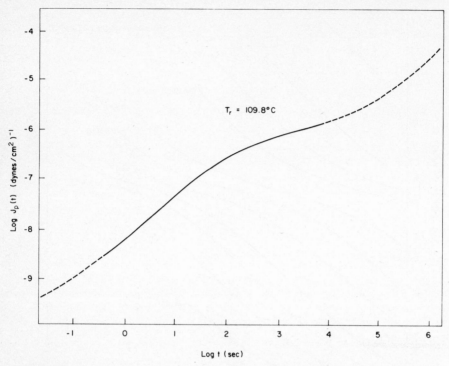

Fig. VI.18 Master curve for the creep compliance $J_0(t)$ of polystyrene, constructed from Fig. VI.17.

temperature. Using the data, and shifting the curves below $100°C$ until they coincide with the dotted master curve, we have

$$\log a(23°C) = 4.8$$
$$\log a(40°C) = 3.5 \qquad \text{(a)}$$
$$\log a(65°C) = 2.5$$

We find $\log a(60°C) = 2.8$ (by interpolation). Therefore, the point at $\log t = 4.7$ on the $100°C$ master curve represents $\log t = 9.5$ at $23°C$. The relaxation modulus $E_r(100 \text{ years}, 23°C)$ can therefore be read from the graph as $E_r(100 \text{ years}, 23°C) = .7 \times 10^{10}$ dynes/cm^2 = 10^5 psi.

The stress in the thin-walled tube is related to the internal pressure and the modulus by

$$\sigma = \frac{pr}{t} = E_r \epsilon$$

$$\frac{r}{t} \leq \frac{E_r \epsilon_{\max}}{p} = \frac{10^5(10^{-2})}{50} = 20 \qquad \text{(b)}$$

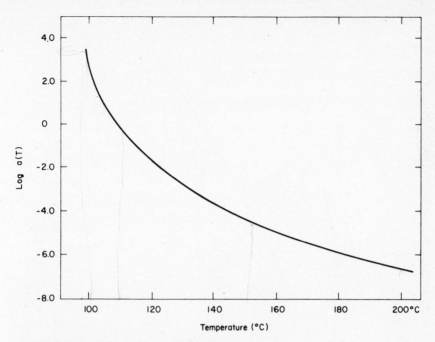

Fig. VI.19 Shift factor $a(T)$ for polystyrene, used to construct the master curve of the creep compliance in Fig. VI.18 from the data in Fig. VI.17.

$$\frac{r}{t}\,(23°\text{C}) \leq 20 \qquad\qquad \begin{array}{c}\textbf{(b)}\\ \textbf{(Con't)}\end{array}$$

which is practical.

For $60°\text{C}$, the shifted curve only extends to log $t = 7.8$, which is only slightly more than a year. When one attempts to extend the master curve to longer times by using the data for $110°\text{C}$ and above, it is found that the curves for these higher temperatures cannot be made to superimpose on the master curve by shifting them. From a straight-line extrapolation of the master curve, $E_r(100 \text{ years}, 60°\text{C}) = .24 \times 10^{10}$ dynes/cm^2 = 35,000 psi, for which we get $r/t \leq 7$. However, the modulus may fall off more rapidly than the straight-line extrapolation. Alternatively, we could extrapolate the master curve with a slope parallel to the $110°\text{C}$ data. This gives $E_r(100 \text{ years}, 60°\text{C}) = .1 \times 10^{10}$ dynes/cm^2 = 14,500 psi, for which we get $r/t \leq 2.9$. This figure is unacceptable. For this reason one might consider using cold water pipes made from polycarbonate but, unless more conclusive data are available, polycarbonate should not be used for hot water pipes.

VI-8 STRESS-STRAIN RELATIONS OF POLYMERS

Because of the rate and temperature sensitivity of polymers, it is difficult (and sometimes meaningless) to discuss the stress-strain

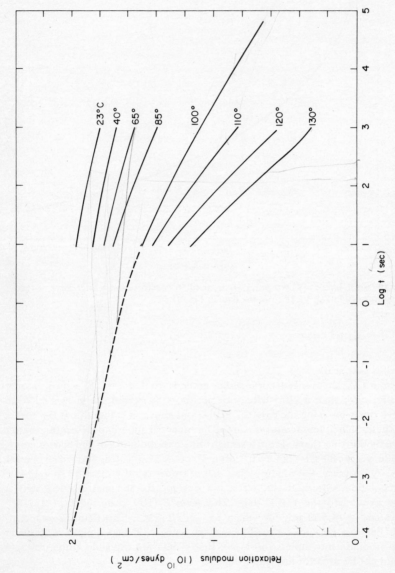

Fig. VI.20 Relaxation modulus of polycarbonate. (*From I. V. Yannas, "The Range of Validity of Linear Viscoelastic Theory," unpublished paper, March 1970.*)

Relaxation modulus (10^{10} dynes/cm^2)

Log t (sec)

23°C
40°
65°
85°
100°
110°
120°
130°

relations. Not only are the elongation, yield stress, and ultimate stress dependent on the temperature and strain rate, but so are the shapes of the stress-strain curves. However, certain generalizations may be made about the stress-strain relationship for various materials undergoing elongation at a given strain rate. Figure VI.21 shows several stress-strain curves at a strain rate of about 0.15 sec^{-1}. Hard and rigid thermosetting plastics typically have stress-strain curves as shown in Fig. VI.21(a). The maximum strain is only a few percent. This type of material is least sensitive to strain rate and temperature. Glassy thermoplastics such as polystyrene (unmodified) and poly-methylmethacrylate have stress-strain curves typical of those shown in Fig. VI.21(b). At room temperature, the maximum strain is less than 25%; this strain is very sensitive to loading rate and tempera-ture. Certain thermoplastics, such as nylon and PVC, usually neck and draw, starting at the weakest section. This type of material elongates with a final strain as much as 700%. The necked region does not break immediately since the flow strength of the necked region increases a great deal due to orientation of the molecules. Typical stress-strain curves of this type are shown in Fig. VI.21(c).

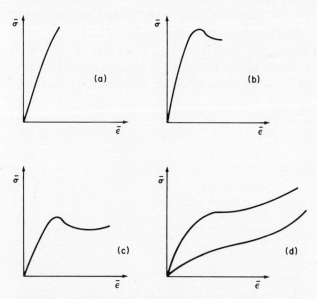

Fig. VI.21 Stress-strain curves typical of plastic materials. (a) rigid thermosetting plastic; (b) glassy thermoplastics. The plastic strain is produced by crazing. (c) crystalline polymers which can be drawn; (d) rubbery polymers.

Before fracture, the entire specimen necks down to a uniform cross section. Figure VI.21(d) illustrates the typical stress-strain curves of rubbers and thermoplastics that do not neck. The upper curve is a typical curve for Teflon, whereas the lower curve is typical of low density polyethylene, plasticized PVC, and rubbers.

The fact that the strain rate and temperature must be carefully specified in discussing the stress-strain behavior of polymers is illustrated in Fig. VI.22. The dotted line shows the locus of rupture or fracture points and the solid lines show the stress-strain curves at a constant strain rate but at different temperatures, or conversely, at a constant temperature but at different strain rates. It can be seen that, at either very high strain rates or very low temperatures, all plastics behave in a glassy manner.

The plastic deformation of polymeric materials depends on the hydrostatic stress. Unlike metals, which have nearly constant yield and flow stresses, the flow and yield stresses of polymers increase with an increase in the hydrostatic compressive stress. Therefore, whereas at yielding in metals the Mohr's circle always has a constant radius, the radius of the Mohr's circle at yielding in polymers changes as a function of the hydrostatic stress, as shown in Fig. VI.23. As discussed in Chapter IV [Eq. (IV.13)], the Mohr-Coulomb yield criterion describes the yield envelope shown in Fig. VI.23.

The toughness of glassy polymers [see Fig. VI.21(b)] is influenced by the formation of *crazes* in the material. A craze is a region which has undergone a type of permanent deformation which increases the

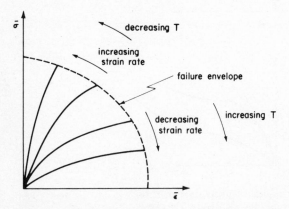

Fig. VI.22 Effects of temperature and strain rate on the stress-strain curve of a polymer.

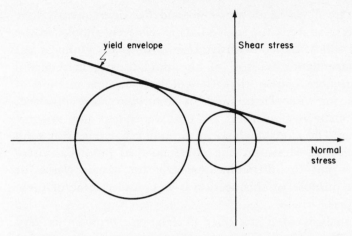

Fig. VI.23 Mohr's circle representation of a pressure-dependent yield condition, such as that for polymers and granular materials.

volume of the material by a large percentage. In many cases this increase in volume is associated with the formation of many small voids in the crazed material. The large local strains which occur during craze formation cause substantial changes in the material properties. In addition to the decrease in density of crazed material, its elastic moduli are also reduced and the molecular orientation which occurs during the deformation causes the strength of the material to be increased. Crazing contributes to the toughness of polymers, because the formation of a craze at the tip of a growing crack dissipates a large amount of energy and also reduces the stress concentration caused by the crack. As will be discussed in Chapter VIII, this increases the applied stress necessary for causing the crack to grow. The toughening effect of copolymerizing glassy polymers with rubber results partially because the rubber phase induces the formation of crazes.

VI-9 STRESS ANALYSIS FOR LINEAR VISCOELASTIC SOLIDS

As emphasized in Chapter II, three basic conditions must be satisfied by the exact stress distribution in a continuum: equilibrium; geometric compatibility; and constitutive relations. The first two conditions are independent of material properties and hence must be satisfied by all continua. The constitutive relations for Hookeian elastic solids are linear and time-independent. Consequently, the method of stress analysis for such solids is relatively simple. In the case of plastic solids, the constitutive relationships are nonlinear but still time independent. Because of the nonlinearity, incremental strains are considered rather than the total strain. The method of stress analysis for viscoelastic solids is somewhat more involved, because the constitutive relations are time dependent and, in some cases, nonlinear. The stress analysis for a nonlinear viscoelastic solid is very complicated. However, the stress analysis for linear visco-elastic solids is not too difficult, especially for certain classes of problems. The purpose of this section is to consider some of these simple techniques.

The stress analysis of a *statically determinate problem* is very simple for all materials, because the solutions can be determined directly from the equilibrium condition without using the material properties. An example of the statically determinate problem is the determination of the hoop stress in the thin walled garden hose discussed in Example VI.1.

Another class of problems which can readily be solved consists of plane stress problems where the boundary conditions are specified in terms of stress. *In this case, the stress distribution in a linear viscoelastic solid is exactly identical to that of the corresponding elastic solid*, as illustrated in Examples VI.4 and VI.5.

EXAMPLE VI.4—Stress Distribution and Deflection
of a Viscoelastic Beam

Consider a cantilever beam, made of Delrin, which is loaded at its free end by a concentrated load $F = 10$ lb. The time-dependent relaxation modulus is given in Fig. VI.24. Determine the stress distribution in the beam and the maximum deflection of the beam 1 year after loading. The temperature of the beam is maintained at $73°$ F.

Solution:

Figure VI.25 shows the dimensions of the beam and the coordinate system used in this analysis. By isolating a part of the cantilever beam, the bending moment at any cross section is obtained as

$$M_3 = -F(L - x_1) \tag{a}$$

The bending moment is related to the stress distribution by

$$M_3 = -b \int_{-h}^{h} \sigma_{11} x_2 \, dx_2 \tag{b}$$

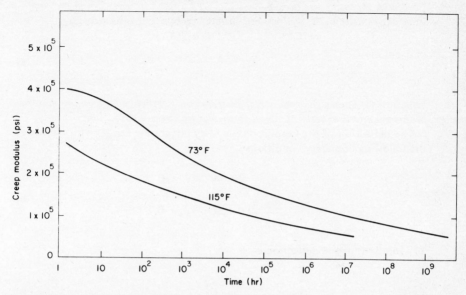

Fig. VI.24 Master curves for the creep modulus for duPont Delrin 500. (*Constructed from Fig. VI.13.*)

The stress at any part of the cross section is related to the strain by

$$\sigma_{11} = E_r(t)\epsilon_{11} \quad \text{or} \quad \bar{\sigma} = E_r(t)\bar{\epsilon} \tag{c}$$

The strain is related to the radius of curvature R of the beam at that point and the distance from the neutral axis as

$$\epsilon_{11} = -\frac{x_2}{R} \tag{d}$$

Substituting Eqs. (c) and (d) into Eq. (b), gives

$$M_3 = b \int_{-h}^{h} \frac{E_r(t)x_2^2}{R} \, dx_2 \tag{e}$$

Since $E_r(t)$ is a function only of time for a linear viscoelastic solid, Eq. (e) may be written as

$$M_3 = \frac{bE_r(t)}{R} \int_{-h}^{h} x_2^2 \, dx_2 = \frac{2bh^3 E_r(t)}{3R} = \frac{E_r(t)I_{22}}{R}$$

$$= E_r(t)I_{22} \frac{d^2 u_2}{dx_1^2} \tag{f}$$

Since $\sigma_{11} = -E_r(t)x_2/R$, the bending moment is related to the stress by

$$\sigma_{11} = -\frac{M_3 x_2}{I_{22}} \tag{g}$$

Equation (g) shows that the *stress distribution is independent of the material properties and is exactly identical to the elastic stress distribution.*

The deflection of the beam is obtained by integrating Eq. (f) with the use of the following boundary conditions

$$u_2 = 0 \quad \text{at} \quad x_1 = 0$$
$$\frac{du_2}{dx_1} = 0 \quad \text{at} \quad x_1 = 0 \tag{h}$$

The solution for the deflection u_2 is

$$u_2 = \frac{-F}{E_r(t)I_{22}} \left(\frac{Lx_1^2}{2} - \frac{x_1^3}{6} \right) \tag{i}$$

The deflection after 1 year is obtained by substituting the numerical value for $E_r(1 \text{ year})$ from the creep modulus given in Fig. VI.24, which is $E_r(8.76 \times 10^3 \text{ hr}) = 2.2 \times 10^5$ psi. The maximum deflection occurs at $x_1 = L$.

EXAMPLE VI.5—Creep of a Torsion Rod

A torsion bar is made of polymethymethacrylate as part of a display to be used at a Florida amusement park. The relaxation modulus is given by the data of Fig. VI.14. The diameter of the bar is 1 in. and the length is 36 in. A constant moment of 20 in.-lb acts on the bar. The temperature is approximated as cycling between $40°C$ during the night and $60°C$ during the day (12 hr at each temperature). Determine the deflection of the bar after 5 years. Assume the time-dependent shear modulus to be $E_r(t)/3$. Be clear as to the

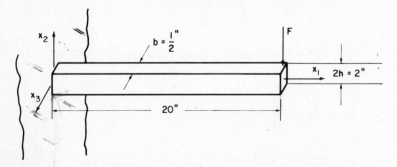

Fig. VI.25 A viscoelastic cantilever beam loaded by a concentrated end load.

assumptions used in relating constant temperature data to this case where the temperature is cycling.

Solution:

The relationship between the applied moment M and the angle of twist, ϕ, will be first derived. The angle of twist is shown in Fig. VI.26. Since the moment is the same everywhere, it is reasonable to assume a constant angle of twist per unit length of the rod, i.e.,

$$\frac{d\phi}{dz} = \text{constant} \qquad (a)$$

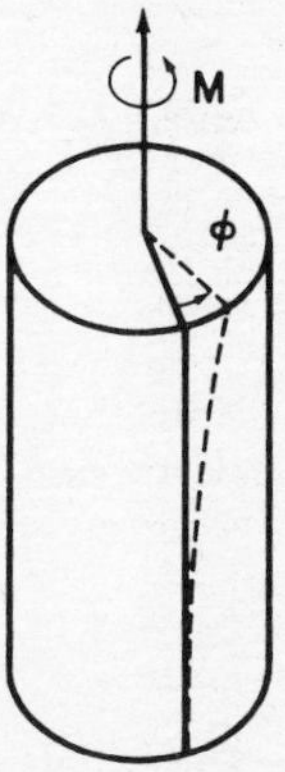

The displacements are

$$u_z = u_r = 0$$
$$u_\theta = \left(\frac{d\phi}{dz} z\right) r \qquad (b)$$

The strains are then

$$\epsilon_{\theta z} = \frac{1}{2}\left(\frac{\partial u_\theta}{\partial z} + \frac{\partial u_z}{\partial \theta}\right) = \frac{1}{2}\frac{d\phi}{dz} r \qquad (c)$$

Fig. VI.26 A viscoelastic torsion rod subjected to a constant twisting moment.

From the stress-strain relationship given by Eq. (VI.11), one obtains

$$d\epsilon_{\theta z}^{\ v} = \frac{3}{2}\frac{d\bar{\epsilon}^v}{\bar{\sigma}}\sigma_{\theta z} \qquad (d)$$

The superscript v may be eliminated from Eq. (d), since the elastic deformation is much smaller than the viscous deformation. The equivalent stress is related to $\sigma_{\theta z}$ as

$$\bar{\sigma} = \sqrt{3}\,\sigma_{\theta z} \qquad (e)$$

Upon substitution of Eq. (e) into Eq. (d),

$$d\dot{\epsilon}_{\theta z} = \frac{\sqrt{3}}{2}d\bar{\epsilon} \quad \text{or} \quad \epsilon_{\theta z} = \frac{\sqrt{3}}{2}\bar{\epsilon} \qquad (f)$$

Since $\bar{\epsilon} = \bar{\sigma}/E_r(t)$, $\sigma_{\theta z}$ may be related to $\epsilon_{\theta z}$ as

$$\sigma_{\theta z} = \frac{\bar{\sigma}}{\sqrt{3}} = \frac{E_r(t)\bar{\epsilon}}{\sqrt{3}} = \frac{2}{3}E_r(t)\epsilon_{\theta z} \qquad (g)$$

The applied moment is related to the stress distribution by

$$M = \int_0^{r_0} \sigma_{\theta z}(2\pi r^2)\,dr = \frac{2\pi}{3}\,E_r(t)\,\frac{d\phi}{dz}\int_0^{r_0} r^3\,dr$$

$$= G_r(t)\,I_p\,\frac{d\phi}{dz}$$

(h)

where $I_p = \pi r_0^4/2$. The expression for stress becomes

$$\sigma_{\theta z} = \frac{rM}{I_p}$$

(i)

Note that the expression for stress distribution, Eq. (i), is the same as in the elastic case. Solving Eq. (h) for ϕ by integrating it over the length gives

$$\phi = \frac{Ml}{G_r(t)\,I_p}$$

(j)

The total deflection ϕ can be determined, if $G_r(t)$ is known for the temperature cycle given at $t = 5$ years.

Since the display will be exposed for 2-1/2 years at $60°C$ and 2-1/2 years at $40°C$, it is necessary to convert the loading cycle to a reference temperature. If we use $60°C$ as the reference temperature, the equivalent time at $60°C$ is

$$(t)_{60°C} = 2\frac{1}{2} + \frac{(t)_{40°C}}{a(40°C)} = 2.5\left(1 + \frac{1}{a(40°C)}\right)$$

(k)

From Fig. VI.14, a master curve may be constructed for a reference temperature of $60°C$. From this it can be estimated that the logarithm of the shift factor, $\log a(40°C)$, is 3.5. Therefore, it is seen that the equivalent time at $60°C$ is only slightly more than 2.5 years. The expression $\log E_r(t)$, or $\log 3G_r(t)$, at 21,360 hr (2.5 years) is 8.6, when $E_r(t)$ is expressed in dynes/cm^2, which gives $G_r(t) = 1.925 \times 10^3$ psi.

Substituting appropriate values into Eq. (j) gives

$$\phi = 3.81 \text{ radians} = 218°$$

(l)

There are more general approaches developed which are based on the generalized constitutive relation Eq. (VI.5). These techniques are based on analogies between elastic and viscoelastic constitutive relations. The difference between elastic and viscoelastic constitutive relations is that the viscoelastic constitutive relations have time as an independent variable and are of higher order. In these techniques, the solution to viscoelastic problems are obtained by eliminating the time variable in the linear viscoelastic constitutive relations, using integral transforms such as Laplace and Fourier transforms. When the equilibrium and geometric compatibility equations are also transformed, the governing equations for viscoelastic solids become identical to those for the corresponding elastic problem with the same boundary conditions. For further discussion of various techniques and their limitations, readers are referred to Refs. 1 and 9. The general result of these techniques is that, for any elastic problem where the elastic constants do not appear in the expressions for the stresses, the stress distribution for the equivalent linear viscoelastic problem will be identical. As mentioned earlier, this class of problems includes many problems of engineering importance, such as beams, circular shafts in torsion, and thin-walled vessels with internal pressure.

VI-10 VISCOUS BEHAVIOR OF POLYMERS

The processing of thermoplastics, such as by extrusion and injection molding, is done at a temperature higher than the melting point of crystalline plastics and much higher than the second-order transition temperature of amorphous plastics. At these temperatures, thermoplastics behave in a viscous manner with negligible elasticity. In this section, we will discuss the viscous behavior of molten thermoplastics at high temperatures and the flow characteristics of unreacted thermosetting plastics. In the analysis of the viscous behavior of these plastics, energy equations must always be considered in conjunction with other governing equations, because the viscosity of these plastics is a very sensitive function of temperature. This is in contrast to the elastic-plastic analysis done for metals, in which it is assumed that the flow stress of metals is not sensitively dependent on temperature.

The viscous behavior of materials is characterized in terms of viscosity,* which is defined as

$$\eta = \frac{\text{shear stress}}{\text{shear strain rate}} = \frac{\tau}{\dot{\gamma}} \qquad \text{(VI.18)}$$

*The unit of viscosity (sometimes referred to as absolute viscosity) is dynes sec/cm². Kinematic viscosity is defined as the ratio of the absolute viscosity to the mass density of the fluid. Its units are cm²/sec. 1 dynes sec/cm² = 1 poise and 1 cm²/sec = 1 stoke.

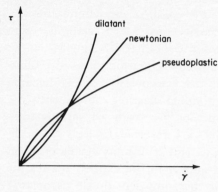

Fig. VI.27 Schematic showing the classification of the viscous behavior of polymers based on strain-rate sensitivity.

It is a measure of power dissipated in a fluid during shear deformation. In a Newtonian fluid, the viscosity is independent of the shear strain rate; this approximates the behavior of many low molecular weight fluids. Most plastics are, on the other hand, non-Newtonian in that the viscosity is sensitive to strain rate. The viscous behavior of a non-Newtonian fluid is classified in terms of its sensitivity to variations in either the strain rate or in the duration of loading at a given strain rate.

The sensitivity to shear strain rate is illustrated in Fig. VI.27. The shear stress of Newtonian fluids increases linearly with shear strain rate. However, most thermoplastics behave such that their viscosity decreases with an increase in shear strain rate. This type of behavior is termed pseudoplastic. On the other hand, some materials, such as plastisol* exhibit dilatant behavior, i.e., the viscosity increases with shear strain rate.

Although there are many models (Refs. 3, 4, 6) that purport to describe the viscous behavior of polymers at high temperatures, the most convenient one, due to its simplicity, is the power law, i.e.,

$$\tau = (\eta_0 \dot{\gamma}^{n-1})\dot{\gamma} = \eta\dot{\gamma} \qquad (\text{VI.19})$$

where $\eta = \eta_0 \dot{\gamma}^{n-1}$ is the strain rate dependent viscosity. For Newtonian fluids, $n = 1$. For most plastics, n is less than 1, i.e., they are pseudoplastic (see Fig. VI.27). The power law is reasonably accurate in the intermediate strain rate range, but at very low strain rates and at very high strain rates it does not predict the correct viscosity. For a pseudoplastic material, it predicts an infinite

*Plastisol is a suspension of PVC resin powders in a plasticizer.

viscosity at zero shear strain rate, which is not correct. It also predicts higher viscosity than in the actual case at high strain rates. The viscosity as defined by Eq. (VI.19), i.e., $\eta_0 \dot{\gamma}^{n-1}$, involves units that are awkward to use. In view of the shortcomings pointed out, it is suggested that the validity of the assumed constitutive relationship for the problem at hand be investigated before actual use. Sometimes it may be necessary to use different constitutive relations in different regions of flow.

The viscosity of a polymer solution and of a colloidal emulsion or dispersion also varies as a function of the loading time at a given strain rate, as shown in Fig. VI. 28. When the viscosity decreases with loading time, the behavior is called *thixotropy*, and when it increases, it is called *rheopexy*. Thixotropy is a very desirable property in a paint. When the paint is stored in a can, it is desirable to have high viscosity, so that the pigments remain suspended in the liquid carrier or solvent. When the paint is being applied by brush, low viscosity will make the painting job easier. However, immediately after application, the paint should not drip. Thixotropy is caused by separation of weak secondary bonds between polar groups when the fluid is stirred. Even normally thixotropic materials exhibit rheopexy when the liquid is stirred very slowly. In this case, slow stirring promotes establishment of these bonds by increasing the mobility of polar groups.

The effect of temperature on the flow stress is great. This dependence is very important in polymer processing, since it is the easiest parameter to control in actual operation. The dependence of viscosity on temperature is shown in Fig. VI.29. Viscous deformation of polymers at high temperatures is also governed by a thermally activated process. It may be expressed in an Arrhenius form as

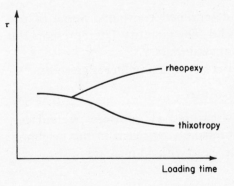

Fig. VI.28 Classification of the viscous behavior of polymers in terms of viscosity as a function of loading time at constant strain rate.

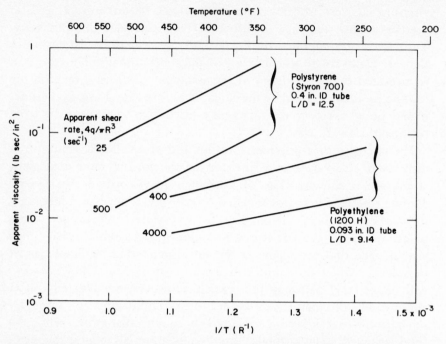

Fig. VI.29 Dependence of the limiting Newtonian viscosity of polymers on temperature. (*From E. C. Bernhardt [ed]: "Processing of Thermoplastic Materials," pp. 593 and 635, Reinhold, New York, 1965.*)

$$\eta = A \exp\left(-\frac{\Delta E}{RT}\right) = \eta_0 \dot{\gamma}^{n-1} \exp\left(-\frac{\Delta E}{RT}\right) \qquad (VI.20)$$

where ΔE is the activation energy. The activation energy of non-Newtonian fluids depends on temperature, in addition to stress and strain rate. Equation (VI.20) is sometimes simplified as

$$\eta = a \exp(-bT) \qquad (VI.21)$$

where a and b are empirically determined constants. Some of the values for b are given in Table VI.3. The quantity $1/b$ represents the temperature sensitivity at constant strain rate $\dot{\gamma}$.

The effect of pressure on the viscosity of a polymer is given by

$$\eta_p = \eta_{p_0} \exp[h(p - p_0)] \qquad (VI.22)$$

where η_p and η_{p_0} are the viscosities at pressures p and p_0, respectively. The factor h is a pressure coefficient in the neighborhood of 3×10^{-15} per psi. It decreases with an increase in shear rate and temperature.

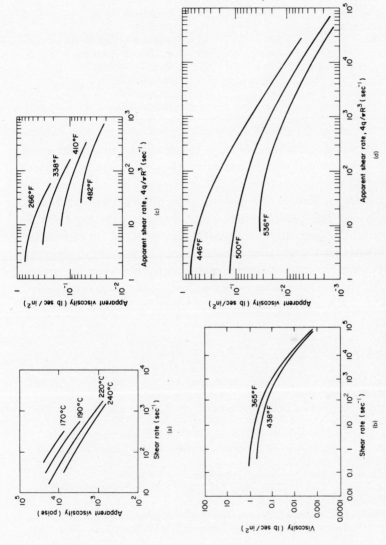

Fig. VI.30 Dependence of the limiting Newtonian viscosity of polymers on the shear rate. (a) Polystyrene; (b) Delrin 100 acetal resin; (c) Polyethylene; (d) Polymethylmethacrylate. ([a] and [b] from Z. Tadmor and I. Klein, "Engineering Principles of Plasticating Extrusion," pp. 476 and 479, Van Nostrand Reinhold, New York, 1970; [c] and [d] from E. C. Bernhardt [ed.], "Processing of Thermoplastic Materials," pp. 595 and 559, Reinhold, New York, 1965.)

TABLE VI.3 Temperature Dependence of the Viscosity of Some Common Thermoplastics at Constant Shear Rate (From Ref. 5, p. 43)

Material	Shear rate (sec^{-1})	$1/b_{\dot{\gamma}}{}^{\circ}C$	Tradename
Polymethylmethacrylate	100	24	Lucite 140
Polymethylmethacrylate	27	18	Plexiglas
Cellulose acetate	100	32	Tenite Acetate 036–H2
Nylon 6	100	60	Plasticon Nylon 8206
Nylon 66	100	56	Zytel 101 NC10
Polyethylene	100	85	Bakelite DYNH
Polystyrene	100	73	Styron 475
PVC	40	51	Geon 8750

Viscosity is often measured using rotational viscometers, which measure torque and rotational velocity. It can also be measured using a capillary viscometer, which measures the flow rate at a given pressure drop.

The viscosities of liquids are diverse, e.g., polyethylene at 340°F has a viscosity of 10^5 poises, whereas the viscosity of water at 68°F is 10^{-2} poise. The viscosity of a plastic is dependent on molecular weight, pressure, and strain rate, in addition to temperature, whereas the viscosity of a Newtonian fluid is mainly sensitive to temperature. Figure VI.30 shows the viscosities of polystyrene, polyacetal, polyethylene, and polymethylmethacrylate as functions of strain rate and temperature.

REFERENCES

1. McClintock, F. A., and A. S. Argon: "Mechanical Behavior of, Materials," Addison-Wesley, Reading, Mass., 1966.
2. Bäer, E. (ed.): "Engineering Design for Plastics," Van Nostrand Reinhold, New York, 1964.
3. Rodriguez, F.: "Principles of Polymer Systems," McGraw-Hill, New York, 1970.
4. Schmidt, A. X., and C. A. Marlies: "Principles of High Polymer Theory and Practice," McGraw-Hill, New York, 1948.
5. McKelvey, J. M.: "Polymer Processing," John Wiley, New York, 1962.
6. Tadmor, Z., and I. Klein: "Engineering Principles of Plasticating Extrusion," Van Nostrand Reinhold, New York, 1970.
7. "Modern Plastics Encyclopedia," McGraw-Hill, New York, latest issue.
8. Alfrey, T.: "Mechanical Behavior of High Polymers," Interscience Publ., New York, 1948.
9. Flügge, W.: "Viscoelasticity," Blaisdell, Waltham, Mass., 1967.
10. Ward, I. M.: "Mechanical Properties of Solid Polymers," John Wiley, New York, 1971.
11. Lee, H., P. Stoffey, and K. Neville: "New Linear Polymers," McGraw-Hill, New York, 1967.
12. Bernhardt, E. C. (ed.): "Processing of Thermoplastic Materials," Reinhold, New York, 1965.

PROBLEMS

VI.1   Acetal, whose duPont trade name is Delrin, is used to make cams, gears, and bearings because of its strength, good frictional properties, and abrasion resistance. It also has good resistance to attack by solvents, but is attacked by strong acids and also by sunlight.

Acme Corporation of Worcester, Massachusetts, a manufacturer of industrial robots, is interested in manufacturing their pneumatic actuators by molding the components with Delrin. One of the parts is a thin-walled cylinder with closed ends. It will be subjected to compressed air at 100 psig. The operating temperature ranges from 40°F to 170°F. These products must last for 10 years.

a) Construct a master curve and determine the shift factor $a(T)$ as a function of temperature (see Fig. VI.13).

b) If the maximum allowable hoop strain is 0.01, determine the maximum r/t ratio for the cylinder.

c) Determine the r/t ratio of the cylinder if the part is subjected to 40°F for 8 hr/day and to 170°F for 16 hr/day for 10 years.

VI.2   The manager of the Stirling-Greswold Book Store is evaluating book shelves made of Delrin for possible display use in his store (data given by Fig. VI.13). The shelves are manufactured by the Zorobo Company of Wilmington, Delaware. The dimensions of the shelf and the manner of support are shown.

As the criterion of his evaluation, the manager has arbitrarily decided that the maximum tolerable sagging of the shelf is 1/16 in. over a period of 10 years. An average book weighs 2 lb per inch of its thickness. The worst possible temperature cycle the shelves may be subjected to is shown in the figures.

a) What is the stress distribution at the midspan of the book shelf?

b) Determine if the book shelves meet the criterion of the store manager by determining the maximum sagging in 10 years of service.

c) If the shelf material was not subjected to cycling temperature, but instead to a constant temperature environment, what would be the maximum temperature to which the shelf could be subjected and still just meet the manager's performance criterion?

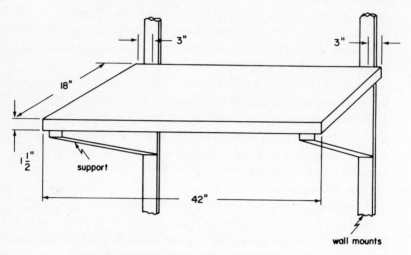

[The shelf is simply placed on top of the supports without the use of fasteners.]

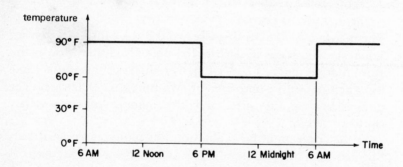

VI.3   Consider Example VI.1 given in the book. If the ends of the garden hose are closed, what is the maximum internal pressure?

VI.4   Consider any piece made of PMMA (polymethylmethacrylate). The deformation is assumed to be due to creep rather than to elastic or plastic phenomena. At an operating temperature of $110°C$, it is noticed that the extent of deformation is too large during its (the part's) expected lifetime of 1 year. To meet specifications, it is calculated that the deformation must be reduced by a factor of 20 for the same lifetime. Due to equipment costs, it is not practical to alter the dimensions of the part. Quantitatively, what is your recommendation?

VI.5 A transparent rectangular duct is to be used in a fluid mechanics demonstration. The duct is built into a wall at one of its ends and the other is essentially free. Water is flowing through the duct at a constant flow rate. The temperature of the water varies cyclically as shown in the figures. Someone suggests that polymethylmethacrylate (PMMA) be used to make the duct. It may be assumed that the duct is loaded by a uniformly distributed load of $w = 4$ lb/in. of the duct length. A master curve and the shift factor vs. temperature curve for PMMA are to be derived from Fig. VI.14.

a) Determine the stress distribution in the duct. If you can write down the expression for the stress distribution directly, it is not necessary to derive it.

b) If the maximum allowable deflection is 0.1 in., determine the service life of the duct.

To convert from lb/in.^{2} to dynes/cm^2, multiply by 6.9 x 10^4. Write all assumptions clearly.

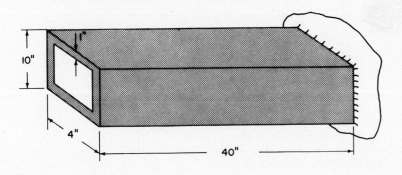

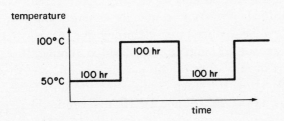

VI.6 A composite shaft is made by plating nickel on Delrin (polyacetal) as shown below. The purpose of the nickel layer is to protect the plastic core from eroding when solid particles are impinging perpendicular to the axis of the shaft. The aerodynamic drag on the shaft and the loading due to the impingement particles are negligible. The shaft is subjected to a constant twisting moment of 500 in.-lb at 73°F.
a) Determine the deflection of the shaft upon loading (i.e., at $t = 0$) and after 100 years in service.
b) Determine the stress distribution in the nickel layer at the onset of loading and after 100 years.

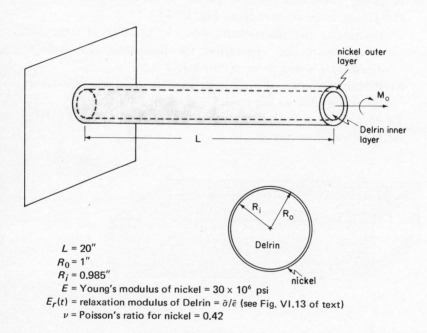

$L = 20''$
$R_0 = 1''$
$R_i = 0.985''$
E = Young's modulus of nickel = 30 x 10⁶ psi
$E_r(t)$ = relaxation modulus of Delrin = $\bar{\sigma}/\bar{\epsilon}$ (see Fig. VI.13 of text)
ν = Poisson's ratio for nickel = 0.42

VI.7 Due to nonuniform cooling from the molding temperature, injection-molded polystyrene panels emerge warped. The warping is to be reduced by clamping the panels flat and holding them for 15 min at an elevated temperature. The initial radius of curvature of the panels is 50 in., and they are 1/2 in. thick. Using the creep compliance and shift factor data given in Figs. IV.17, IV.18, and IV.19, determine the required temperature to achieve a factor of 50 increase in the radius of curvature during the 15-min annealing time.

VI.8 An engineer at the Eastern Electric Company suggested that a microwave guide between Boston and New York City be made by extruding a long rectangular polycarbonate channel. His suggestion calls for plating the inner surfaces of the extruded channel with a continuous layer of nickel, which is then closed by bonding a thin brass sheet to the channel. One of the reasons for attaching the brass sheet is to minimize sagging of the waveguide, which will be simply-supported, as shown in the figure. The only loading is due to the weight of the waveguide. The waveguide will be placed in an underground tunnel in which the temperature fluctuates between 40°F at night (12 hr) and 60°F during the rest of the day.

Determine the maximum deflection of the waveguide after 10 years. Neglect the effect of the thin nickel plating. The modulus of the brass is 16×10^6 psi.

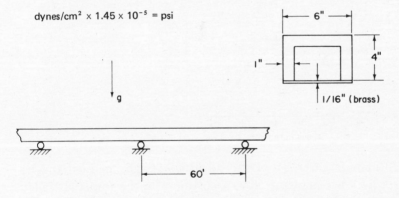

dynes/cm^2 × 1.45 × 10^{-5} = psi

VI.9 Two parts of the casing of an appliance are to be held together by a springlike molding made of polymethyl-methacrylate (PMMA) (see sketch below). The edge of the molding is spread 1/16 in. when it is installed. A clamping force of 1/4 lb/in. of length of molding must be maintained to give satisfactory performance.

A master curve for PMMA at 115°C is given in Fig. VI.16, and shift factor data are given below. Assume that the thick part of the molding is essentially rigid and that the arms deflect as beams.

a) What thickness t is required for the two arms of the molding, if it is to have a lifetime of 10 years, assuming that it operates continuously at its maximum service temperature of 40°C?

b) What is the maximum bending stress in the arm of the molding when it is first installed?

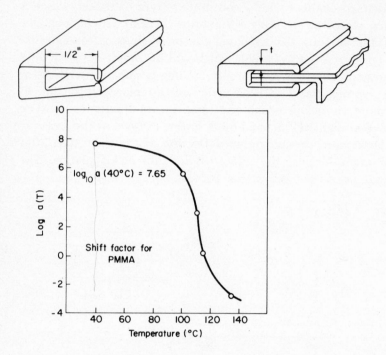

VI.10 A manufacturer of plastic advertising signs in Los Angeles receives a complaint from a customer in the Imperial Valley near the Mexican border, who claims that several signs installed only 3 months earlier are badly distorted and about to fall down. The manufacturer, who has had similar signs in Los Angeles in use for 3 years, is suspicious of the customer's claim. The signs are made from polymethylmethacrylate (PMMA). The mean summer daytime temperature in Los Angeles is 80°F, with less than ten days above 95°F, and the mean daytime temperature during the 3 months in the Imperial Valley has been 105°F, with 15 days above 115°F, and maximum temperatures of 120°F. Using the data given in Figs. VI.14 and VI.16 for the modulus and the shift factor

given in Problem VI.9 for PMMA, determine whether the customer's claim is reasonable. Your discussion should be as *quantitative* as possible. You may assume that the loads are the same in both cases.

VI.11 In order to observe the mixing process of two liquid components in a mixing machine, a PMMA tube is used as an outer cylinder. However, it has been found that because of the high temperature and pressure involved, radial creep of the tube is excessive. In order to overcome this problem, CASING (Cross Linking by Activated Species (ionized) of Inert Gas) was used to cross link the outer layer of PMMA. The cross-linked surface layer has a time-independent modulus of 10 dynes/cm². If the service temperature is 135°C and the internal pressure is 60 psi, determine the expansion of the tube after 1/2 year of service. The thickness of the outer layer is 0.070 in.

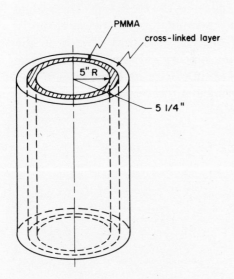

VI.12 An injection-molded container of translucent, high-density polyethylene is used to store glycol. The temperature of glycol fluctuates between 20°C and 40°C every 30 min. The fluid is at 30 psig. The inlet and outlet of the container are reinforced to eliminate the problems associated with stress concentration. Determine the service life of the container. The tensile strength of the plastic under uniaxial loading at

40°C is 2,500 psi. (The creep data for polyethylene are in the table on pp. 354–357.)

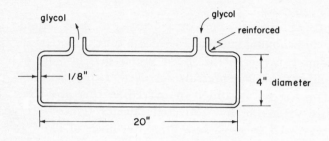

VI.13 For some unknown reason, the container for the glycol of Problem VI.12 became transparent and ruptured. What do you think happened?

VI.14 Polytetrafluoroethylene (PTFE) is used as a gasket material between two Pyrex glass tubes, as shown below. The sealing force is 100 lb. The service temperature fluctuates abruptly between 32°F and 150°F every 2 hr. Leaks develop in the system elsewhere when the spacing between the tube ends changes by more than 1/32 in. Determine the life of the gasket. (The data for the particular plastic used [Teflon®] are in the table on pp. 358–359.)

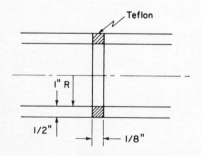

VI.15 Electric insulation and a lap joint between two copper sheets are provided by using copper/epoxy/polytetrafluoroethylene/ epoxy/copper. The thickness of the epoxy is negligible. Assuming that the stress is uniformly distributed, determine the life of the part, if it stays for 2 hr at 73°F and 4 hr at 90°F cyclically. The normal and tangential loads are 50 and 200 lb,

respectively. The polytetrafluoroethylene is the same as that used in Prob. VI.14

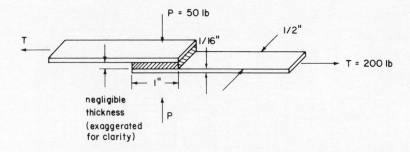

VI.16 One source of error in strain measurements made with strain gages is the creep of adhesives. This ordinarily manifests itself by a loss of output, or a continual drift in the reading after a load is applied to the gaged structure. Assuming that the adhesive used to bond a wire strain gage has the property given in the table for polyester (Arnite 151), estimate the error involved in measuring a strain of 0.1% magnitude (the gage remains within its elastic regime) at 74°F after 2000 hr. How would you overcome the problem? (The creep data for polyester are in the table on pp. 360–361.)

The properties and dimensions of the constantan alloy gage are:

Cross-sectional area	$\cong 10^{-6}$ in.^{2}
Gage length	= .125 in.
Gage width	= .090 in.
Number of parallel wires	= 15
Young's modulus for constantan	= 22×10^6 psi

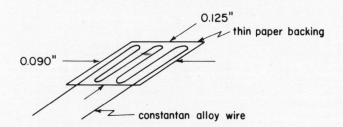

Creep Chart

Generic name	Trade name and grade designation	Grade description	Supplier	ASTM, military or other specification Classification	Specific gravity (density) by ASTM D 792 Method A	Melt-flow property Value	Melt-flow property Test method	Specimen fabrication process	Specimen type or shape	Principal specimen dimensions, in.	
Polyethylene	Alkathene® WJG II	Low density GP molding	13		0.922			Compression molded	Tensile bar	5 × 1/32 × 0.392	①
					0.921			Compression molded	Tensile bar	5 × 1/32 × 0.392	②
	Plaskon® 3005	High density Wire and cable extrusion	1	ASTM D 1248 Type III Grade 5	0.941	0.3 MI	ASTM D 1238 Cond. E	Compression molded per ASTM D 1928	ASTM D 638 Type II tensile bar	1/4 × 1/16	③
	Plaskon® AA50-003	High density Blow molding	1	ASTM D 1248 Type III Grade 5	0.950	0.3 MI	ASTM D 1238 Cond. E	Compression molded per ASTM D 1928	ASTM D 638 Type II tensile bar	1/4 × 1/16	④
	Plaskon® AA60-003	High density Blow molding and extrusion	1	ASTM D 1248 Type III Grade 5	0.960	0.3 MI	ASTM D 1238 Cond. E	Compression molded per ASTM D 1928	ASTM D 638 Type II tensile bar	1/4 × 1/16	⑤
	Bakelite® DMD-7014	High density GP molding	20	ASTM D 1248 Type III	0.96	3 MI	ASTM D 1238 Cond. E	Compression molded	ASTM D 638 Type I tensile bar	1/2 × 1/8	⑥
	Marlex® 6050	High density GP Inj. molding	17	ASTM D 1248 Type III, Grade 3	0.96 ASTM D 1505	5 MI	ASTM D 1238 Cond. E	Compression molded	ASTM D 412 Die C tensile bar	1/4 × 0.062	⑦

Source: "Modern Plastics Encyclopedia," vol. 49, pp. 190–193, 198–199, McGraw-Hill, New York, 1972.

354

#	Type of load	Method of measuring strain or deflection	Special specimen conditioning	Test temp. °F	Initial applied stress psi	Creep (apparent) modulus, thousand psi (before rupture and onset of yielding) Calculated from total creep strain or deflection at the following test times:							Time at longest test point, hr	Time at rupture or onset of yielding in air, hr
						1 hr	10 hr	30 hr	100 hr	300 hr	1,000 hr	@ longest test point		
1	Tension	Strain in red. section	Annealed 3 hr @ 111°C cooled @ 5°C/hr Stored 7 days @ 68° F	68	290	18.0	15.7	14.7	13.8			13.6	139	
					580	15.4	12.4	11.4	10.4			10.1	167	
2	Tension	Strain in red. section	Annealed 2 hr @ 145°C cooled @ 5°C/hr Stored 420 days @ 68° F	140	72.5	7.4	6.6	6.4	6.0	5.8	5.5	5.3	2,780	
					145	6.8	6.0	5.7	5.4	5.1		5.0	416	
					212	5.8	5.2	4.9	4.6	4.3		4.3	330	
3	Tension	Strain in red. section		73	1,250	38	26	22	19	16	15	13	2,500	
4	Tension	Strain in red. section		73	1,250	55	36	31	26	24	22	21	2,500	
5	Tension	Strain in red. section		73	1,250	60	42	37	32	30		27	800	800
6	Tension	Strain in red. section		73	250	—	139	131	109	93	74	54	3,400	
					500	—	83	77	68	60	54	46	3,400	
					750	—	68	65	56	47	42	38	3,400	
					1,000	—	60	56	49	44	40	34	3,400	
				105	250	—	46	45	41	40	38	34	3,400	
					500	—	34	33	32	31	30	28	3,400	
					750	—	32	31	29	28	28	25	3,400	
					1,000	—	25	25	24	21		21	350	350 (rupture)
7	Tension	Strain in red. section, 1-in. gage		75	1,000	48	36	32	28	26	24			1,600 (rupture)
					1,250	42	30	25	21	18.8				400 (rupture)
					1,500	30	16	11	7.5					150

Creep Chart (*Continued*)

Generic name	Trade name and grade designation	Grade description	Supplier	ASTM, military or other specification Classification	Specific gravity (density) by ASTM D 792 Method A	Melt-flow property Value	Melt-flow property Test method	Specimen fabrication process	Specimen type or shape	Principal specimen dimensions, in.	
Polyethylene (Continued)	Marlex® 6002	High density GP extrusion	17	ASTM D 1248 Type III, Grade 5	0.96 ASTM D 1505	0.2 MI	ASTM D 1238 Cond. E	Compression molded	ASTM D 412 Die C tensile bar	1/4 × 0.062	⑧
	Amoco® 30-670 B4	High density Inj. molding	2	ASTM D 1248 Type IV, Grade 3	0.96 ASTM D 1505	9.0	ASTM D 1238 Cond. E	Compression molded	ASTM D 638 Type I tensile bar	1/2 × 1/8	⑨
	Marlex® 5003	Butene-1 copolymer Stress crack resistant Blow molding	17	ASTM D 1248 Type III, Grade 5	0.95 ASTM D 1505	0.3 MI	ASTM D 1238 Cond. E	Compression molded	ASTM D 412 Die C tensile bar	1/4 × 0.062	⑩
	Ethofil® G 90/20	20% glass fiber reinforced Inj. molding	8		1.10			Inj. molded	Rectang. bar	5 × 1/2 × 1/16	⑪
	LNP® Thermocomp® FF-1004	20% glass fiber reinforced Inj. molding	14		1.10			Inj. molded	Rectang. bar	5 × 1/2 × 1/8	⑫
	LNP® FF 1006	30% glass fiber reinforced Inj. molding	14		1.17			Inj. molded	Rectang. bar	5 × 1/2 × 1/2	⑬
	Ethofil® G 90/40	40% glass fiber reinforced Inj. molding	8		1.28			Inj. molded	Rectang. bar	5 × 1/2 × 1/16	⑭

356

Creep test conditions | Creep test data

No.	Type of load	Method of measuring strain or deflection	Special specimen conditioning	Test temp. °F	Initial applied stress psi	Creep (apparent) modulus, thousand psi (before rupture and onset of yielding) Calculated from total creep strain of deflection at the following test times:							Time at longest test point, hr	Time at rupture or onset of yielding in air, hr
						1 hr	10 hr	30 hr	100 hr	300 hr	1,000 hr	@ longest test point		
⑧	Tension	Strain in red. section, 1-in. gage		75	750	60	45	39	34	29	26	25	21,000	2,000
					1,000	50	37	32	27	24	22	20	21,000	120
					1,250	42	30	23	17.9	13.6	9.5			
					1,500	30	17.6	12.2	7.8					
⑨	Tension	Grip separation		75	1,000	123	62	47	36				100	
⑩	Tension	Strain in red. section, 1-in. gage		75	750	46	34	29	26	25	24	22	16,600	140
					1,000	37	27	24	21	20	24	17	16,600	
					1,250	30	22	18.8	17	14	14.1	12.5	16,600	
					1,500	25	16.3	12.3	8.8					
⑪	Simple beam bending, 2-in. span, Load at center	Deflection at center		73	2,000	370	340	325	310				100	
				100	2,000	280	250	240	230				100	
⑫	Simple beam bending, 4-in. span, Load at center	Deflection at center	Equilib. with 50% rel. hum.	75	2,000	–	345	330	310	270	260		1,000	
⑬	Simple beam bending, 4-in. span, Load at center	Deflection at center		73	500	–	695	625	610	595	595		1,000	
					1,500	–	550	500	470	440	420		1,000	
⑭	Simple beam bending, 2-in. span, Load at center	Deflection at center		73	2,000	840	740	700	680				100	
				100	2,000	690	620	580	570				100	
				140	2,000	580	510	490	480				100	
				180	2,000	500	480	460	420				100	

Creep Chart (*Continued*)

Generic name	Material trade name and pertinent descriptive information							Test specimen		
						Melt-flow property			Test specimen	
	Trade name and grade designation	Grade description	Supplier	ASTM, military or other specification Classification	Specific gravity (density) by ASTM D 792 Method A	Value	Test method	Specimen fabrication process	Specimen type or shape	Principal specimen dimensions, in.
Polytetrafluoroethylene (PTFE)	Teflon® 7	High mechanical performance, GP molding resin	6	ASTM D 1457 Type IV, Grade 1	2.13–2.18			Compression molded and free sintered @ 720° F; cooled @ 180° F/hr	Rectang. strip	1/4 × 1/16 and 1/2 × 1/16
								Compression molded and free sintered @ 720° F; cooled @ 300° F/hr	Cylindrical rod	1/2 diam. × 1 high

⑮

⑯

Type of load	Method of measuring strain or deflection	Special specimen conditioning	Test temp. °F	Initial applied stress psi	Creep (apparent) modulus, thousand psi (before rupture and onset of yielding) Calculated from total creep strain or deflection at the following test times:							Time at longest test point, hr	Time at rupture or onset of yielding in air, hr
					1 hr	10 hr	30 hr	100 hr	300 hr	1,000 hr	@ longest test point		
Tension	Grip separation		-65	1,000	178	178	178	178			178	115	
				2,000	95	89	82	73			70	115	
				3,000	60	47	38	25			24	115	
			73	500	60	44	38	32			30	160	
				1,000	17.5	10.5	8.1	6.3			5.6	160	
			212	200	16.6	14.8	13.8	12.9			12.9	100	
				500	6.2	5	4.4	4.3			4.2	120	
				580	4	3.1	2.8	2.4			2.2	180	
			392	100	7.1	5.7	5.3	5			4.9	140	
				200	5.9	4.6	4.2	3.6			3.5	140	
				300	2.5	2.1	1.9	1.8			1.7	140	
Compression	Reduction in height		73	500	50	42	40	37	35		34	330	
				1,000	40	33	30	28			27	140	
				1,750	21	17.5	15.9				15.2	70	
			212	200	25	19	16.7	14.8			14.3	125	
				500	13.9	12.2	11.1	10.2			10	160	
				750	11.5	9.9	9.3				8.8	50	

⑮

⑯

Creep Chart (*Continued*)

| Generic name | Material trade name and pertinent descriptive information | | | | | Melt-flow property | | Test specimen | | |
	Trade name and grade designation	Grade description	Supplier	ASTM, military or other specification Classification	Specific gravity (density) by ASTM D 792 Method A	Value	Test method	Specimen fabrication process	Specimen type or shape	Principal specimen dimensions, in.
Polyester	Arnite® A151 ⑰	GP Inj. molding	7		1.38			Inj. molded Mold temp. 265° F	ASTM D 638 Type 1 tensile bar	1/2 × 1/8
	Arnite® A300 ⑱	18% glass fiber reinforced Inj. molding	7		1.50			Inj. molded Mold temp. 265° F	ASTM D 638 Type 1 tensile bar	1/2 × 1/8
	LNP® Thermocomp® WFL-4036 ⑲	30% glass fiber reinforced 15% PTFE lub. Inj. molding	14		1.70			Inj. molded	Rectang. bar	5 × 1/2 × 1/8
	LNP® WF-1008 ⑳	40% glass fiber reinforced Inj. molding	14		1.68			Inj. molded	Rectang. bar	5 × 1/2 × 1/8

| | Creep test conditions | | | | | Creep test data | | | | | | | | | |
| | | | | | | Creep (apparent) modulus, thousand psi (before rupture and onset of yielding) Calculated from total creep strain or deflection at the following test times: | | | | | | | Time at longest test point, hr | Time at rupture or onset of yielding in air, hr |
	Type of load	Method of measuring strain or deflection	Special specimen conditioning	Test temp. °F	Initial applied stress psi	1 hr	10 hr	30 hr	100 hr	300 hr	1,000 hr	@ longest test point		
(17)	Tension			50	1,060	440	415	408	400	385	378		1,000	
					2,130	440	415	408	400	385	378		1,000	
					3,540	440	415	408	400	383	376		1,000	
				104	1,060	393	348	331	311	286	240		1,000	
					1,422	393	348	330	310	285	238		1,000	
					2,130	393	348	328	300	276	225		1,000	
(18)	Tension			50	2,845	950	890	860	840	790	750		1,000	
					5,690	950	890	860	840	790	750		1,000	
					8,534	950	890	860	840	790	750		1,000	
				104	2,845	890	800	760	750	730	690		1,000	
					5,690	860	770	740	730	710	580		1,000	
					8,534	815	710	630	580	530	475		1,000	
				158	1,422	490	375	330	285	255	220		1,000	
					2,845	440	340	295	260	235	195		1,000	
					4,266	390	305	265	235	210	180		1,000	
(19)	Simple beam bending, 4-in. span Load at center	Deflection at center		75	2,000	–	1,100	910	800	750	720		1,000	
(20)	Simple beam bending, 4-in. span Load at center	Deflection at center		75	5,000	–	1,830	1,800	1,760	1,730	1,680		1,000	
					10,000	–	1,700	1,650	1,600	1,575	1,560		1,000	

CHAPTER SEVEN

Time-dependent Plastic Deformation in Metals— Creep

VII-1 INTRODUCTION

The time-independent theory of plastic deformation presented in Chapter IV assumes that the stress required to cause plastic deformation is independent of the rate of deformation. The justification for this approximation is that, for most structural metals, the tensile stress-strain curves measured at room temperature are changed by only a few percent when the strain rate is changed by an order of magnitude. Under static loading conditions, the strain rate is imperceptible and of no engineering significance when a static stress is applied which is a few percent less than the yield stress measured in a short-time tensile test. This type of behavior is typical of a metal at temperatures below about one-third of its melting point on an absolute temperature scale. In this temperature regime, the concept of a rate- and time-independent yield locus with purely elastic behavior for states of stress inside the yield locus gives reasonable results.

As the temperature is raised above one-third of the melting temperature, these approximations become increasingly inaccurate. In general the effects of increasing the temperature on the mechanical properties of metals are to decrease both the yield strength and the modulus, and to increase the strain rate dependence of the stress-strain curves. Definition of a general yield strength becomes increasingly difficult, and standard strain rates must be used in short-time tensile tests to ensure reproducibility of the results. In terms of behavior under static loading, it is found that loads well below the short-time yield strength will cause straining at appreciable rates, and that maintaining these loads for long periods of time may eventually cause fracture. This phenomenon of slow straining at constant stress at elevated temperature is called *creep*. Creep is a limiting design consideration for nearly all metal parts used at elevated temperatures. Since accumulated strain in creep depends on the time under stress, design criteria at elevated temperatures must be based on finite lifetimes. The emphasis in creep analysis is therefore on the prediction of strain rates and time-to-fracture for various conditions of stress and temperature.

The importance of the elevated temperature performance of materials is readily apparent from the relationship between the maximum temperature and the thermal efficiency of a power cycle. Since the thermal efficiency and consequently the overall efficiency of an energy conversion process increases as the maximum temperature in the cycle increases (the low temperature in the cycle is always limited by the ambient temperature), considerable economic advantage can be gained by raising the maximum operating temperature. This maximum temperature is limited by the properties of the material from which the hot part of the device is constructed, as, for example, the boiler tubes of a steam power plant, the turbine section of a gas turbine, or the parts of the core of a nuclear reactor. Development of improved materials and a better understanding of the behavior of existing materials, which will allow less conservative design without sacrificing safety, offer promise of substantial gains in power conversion efficiency. As the supplies of fossil fuels become depleted and the requirements of environmental protection become more severe, these economic incentives for improvements in high-temperature materials can be expected to increase.

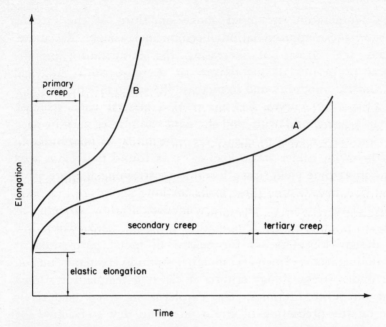

Fig. VII.1 Schematic representation of the typical creep behavior of metals. (Curve A) at relatively low temperatures and low stresses; (Curve B) at relatively high temperatures or high stresses.

VII-2 MECHANISMS OF CREEP

The result of a typical high-temperature creep experiment, where a constant load is applied to the specimen and the strain is measured as a function of time, is illustrated schematically in Fig. VII.1.* A small amount of elastic strain accompanies the application of the load. This elastic component of strain is present as long as the load is maintained and is recoverable by unloading at any stage of the experiment. As the time under load increases, the strain in the specimen increases continuously as shown. This additional strain is permanent.

The creep curve A in Fig. VII.1 is traditionally divided into three parts, as indicated. The first stage, which is called transient or

---

*Creep experiments are nearly always performed under constant load or constant engineering stress. Strain is given as engineering strain. At small strains, the results are approximately the same as they would be if the experiment were done at constant true stress. However the behavior at large strains reflects the fact that the cross-sectional area is changing and the true stress is increasing.

primary creep, is characterized by a relatively high strain rate which decreases with time. The strain rate eventually reaches a minimum value which may remain constant for some period of time. This stage is called steady-state or secondary creep. At some point, another transition occurs and the creep rate again accelerates. This final stage which is terminated by the fracture of the sample is called tertiary creep. The relative durations of the three stages depend on the temperature, stress, and prior history of the sample. At high stress or high temperatures, the secondary stage may be severely restricted or absent completely. A creep curve for such a situation where there is only an inflection point between the decelerating primary region and the accelerating tertiary region is shown as curve B in Fig. VII.1.

There are a number of strain-producing processes which can occur only at elevated temperature and contribute to creep. For the temperature and stress conditions of engineering importance, the mechanism which limits the rate of creep is probably diffusion-controlled climb of dislocations. However, thermally activated glide of dislocations, grain boundary sliding, and direct mass transfer by diffusion may also contribute to the creep process. Some of the mechanisms of creep are illustrated schematically in Fig. VII.2. These processes usually go on simultaneously and may be combined in various ways, but one mechanism is usually rate controlling for a particular set of conditions. At relatively low temperatures close to one-third of the melting temperature, diffusion rates are small and direct thermal activation of dislocations probably accounts for most of the creep strain. At very high temperatures and low stresses, where dislocation processes are slow, the rate of diffusion may become sufficiently large that direct mass transport by diffusion becomes the rate controlling process. In this mechanism, diffusion of atoms from the sides of the grain which are under compression to the sides of the grain which are under tension causes the grains and the whole body to elongate in the tensile direction. The atom transport can take place by bulk diffusion through the grain,* in which case it is called Nabarro-Herring creep, or along the grain boundaries, when it is called Coble creep. Grain boundary sliding contributes to creep strain in a polycrystal but by itself cannot produce very much strain. Since grain boundaries do not extend on a single plane across a sample,

---

*Bulk diffusion in a crystal lattice actually takes place by the diffusion of lattice vacancies in the direction opposite to the mass flow.

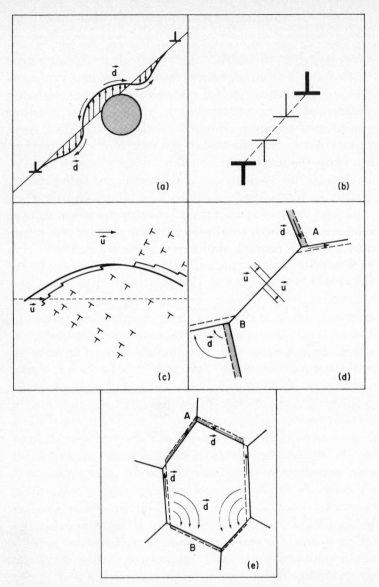

Fig. VII.2 Mechanisms of creep. (a) A dislocation can climb to get around a particle which intersects its slip plane. Here climb is taking place by diffusion of atoms along the dislocation, as indicated by **d**. Vertical arrows are the climb displacement of different parts of the dislocation. (b) The two edge dislocations of a dipole can climb together along the dashed line and annihilate each other. (c) At an irregularity on a grain boundary which has a general direction indicated by the dashed line, grain boundary sliding through a distance **u** requires that the bump of the irregularity deform to accommodate the motion. Here accommodation is occurring by dislocation motion. (d) The interference to grain boundary sliding at triple points of the grain boundaries can be accommodated by diffusion of atoms around the corners, as indicated by the vectors **d**. Diffusion can be either through the grain (B) or along the grain boundary (A). (e) Creep can occur directly by diffusional flow of atoms from one part of the grain to another. In Nabarro-Herring creep, the flow is through the grain; in Coble creep, the flow is along the grain boundary.

sliding of the grain boundary is limited by the incompatibility of the motion at grain boundary corners and irregularities. Further sliding can occur only if compatibility is maintained by deformation of the interior of the grain. The accommodating deformation can occur by either dislocation motion or diffusion, as shown in Fig. VII.2. Since grain boundary sliding can take place very rapidly, the accommodation process limits the rate of creep.

At temperatures of engineering importance, the predominant mechanism of creep is dislocation climb. The ability of a dislocation to move by climb when atoms diffuse to or away from it adds an additional degree of freedom to its motion. Thus, a dislocation which is being held up by an obstacle on its glide plane can climb to a new glide plane which is not intersected by the obstacle. This releases the dislocation so that it can undergo additional glide, producing more strain. Climb also allows dislocations of opposite sign on noninter-secting glide planes to move together to annihilate each other. This reduces the dislocation density and decreases the number of obstacles to dislocation motion. These climb processes may occur by bulk diffusion, where vacancies diffuse to and from the dislocation through the lattice, or by core diffusion, where the atoms simply diffuse along the dislocation. On a single dislocation, core diffusion causes different parts of the dislocation to climb in opposite directions and can release a dislocation from a discrete obstacle like a particle, as shown in Fig. VII.2. Core diffusion can also be an important mechanism when two or more dislocations have inter-sected and diffusion of atoms from one dislocation to the other causes cooperative motion. Such cooperative climb of intersecting dislocations by core diffusion may be very important to the motion of complex dislocation networks which form during creep.

The differences in the various stages of creep can be related to the creep mechanisms discussed above and to the changes in the dislocation structure which are occurring. In any metal at low temperature and at a state of stress inside the yield locus, the dislocations in the material are in a state of quasi-equilibrium. This means that they are in positions of local equilibrium with respect to the glide forces which act on them. However, many of the dislocations could reach lower energy configurations if they could overcome short-range obstacles or if they could move by climb. At high temperatures, thermal activation and climb can take place, allowing the dislocation structure to move toward a lower energy state. Initially when the material is first heated to a high temperature, there is a large distribution of the energies required to activate different dislocations. The dislocations requiring the minimum energy or in the case of climb those which can be freed by the least climb tend to be released first. Since the activation energies and the climb distances are small, the strain rate is initially quite high. As the dislocation structure changes with increasing creep strain, the favorable sites are depleted, and the strain rate decreases. This accounts for the decelerating strain rate of primary creep. Since the initial dislocation structure depends on the prior history of the material, primary creep behavior is expected to be quite history dependent.

Eventually the internal structure of the material approaches a steady state which is characteristic of secondary creep. During this part of the creep process, the rate of work hardening by such processes as immobilization of dislocations by obstacles, creation of dislocation structures which impede dislocation motion, and increase in dislocation density by dislocation generation, is balanced by the rate of recovery related to activation of dislocations past obstacles, annealing of dislocation structures, and the annihilation of dislocations by recombination of those of opposite sign. The rate of these processes is limited by the rate of dislocation climb, which is in turn determined by the rate of diffusion of atoms and vacancies to or along dislocations. The dislocation structure which has developed by the beginning of the second stage of creep is usually independent of the history of the material prior to its being subjected to creep conditions. The primary creep stage has served to erase the history effects. Because of this, the structure of the material and the strain rate experienced during secondary creep depend only on the stress and temperature.

Since the average internal structure of the metal is not changing during second-stage creep, one might expect this type of behavior to continue until the cross-sectional area of the sample had been reduced to zero. In such cases, the region of tertiary creep would be associated with the familiar phenomenon of necking. This is nearly true for a limited number of materials when creep takes place at very slow rates. However, for most materials, the beginning of tertiary creep is associated with the formation of internal flaws or holes before the formation of an external neck. These holes usually begin at grain boundaries or precipitate particles where the creep process has produced strain concentrations. During the early stages of creep, these defects are of insufficient size and number to affect the average strain rate, so that the steady-state nature of secondary creep is undisturbed. At the beginning of tertiary creep, however, these defects have become of sufficient severity to cause localization of the flow. During tertiary creep, these cracks grow and link up to form the final fracture.

VII-3 EVALUATION OF CREEP DATA FOR DESIGN

Since many applications require the use of metals at temperatures where creep is significant, a theory is needed to allow the use of experimental data to predict the service behavior of parts. The requirements of such a theory are that it should allow not only extrapolation of the short-term tests to predict longer-term behavior, but also the interpolation of measurements to predict behavior at temperatures and stresses intermediate to those measured. The ideas to be presented will first be illustrated by the simple case of uniaxial stress. The results will then be generalized to more complex states of stress, using the ideas presented in Chapter IV.

Primary creep usually obeys the functional relation

$$\epsilon_{11} = \beta t^{1/3} \tag{VII.1}$$

where t is the time of loading, and β is a function of the stress, temperature, and prior strain history. Because of this prior history dependence, consistent data for the evaluation of β and its dependence on stress and temperature are rarely available. However, for engineering purposes involving materials with a uniform prior history, the necessary information can be obtained from a series of short-term experiments. Furthermore, creep is generally a severe problem only in parts designed for long lives at elevated temperature. Under these circumstances the stress must be kept low in order to keep the second-stage creep rates at an acceptable level. In such cases the transient creep is often small compared to the secondary creep, so that design may be based exclusively on the second-stage behavior.

During steady-state creep, the creep rate should be independent of the strain and relatively insensitive to prior history. Since the above discussion indicates that the rate limiting process in creep is thermally activated, a typical Arrhenius type relationship is expected.*

$$\dot{\epsilon}_{11} = A\sigma_{11}{}^{m} \exp\left[-\frac{H(\sigma_{11})}{kT}\right] \qquad \text{(VII.2)}^{\dagger}$$

where the term $\sigma_{11}{}^{m}$ is inserted to account for the effects of stress on the steady-state dislocation structure, and the stress dependence of the activation energy H reflects the fact that thermal activation is assisted by the applied stress. Over the range of interest, A and m can be considered to be constant and H to be a function of stress only; k is Boltzmann's constant. Complete description of the creep process, therefore, requires determination of the two constants, A and m, and the stress-dependent function H from experimental data.

The procedure for evaluating all of the constants will be discussed first. This procedure can be used when the experimental data available is sufficient to justify this detailed analysis. In many cases the creep behavior can be adequately described by one of the more simplified formulations which will be discussed later. It is convenient to begin by writing the creep equation in dimensionless form‡

$$\frac{\dot{\epsilon}_{11}}{\dot{\epsilon}_0} = \left(\frac{\sigma_{11}}{\sigma_0}\right)^{m} \exp\left[-\frac{H(\sigma_{11})}{kT}\right] \qquad \text{(VII.3)}$$

where $\dot{\epsilon}_0$ is an arbitrarily chosen strain rate (usually unity in convenient units such as $\sec^{-1}$ or hr^{-1}), and σ_0 is an experimentally determined constant which replaces A in Eq. (VII.2). Now that the terms are dimensionless, the equation can be rewritten in logarithmic form:

*From physical considerations, the stress and strain in this expression should be true stress and true strain. In actual practice, the parameters are usually determined from engineering stress-engineering strain curves. Since the linear portion of the creep curve occurs at small strain, the approximation should be satisfactory.

†In this chapter $\dot{\epsilon}_{ij}$ and $\dot{\epsilon}$ refer to creep strain rates since under constant stresses the elastic strain rates are zero.

‡Either $\dot{\epsilon}_0$ or σ_0 may be chosen arbitrarily and the other determined from the data. The value of the arbitrarily chosen one will affect the value determined for the other. The nondimensionalization is carried out simply to avoid the awkwardness of taking logarithms of dimensioned quantities.

$$\ln\left(\frac{\dot{\epsilon}_{11}}{\dot{\epsilon}_0}\right) = m \ln\left(\frac{\sigma_{11}}{\sigma_0}\right) - \frac{H(\sigma_{11})}{kT} \tag{VII.4}$$

The first term on the right contains only quantities which are assumed to be independent of temperature. Therefore, the activation energy $H(\sigma_{11})$ can be determined from the derivative of the strain rate at constant stress with respect to the reciprocal of the temperature.

$$H(\sigma_{11}) = -k\left[\frac{\partial \ln(\dot{\epsilon}_{11}/\dot{\epsilon}_0)}{\partial(1/T)}\right]_{\sigma_{11}=\text{const.}} \tag{VII.5}$$

This derivative is usually determined graphically from the experimental data as outlined in the following example. Once $H(\sigma_{11})$ has been determined, the stress exponent m can be determined from the derivative of $[\ln(\dot{\epsilon}_{11}/\dot{\epsilon}_0) + H(\sigma_{11})/kT]$ with respect to $\ln \sigma_{11}$ at constant temperature.

$$m = \left\{\frac{\partial[\ln(\dot{\epsilon}_{11}/\dot{\epsilon}_0) + H(\sigma_{11})/kT]}{\partial \ln(\sigma_{11}/\sigma_0)}\right\}_{T=\text{const.}} \tag{VII.6}$$

Once again this is done graphically, as described in Example VII.1. At this point, σ_0 is the only unknown parameter so that Eq. (VII.4) can be solved for σ_0 and it can be directly evaluated from the experimental data.

EXAMPLE VII.1—Creep of a Support Rod in a Boiler

A support rod in a boiler carries a constant tensile stress of 10,000 psi. If the rod is made from the medium-carbon steel for which creep data are given in Fig. VII.3, what is the expected lifetime for the rod before it elongates 10%? The operating temperature of the boiler is $1000°$ F.

Solution:

It will be assumed that the creep strain comes exclusively from second-stage creep. The first step in the solution is to determine, from the experimental data presented in Fig. VII.3, the values of the parameters m, σ_0, and $H(\sigma)$ which appear in Eq. (VII.3). From Eq. (VII.5) it can be seen that the slopes of the lines in Fig. VII.3 are proportional to the activation energies $H(\sigma)$.

$$H = -k\left(\frac{\Delta \ln \dot{\epsilon}_{11}}{\Delta 1/T}\right) = -2.3k\left(\frac{\Delta \log_{10} \dot{\epsilon}_{11}}{\Delta 1/T}\right) \tag{a}$$

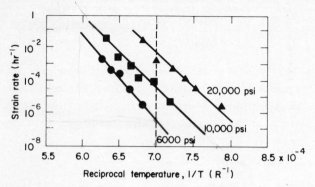

Fig. VII.3 Creep rate for medium-carbon forged steel. (*Data from Ref. 7, p. 636.*)

TABLE VII.1 Boltzmann's Constant

$k = 1.38 \times 10^{-16}$ erg/$°$K $= 1.38 \times 10^{-23}$ Joule/$°$K
$= 8.61 \times 10^{-5}$ eV/$°$K $= 6.79 \times 10^{-23}$ in. lb/$°$R

Considering the line for $\sigma_{11} = 10{,}000$ psi, it is found that when $1/T$ changes by $1.82 \times 10^{-4}\ °R^{-1}$ from $6.0 \times 10^{-4}\ °R^{-1}$ to $7.82 \times 10^{-4}\ °R^{-1}$, $\dot{\epsilon}_{11}$ goes from 1 to 10^{-8}/hr. That is $\Delta\ 1/T = 1.82 \times 10^{-4}\ °R^{-1}$ and $\Delta \log_{10} \dot{\epsilon}_{11}/\dot{\epsilon}_0 = -8$. Values for Boltzmann's constant for various units are given in Table VII.1. Therefore, from Eq. (VII.5)

$$H\ (10{,}000\ \text{psi}) = -2.3\,(6.79 \times 10^{-23})\left(\frac{-8}{1.82 \times 10^{-4}}\right) \text{in. lb}$$

$$H\ (10{,}000\ \text{psi}) = 6.9 \times 10^{-18}\ \text{in. lb}$$

(b)

Similarly the activation energies for 6,000 psi and 20,000 psi are determined to be

$$H\ (6{,}000\ \text{psi}) = 7.25 \times 10^{-18}\ \text{in. lb}$$
$$H\ (20{,}000\ \text{psi}) = 6.50 \times 10^{-18}\ \text{in. lb}$$

(c)

The exponent m is to be determined from the relationship in Eq. (VII.6). Therefore, a graph of $[\ln (\dot{\epsilon}_{11}/\dot{\epsilon}_0) + H(\sigma_{11})/kT]$ vs. $\log_{10} \sigma_{11}$ will be construed using the values of H in Eqs. (b) and (c). A suitable temperature is chosen, in this case, $1/T = 7 \times 10^{-4}\ °R^{-1}$ ($T = 1430°R = 970°F$), as indicated by the dashed line in Fig. VII.3. The points calculated for the graph of Fig. VII.4 are

σ_{11} (psi)	$\ln \dot{\epsilon}_{11}/\dot{\epsilon}_0$	H/kT	$\ln (\dot{\epsilon}_{11}/\dot{\epsilon}_0) + H/kT$
6,000	−15.0	74.6	59.6
10,000	−10.1	70.7	60.6
20,000	−5.0	67.0	62.0

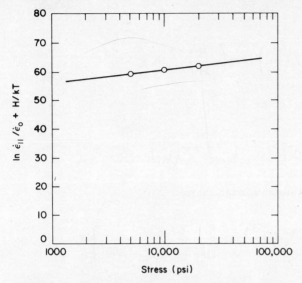

Fig. VII.4 Graph for the determination of the stress exponent m.
(*Prepared from Fig. VII.3.*)

from which m is determined using

$$m = \frac{\Delta\left[\ln(\dot{\epsilon}_{11}/\dot{\epsilon}_0) + H/kT\right]}{2.3\Delta\,\log_{10}\sigma_{11}} = \frac{62.0 - 59.6}{2.3\left[\log_{10}(2\times10^4) - \log_{10}(6\times10^3)\right]} \quad (d)$$

$$m = 2.0$$

The constant σ_0 is determined by solving Eq. (VII.3) for σ_0 and using the values for one point from the data of Fig. VII.3 to evaluate the expression

$$\sigma_0 = \sigma_{11}\left(\frac{\dot{\epsilon}_0}{\dot{\epsilon}_{11}}\right)^{1/m} \exp\left(-\frac{H}{mkT}\right) \quad (e)$$

Using $\dot{\epsilon}_{11} = 4\times10^{-5}$ hr^{-1}, $\sigma_{11} = 10{,}000$ psi, $T = 1430^\circ$R as the reference point, one obtains

$$\sigma_0 = 6.3\times10^{-10}\ \text{psi} \quad (f)$$

The remainder of the solution to the stated problem is straightforward. The state of stress is simple uniaxial tension. It is assumed that the entire 10% strain is from second-stage creep. Therefore, the strain rate is constant and the time t is given by

$$t = \frac{\epsilon_{11}}{\dot{\epsilon}_{11}} = \frac{0.1}{\dot{\epsilon}_{11}} \quad (g)$$

From Eq. (VII.3)

$$\dot{\epsilon}_{11} = \dot{\epsilon}_0 \left(\frac{\sigma_{11}}{\sigma_0}\right)^m \exp\left(-\frac{H}{kT}\right); \quad (\dot{\epsilon}_0 = 1 \text{ hr}^{-1}) \tag{h}$$

Substituting Eq. (g) into Eq. (h)

$$t = \frac{0.1}{\dot{\epsilon}_0} \left(\frac{\sigma_0}{\sigma_{11}}\right)^m \exp\left(\frac{H}{kT}\right) \tag{i}$$

Using the values of the constants found above

$$t = 0.1 \left(\frac{6.3 \times 10^{-10}}{10^4}\right)^2 \exp\left[\frac{6.9 \times 10^{-18}}{(6.79 \times 10^{-23})(1,460)}\right] \text{hr} \tag{j}$$

$$t = 1.59 \times 10^3 \text{ hr}$$

Since the stress of 10,000 psi is the same as for one of the curves in Fig. VII.3, this point could have been obtained directly from the graph. However, the curve for $H(\sigma)$ can be plotted for interpolation to other stresses. Then, using the known values of m and σ_0, the strain rates can be calculated for other values of stress. (See Fig. VII.5 for H vs. σ.)

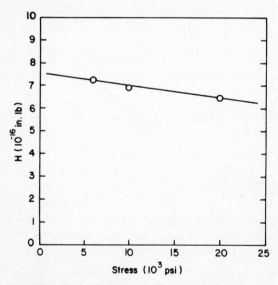

Fig. VII.5 Creep activation energy as a function of stress for medium-carbon forged steel. (*Calculated from the data in Fig. VII.3.*)

In most cases, one must work with less extensive data, and it is generally found that data are not available for creep rates at the same stress for several temperatures. Then, the activation energy cannot be determined by the method described above. In such circumstances it is usually sufficient to work with one of the following simplified forms of the creep rate equation:

$$\frac{\dot{\epsilon}_{11}}{\dot{\epsilon}_0} = \left(\frac{\sigma_{11}}{\sigma_0}\right)^m \exp\left(-\frac{H}{kT}\right) \tag{VII.7}$$

$$\frac{\dot{\epsilon}_{11}}{\dot{\epsilon}_0} = \exp\left[-\frac{H(\sigma_{11})}{kT}\right] \tag{VII.8}$$

The choice between these two equations can be made on the basis of the behavior of the creep rate as a function of stress at a constant temperature, such as is shown in Fig. VII.6. If the slopes of all of the lines of log stress vs. log strain rate for different temperatures are the same, Eq. (VII.7) is appropriate. The lines in the log-log plot can then be extrapolated sufficiently to give the needed information for the determination of the constant activation energy H. The constant σ_0 is determined as before. See the following example for the details of this procedure.

EXAMPLE VII.2—Creep in a Chemical Reaction Chamber

A chemical reaction chamber of the type sketched in Fig. VII.7(a) is made of type 304 stainless steel for which the creep data is given in Fig. VII.6. The cover bolts of the chamber, which are made from the same steel, are pretensioned at room temperature to a stress of 10,000 psi. The service temperature of the chamber is $1100°F$. How long can the chamber be used at this temperature before the stress in the bolts drops below 5,000 psi at the operating temperature? It may be assumed that the deflection of the flanges during tightening is negligible compared to the extension of the bolts. Note that the Young's modulus of 304 SS is 29×10^6 psi at room temperature and 22.3×10^6 psi at $1100°F$.

Solution:

Once again, the first step is to determine the necessary parameters from the data given in Fig. VII.6. Since the lines in the figure can be drawn parallel, the activation energy can be assumed to be stress independent and Eq. (VII.7) can be used as a description of the creep behavior. The slope of the line in Fig. VII.6 gives the exponent m:

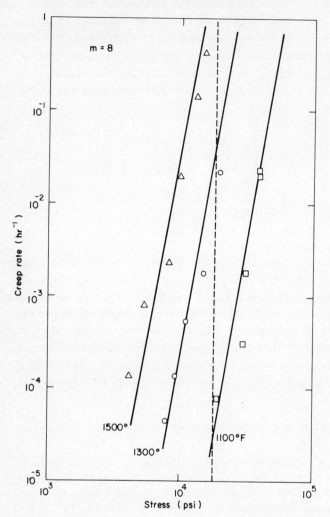

Fig. VII.6 Minimum creep rate of 304 stainless steel, as a function of stress. (*Data from W. F. Simmons and H. C. Cross, The Elevated-Temperature Properties of Stainless Steels*, ASTM, Philadelphia, *p. 60, 1952.*)

$$m = \frac{\Delta \ln \dot{\epsilon}_{11}}{\Delta \ln \sigma_{11}} = \frac{\Delta \log \dot{\epsilon}_{11}}{\Delta \log \sigma_{11}} \tag{a}$$

Taking points at opposite ends of the $1500°$F curve, i.e., $\dot{\epsilon}_{11} = 10^{-5}$ hr^{-1}, $\sigma_{11} = 3,600$ psi and $\dot{\epsilon}_{11} = 1$ hr^{-1}, $\sigma_{11} = 15,000$ psi, one obtains $m = 8$.

If the lines for $1500°$F and $1100°$F are extrapolated for a short distance, the values for strain rate at each of the three temperatures for a stress of 18,000 psi (dashed line) can be determined. The values are found to be

T ($^\circ$F)	σ_{11} (psi)	$\log_{10} \dot{\epsilon}_{11}$
1500	18,000	.4
1300	18,000	−1.33
1100 .	18,000	−4.5

The temperatures can then be converted to $^\circ$R and the graph of $\log_{10}\dot{\epsilon}_{11}$ vs. $1/T$ can be constructed as shown in Fig. VII.7(b). The activation energy H is determined from the slope of the line in Fig. VII.7(b) by

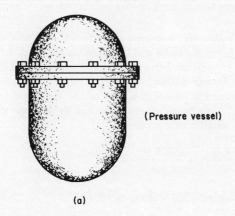

(Pressure vessel)

(a)

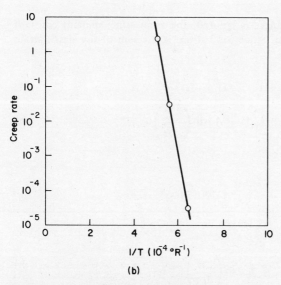

(b)

Fig. VII.7 (a) Pressure vessel with bolted flange. (b) Creep rate as a function of reciprocal temperature at a constant stress of 18,000 psi for 304 stainless steel. (*Constructed from data in Fig. VII.6.*)

$$H = -\frac{k \, \Delta \ln \dot{\epsilon}_{11}}{\Delta 1/T} = -\frac{2.3k \, \Delta \log \dot{\epsilon}_{11}}{\Delta 1/T} = \frac{-2.3k \, [-4.5 - (.4)]}{(1/1,560) - (1/1,960)} \tag{b}$$

which gives $H = 5.85 \times 10^{-18}$ in. lb.

Solving Eq. (VII.7) for σ_0 and using the values for the point $\dot{\epsilon}_{11} = 10^{-4}$ /hr when $\sigma_{11} = 19,500$ psi and $T = 1100°F$, σ_0 is found to be

$$\sigma_0 = \left(\frac{\dot{\epsilon}_0}{\dot{\epsilon}_{11}}\right)^{1/m} \sigma_{11} \exp\left(-\frac{H}{mkT}\right) = 62 \text{ psi} \tag{c}$$

Now that the necessary constants have been determined, the remainder of the problem can be solved. Since the bolts and the chamber are made from the same material, the thermal expansions for the two parts match and the elastic strain in the bolts does not change during heating. The stress in the bolts does change during heating because the elastic modulus of the steel changes. The initial stress at $1100°F$, $\sigma_{11}{}^0$, can be calculated from the initial stress at room temperature in terms of the initial elastic strain, $\epsilon_{11}{}^0$.

$$\epsilon_{11}{}^0 = \frac{\sigma_{11}}{E} = \frac{10,000}{29 \times 10^6} = 3.45 \times 10^{-4} \tag{d}$$

Since the strain does not change during heating, at $1100°F$

$$\sigma_{11}{}^0 = E\epsilon_{11}{}^0 = 22.3 \times 10^6 \, (3.45 \times 10^{-4}) = 7,700 \text{ psi}$$

If the deflection of the flanges is ignored, the total strain remains constant during the subsequent creep. Since total strain is the sum of the elastic strain and the creep strain

$$\epsilon_{11}{}^e + \epsilon_{11}{}^c = \epsilon_{11}{}^0; \quad \dot{\epsilon}_{11}{}^e = -\dot{\epsilon}_{11}{}^c \tag{e}$$

$$\dot{\epsilon}_{11}{}^c = -\dot{\epsilon}_{11}{}^e = A \exp\left(-\frac{H}{kT}\right)(\sigma_{11})^m \tag{f}$$

where $A = \dot{\epsilon}_0 \, (1/\sigma_0)^m$ and

$$\sigma_{11} = E(1100°F)\epsilon_{11}{}^e \tag{g}$$

Substituting Eq. (g) into Eq. (f), we obtain

$$\frac{d\epsilon_{11}{}^e}{dt} = A \exp\left(-\frac{H}{kT}\right) E^m \, (\epsilon_{11}{}^e)^m \tag{h}$$

where E is Young's modulus at $1100°F$. This can be integrated giving

$$\int \frac{d\epsilon_{11}{}^e}{(\epsilon_{11}{}^e)^m} = -A \exp\left(-\frac{H}{kT}\right) E^m t + C \tag{i}$$

$$\frac{(\epsilon_{11}{}^e)^{1-m}}{m-1} = A \exp\left(-\frac{H}{kT}\right) E^m t + C \tag{j}$$

From the initial conditions $\epsilon_{11}{}^e = \epsilon_{11}{}^0$ at time $t = 0$, the integration constant is evaluated as

$$C = -\frac{(\epsilon_{11}{}^0)^{1-m}}{m-1} \tag{k}$$

Substituting Eq. (k) into Eq. (j) and rearranging terms, we get

$$A \exp\left(-\frac{H}{kT}\right) E^m t = \frac{(\epsilon_{11}{}^e)^{1-m} - (\epsilon_{11}{}^0)^{1-m}}{m-1} \tag{l}$$

The substitution of $\epsilon_{11}{}^e = \sigma_{11}/E$ into Eq. (1) yields

$$A \exp\left(-\frac{H}{kT}\right) E^m t = \frac{E^{m-1}}{m-1}[(\sigma_{11})^{1-m} - (\sigma_{11}{}^0)^{1-m}] \tag{m}$$

which may be solved for t giving

$$t = \left(\frac{E^{-1}}{Ae^{-H/kT}}\right)\left(\frac{1}{m-1}\right)[(\sigma_{11})^{1-m} - (\sigma_{11}{}^0)^{1-m}] \tag{n}$$

Factoring out $(\sigma_{11}{}^0)^{1-m}$ gives

$$t = \frac{1}{E(m-1)} \frac{(\sigma_{11}{}^0)^{1-m}}{Ae^{-H/kT}}\left[\left(\frac{\sigma_{11}}{\sigma_{11}{}^0}\right)^{1-m} - 1\right] \tag{o}$$

Substituting for A and rearranging terms gives

$$t = \frac{\sigma_{11}{}^0}{E(m-1)\dot{\epsilon}_0(\sigma_{11}{}^0/\sigma_0)^m e^{-H/kT}}\left[\left(\frac{\sigma_{11}{}^0}{\sigma_{11}}\right)^{m-1} - 1\right] \tag{p}$$

with $\dot{\epsilon}_0 = 1 \text{ hr}^{-1}$. Substituting the values for m, H, and σ_0 determined above, the desired solution for t is obtained:

$$t = 1.7 \times 10^4 \text{ hr} \tag{q}$$

When the lines of log stress vs. log strain rate at constant temperature are not parallel, Eq. (VII.8) is preferable. In this case, a linear dependence of the activation energy on stress is usually adequate.

$$H = H_0 - V\sigma_{11} \tag{VII.9}$$

With this expression, it is more convenient to work with a plot of the logarithm of strain rate vs. stress, as in Fig. VII.8. The slope of the curves gives the constant V.

$$V = kT \frac{\partial \ln(\dot{\epsilon}_{11}/\dot{\epsilon}_0)}{\partial \sigma_{11}} \tag{VII.10}$$

The constant H_0 can be determined from the intercepts of the curves at zero stress, using Fig. VII.9, and from the relation

$$H_0 = k \frac{\partial \ln(\dot{\epsilon}_{11}/\dot{\epsilon}_0)}{\partial 1/T} \tag{VII.11}$$

and $\dot{\epsilon}_0$ can be determined from the intercept of Fig. VII.9 at $1/T$ equals zero.

EXAMPLE VII.3—Maximum Allowable Stress in a Turbine Disk

Determine the maximum stress allowable in a turbine disk operating at $1350°$ F, if it is made from Waspaloy superalloy. The design criterion requires less than 1% strain in 10^4 hr.

Solution:
From the nature of the presentation of the data in Fig. VII.8, a creep law of the following form is assumed to be applicable:

$$\frac{\dot{\epsilon}_{11}}{\dot{\epsilon}_0} = \exp\left(-\frac{H(\sigma_{11})}{kT}\right) \tag{a}$$

which can be written, for H linear in stress, as

$$\ln \frac{\dot{\epsilon}_{11}}{\dot{\epsilon}_0} = -\frac{H - V\sigma_{11}}{kT} \tag{b}$$

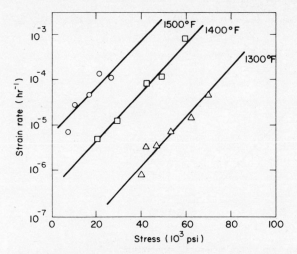

Fig. VII.8 Minimum creep rate for the superalloy Waspaloy as a function of stress. (*Data from D. P. Moon, R. C. Simon, and R. J. Favor, The Elevated-Temperature Properties of Selected Superalloys*, ASTM, Philadelphia, *p. 335, 1968.*)

Therefore,

$$V = kT \frac{\partial \ln(\dot{\epsilon}_{11}/\dot{\epsilon}_0)}{\partial \sigma_{11}} = kT \frac{2.3 \Delta \log \dot{\epsilon}_{11}}{\Delta \sigma_{11}} \tag{c}$$

Using the slopes of the lines in Fig. VII.8, we calculate the values

$$V(1300°F) = 1.48 \times 10^{-23} \text{ in.}^3$$
$$V(1400°F) = 1.55 \times 10^{-23} \text{ in.}^3 \tag{d}$$
$$V(1500°F) = 1.59 \times 10^{-23} \text{ in.}^3$$

Since these values are nearly the same, it may be assumed that V is independent of temperature and equal to

$$V_{ave} = 1.54 \times 10^{-23} \text{ in.}^3 \tag{e}$$

By extrapolating the lines of Fig. VII.8 to zero stress, the points for plotting Fig. VII.9 are obtained. The activation energy at zero stress, H_0, is proportional to the slope of the line in Fig. VII.9.*

$$H_0 = -\frac{k \partial \ln(\dot{\epsilon}_{11}/\dot{\epsilon}_0)}{\partial 1/T} = -2.3k \frac{\Delta \log \dot{\epsilon}_{11}}{\Delta 1/T} \tag{f}$$

*The extrapolation is artificial, since the creep rate must go to zero for zero stress. This means that the true behavior deviates from the behavior described by Eq. (b) at low stress.

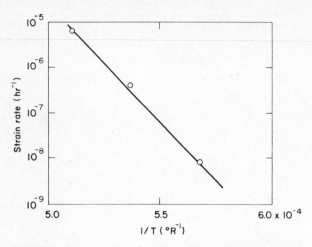

Fig. VII.9 Strain-rate as a function of the reciprocal temperature, extrapolated to zero stress for Waspaloy. (*Constructed from the data in Fig. VII.8.*)

This gives the result

$$H_0 \; = \; 7.9 \times 10^{-18} \text{ in. lb} \tag{g}$$

By extrapolating the line of Fig. VII.9 to $1/T = 0$, we can obtain the value of $\dot{\epsilon}_0$ which is the strain rate when the exponent H/kT is zero:

$$\dot{\epsilon}_0 \; = \; 8 \times 10^{20} \text{ hr}^{-1} \tag{h}$$

Now that the constants have been determined, the stress which will give 1% strain in 10,000 hr can be calculated. For $\epsilon_{11} = 10^{-2}$ in 10^4 hr, $\dot{\epsilon}_{11} = 10^{-6}$ hr^{-1}. Solving Eq. (b) for σ_{11}

$$\sigma_{11} \; = \; \frac{kT \ln(\dot{\epsilon}_{11}/\dot{\epsilon}_0) + H_0}{V} \tag{i}$$

Using the values obtained above for V, H_0 and $\dot{\epsilon}_0$, we find that the maximum $\sigma_{11} = 24,700$ psi.

In all of the methods of interpreting experimental data discussed above and illustrated in the examples, the accuracy of the determinations depends on the completeness of the data. Errors of course increase with increasing distance of extrapolation so that these methods must be applied with care. The results should, however, be sufficiently accurate for interpolating the experimental data to predict behavior at stresses and temperatures intermediate to those measured. One must be particularly aware of the possible inaccuracies which are associated with extrapolation to long times when processes such as corrosion are taking place.

VII-4 CORRELATION OF CREEP-RUPTURE DATA*

Because of the simplicity of the test, creep data are often reported as total time to failure as a function of engineering stress and temperature. For a material which obeys Eq. (VII.7), the strain rate at constant temperature is given by

$$\dot{\epsilon}_{11} = B\sigma_{11}{}^m \qquad\qquad (\text{VII.12})$$

where $\dot{\epsilon}_{11}$ and σ_{11} are the true strain rate and the true stress, respectively,[†] and $B = Ae^{-H/kT}$. In a typical experiment, since creep occurs at constant volume, the following expression can be written:

$$\dot{\epsilon}_{11} = \frac{1}{l}\frac{dl}{dt} = -\frac{1}{A}\frac{dA}{dt}$$

$$\dot{\epsilon}_{11} = B\sigma_{11}{}^m = B\left(\frac{F}{A}\right)^m = -\frac{1}{A}\frac{dA}{dt} \qquad\qquad (\text{VII.13})$$

$$-A^{m-1}dA = BF^m dt$$

This can be integrated to find the time t:

$$-\frac{1}{m}(A_f{}^m - A_0{}^m) = BF^m t$$

$$\frac{1}{m}\left[1 - \left(\frac{A_f}{A_0}\right)^m\right] = B\left(\frac{F}{A_0}\right)^m t = \dot{\epsilon}_i t \qquad\qquad (\text{VII.14})$$

---

*Here, the term rupture is used loosely to include fracture at greater than zero cross-sectional area.

†As discussed earlier, the stress and strain in the creep equations should be true stress and true strain. The justification for determining the parameters from engineering stress-engineering strain data is that the evaluation of the second-stage creep rate is usually made at relatively small strains. Here the creep equation is going to be integrated to large strains so it is necessary to include the cross-sectional area changes in the equation.

$$t = \frac{1}{m\dot{\epsilon}_i}\left[1 - \left(\frac{A_f}{A_0}\right)^m\right] = \frac{1}{m\dot{\epsilon}_0}\left(\frac{\sigma_0}{\sigma_i}\right)^m \exp\left(\frac{H}{kT}\right)\left[1 - \left(\frac{A_f}{A_0}\right)^m\right]$$

$$\text{(VII.14)}$$
$$\text{(Cont.)}$$

The parameter $\dot{\epsilon}_i$ is the initial creep rate given by $\dot{\epsilon}_i = B(F/A_0)^m$. Equation (VII.7) is used to substitute for $\dot{\epsilon}_i$. The parameter σ_i is the initial stress, and σ_0 and ϵ_0 are the constants from Eq. (VII.3).

The reason for expressing the equations in terms of changes in cross-sectional area instead of elongation is that samples which are deforming by necking can also be included in this type of analysis.

The above result indicates that when failure occurs at a particular value of strain, i.e., if A_f/A_0 is independent of the initial stress or initial strain rate and temperature, the time-to-failure should be inversely proportional to the initial strain rate. This would allow the use of creep rupture data to determine the constants m and H of the creep expression. For cases where the reduction of area approaches 50%, the results will not even be very sensitive to the exact value of the strain at failure. In these cases, the expression can be simplified by simply assuming rupture occurs (i.e., the cross-sectional area becomes zero). This gives the expression

$$t = \frac{1}{m\dot{\epsilon}_i} = \frac{1}{m\dot{\epsilon}_0}\left(\frac{\sigma_0}{\sigma_i}\right)^m \exp\left(\frac{H}{kT}\right) \qquad \text{(VII.15)}$$

This analysis assumes that tertiary creep is simply the result of necking which is equivalent to assuming that the creep mechanism remains unchanged all the way to failure. In many alloys this is not the case. In such cases the transition to tertiary creep occurs at a strain which depends on temperature and stress, and begins at some temperatures after a very small strain. In these materials, the transition to tertiary creep is usually associated with fundamental changes in internal structure such as overaging or recrystallization. When these conditions exist, the use of Eq. (VII.15) to extrapolate short-term high-temperature behavior to predict lifetimes at lower temperatures will result in overestimation of the life by a substantial factor. Example VII.4 following Sec. VII-5 illustrates the use of creep rupture data to determine the constants of the creep equation.

VII-5 CREEP UNDER COMBINED STRESS

Since creep is a form of permanent deformation by shear, the ideas introduced in Chapter IV should allow the extension of the uniaxial theory of creep described above to more complex states of stress. Just as the plastic stress-strain curve for a low-temperature tensile test could be interpreted as a special case of a curve of equivalent stress $\bar{\sigma}$ vs. equivalent strain $\bar{\epsilon}$, the tensile creep law given in Eq. (VII.3) can be interpreted as a special case of a multiaxial creep law given by

$$\frac{\dot{\bar{\epsilon}}}{\dot{\epsilon}_0} = \left(\frac{\bar{\sigma}}{\sigma_0}\right)^m \exp\left(-\frac{H(\bar{\sigma})}{kT}\right) \tag{VII.16}$$

Thus, for any state of stress, the equivalent strain rate $\dot{\bar{\epsilon}}$ can be determined from $\bar{\sigma}$. Furthermore, the individual strain rate components can be determined from the strain-rate form of the Prandtl-Reuss equations,

$$\dot{\epsilon}_{11} = \frac{\dot{\bar{\epsilon}}}{\bar{\sigma}}\left[\sigma_{11} - \frac{1}{2}(\sigma_{22} + \sigma_{33})\right]$$

$$\tag{VII.17}$$

$$\dot{\epsilon}_{12} = \frac{3}{2}\frac{\dot{\bar{\epsilon}}}{\bar{\sigma}}\sigma_{12}$$

In terms of deviator strains and deviator stresses, this can be written

$$\dot{\epsilon}'_{ij} = \frac{3}{2}\frac{\dot{\bar{\epsilon}}}{\bar{\sigma}}\sigma'_{ij} \tag{VII.18}$$

The following example illustrates the analysis of creep under multiaxial stress.

EXAMPLE VII.4—Creep of Boiler Tubes

The boiler tubes of the superheater section of a central power station boiler are 2.5 in. in diameter by 0.37 in. wall thickness. They operate at 1100°F. It is

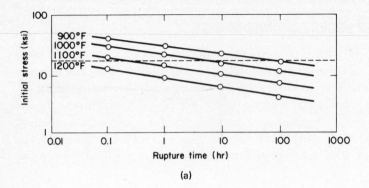

(a)

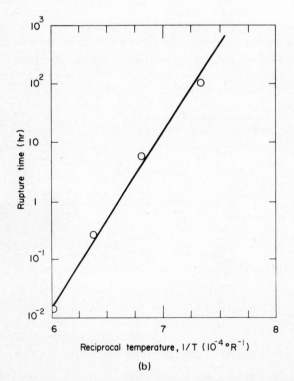

(b)

Fig. VII.10 (a) Creep rupture behavior of 2.25% Cr, 1.0% Mo steel. (b) Rupture time vs. reciprocal temperature for a stress of 16,000 psi for the 2.25%, 1.0% Mo steel of (a). (*Data for [a] from W. F. Simmons and H. C. Cross, The Elevated-Temperature Properties of Chromium-Molybdenum Steels, ASTM, Philadelphia, p. 11, 1953.*)

desired that the radial expansion of the tubes over 10^5 hr be less than 10%. Calculate the maximum allowable internal pressure p which could be used for tubes of a 2.25% Cr, 1.0% Mo steel with the creep rupture behavior shown in Fig. VII.10(a).

Solution:

The curves for rupture time vs. stress at constant temperature are parallel, so the creep behavior should be described by

$$\frac{\dot{\epsilon}}{\dot{\epsilon}_0} = \left(\frac{\bar{\sigma}}{\sigma_0}\right)^m \exp\left(-\frac{H}{kT}\right) \tag{a}$$

where H is independent of stress. For this type of behavior one expects the rupture time to obey the law

$$t = \frac{1}{m\dot{\epsilon}_i} = \frac{\exp(H/kT)}{m(\bar{\sigma}_i/\sigma_0)^m \dot{\epsilon}_0} \qquad (\dot{\epsilon}_0 = 1 \text{ hr}^{-1}) \tag{b}$$

where $\bar{\sigma}_i$ and $\dot{\bar{\epsilon}}_i$ are the initial equivalent stress and equivalent strain rate, respectively. In logarithmic form,

$$\ln(t\dot{\epsilon}_0) = -\ln m - m \ln\left(\frac{\bar{\sigma}_i}{\sigma_0}\right) + \frac{H}{kT} \tag{c}$$

The stress exponent m is determined by differentiating Eq. (c)

$$\frac{\partial \ln(t\dot{\epsilon}_0)}{\partial \ln(\bar{\sigma}_i/\sigma_0)} = -m = \frac{\Delta \log(t\dot{\epsilon}_0)}{\Delta \log(\bar{\sigma}_i/\sigma_0)} \tag{d}$$

From Fig. VII.10(a) we find that $m = 7$. From Eq. (c)

$$H = k\frac{\partial \ln(t\dot{\epsilon}_0)}{\partial 1/T} = 2.3k\frac{\Delta \log t}{\Delta 1/T} \tag{e}$$

holding σ_i constant. Using a stress of 16,000 psi as a base stress [dashed line in Fig. VII.10(a)], points for the graph of rupture time vs. reciprocal temperature shown in Fig. VII.10(b) can be obtained. From the slope of the line in Fig. VII.10(b), using Eq. (e), H is calculated to be 4.65×10^{-18} in. lb. The stress σ_0 is evaluated by solving Eq. (VII.15) for σ_0 and using the values $t = 100$ hr for a temperature of $900°$F and a stress of 16,000 psi.

$$\sigma_0 = (tm\dot{\epsilon}_0)^{1/m} \sigma_i \exp\left(-\frac{H}{mkT}\right) \tag{f}$$

$$\sigma_0 = 31 \text{ psi}$$

For the boiler tube

$$\sigma_{rr} \simeq 0 \qquad \sigma_{\theta\theta} = \frac{pR}{h} \tag{g}$$

$$\sigma_{zz} = \frac{pR}{2h} \qquad \bar{\sigma} = \frac{\sqrt{3}}{2}\frac{pR}{h}$$

where p is the internal pressure, R is the tube radius, and h is the wall thickness. From the strain rate equations

$$\dot{\epsilon}_{\theta\theta} = \frac{\sqrt{3}}{2}\dot{\bar{\epsilon}} \tag{h}$$

Expansion by less than 10% in 10^5 hr implies that

$$\dot{\epsilon}_{\theta\theta} = 10^{-6} \quad \text{or} \quad \dot{\bar{\epsilon}} = 1.16 \times 10^{-6} \tag{i}$$

Solving Eq. (a) for $\bar{\sigma}$

$$\bar{\sigma} = \left(\frac{\dot{\bar{\epsilon}}}{\dot{\epsilon}_0}\right)^{1/m} \exp\left(\frac{H}{mkT}\right)\sigma_0 \tag{j}$$

Substituting values determined above gives

$$\bar{\sigma} = 2{,}300 \text{ psi} = \frac{\sqrt{3}}{2}\frac{pR}{h} \tag{k}$$

$$p = \frac{2{,}300\,(0.37)\,2}{(1.25)\sqrt{3}} \text{ psi} = 795 \text{ psi}$$

VII-6 STRESS ANALYSIS IN CREEP

When a part is first loaded under creep conditions, the stress distribution is elastic. However, the large stress dependence of the creep strain rate causes this stress to change as the creep strain increases. Analysis of the stress distribution during this transient period is difficult to do analytically. However, for problems where the initial elastic stress distribution is known, it should be possible to approach them by numerical methods using small finite time increments.

In most problems where creep is important, the creep strains are large compared to the initial elastic strains. If this is the case, the stress distribution can be determined using the second stage creep equation as the constitutive equation. In the absence of macroscopic changes in geometry, the stress distribution will be constant during second-stage creep after transient effects disappear. Very few problems have been solved for the creep constitutive relation. Problems for which the strain distribution for any period of time is proportional to the initial elastic strain distribution can be solved. The stress distribution can be calculated from the known strain rate distribution. This class of problems includes beams in plane strain bending, and circular shafts in torsion where the strain rates must vary linearly from the neutral axis. Such a problem is illustrated in the following example.

EXAMPLE VII.5–Creep of a Shaft Under Torsion

Determine the stress distribution in a shaft made from 304 stainless steel which has been operating at $1300°F$ for a sufficient time that the second-stage creep strains are large compared to the elastic strains. What is the rate of twist per unit length, $\dot{\phi}$, due to creep? The shaft is 2 in. in diameter and carries a steady torque of 1,500 ft-lb. If the torque is removed quickly, what is the residual stress in the shaft?

Solution:

An *isotropic circular shaft* in torsion which obeys any of the constitutive relations discussed in this book will have a strain distribution given by

$$\epsilon_{z\theta} = \frac{1}{2} r \frac{d\phi}{dz} \tag{a}$$

with all other strain components being zero. The derivative $d\phi/dz$ is the angle of twist per unit length. Therefore, the strain rate is given by $\dot{\epsilon}_{z\theta} = 1/2\, r\, d/dt\, (d\phi/dz)$. From Eq. (VII.17)

$$\dot{\epsilon}_{z\theta} = \frac{3}{2} \frac{\dot{\bar{\epsilon}}}{\bar{\sigma}} \sigma_{z\theta} \tag{b}$$

From the definition of equivalent stress

$$\bar{\sigma} = \sqrt{3}\, \sigma_{z\theta} \tag{c}$$

Therefore,

$$\dot{\epsilon}_{z\theta} = \frac{\sqrt{3}}{2} \dot{\bar{\epsilon}} = \frac{\sqrt{3}}{2} \left(\frac{\bar{\sigma}}{\sigma_0}\right)^m \dot{\epsilon}_0 \exp\left(-\frac{H}{kT}\right) = \frac{r}{2} \frac{d}{dt}\left(\frac{d\phi}{dz}\right) \tag{d}$$

Solving Eq. (d) for $\bar{\sigma}$ and substituting the value into Eq. (c) yields

$$\sigma_{z\theta} = \frac{\sigma_0}{\sqrt{3}} \left[\frac{(d/dt)(d\phi/dz)}{\sqrt{3}\,\dot{\epsilon}_0} \exp\left(\frac{H}{kT}\right)\right]^{1/m} r^{1/m} \tag{e}$$

which is of the form

$$\sigma_{z\theta} = \sigma_{max}\left(\frac{r}{R}\right)^{1/m} \tag{f}$$

where R is the outside radius of the shaft. The total moment is, therefore

$$M = \int_0^R \sigma_{max}\left(\frac{r}{R}\right)^{1/m} r(2\pi r)\, dr$$

$$M = \frac{\sigma_{max}\, 2\pi}{R^{1/m}} \int_0^R r^{2+1/m}\, dr = \frac{R^{3+1/m}}{3 + 1/m} \frac{(2\pi)\sigma_{max}}{R^{1/m}} \tag{g}$$

$$= \frac{2\pi R^3 \sigma_{max}}{3 + 1/m} = 18,000 \text{ in.-lb}$$

$$\sigma_{max} = \frac{(3 + 1/m)(1.8)10^4}{2\pi (1)^3} \text{ lb/in.}^2$$

Using the data for 304 SS from Fig. VII.6, $m = 8$.

$$\sigma_{max} = \frac{(3.12)(1.8)10^4}{2\pi} = 8,960 \text{ psi}$$

$$\bar{\sigma}_{max} = \sqrt{3}\,\sigma_{max} = 15,500 \text{ psi} \tag{h}$$

Using the values of $m = 8$, $H = 5.85 \times 10^{-18}$ in.-lb, and $\sigma_0 = 62$ psi (obtained in Example VII.2) and Eq. (VII.16), $\bar{\epsilon}_{max}$ can be calculated

$$\dot{\bar{\epsilon}}_{max} = 8.4 \times 10^{-3}\,\text{hr}^{-1}$$

$$\left(\dot{\epsilon}_{z\theta}\right)_{max} = 7.3 \times 10^{-3}\,\text{hr}^{-1} = \frac{\dot{\phi}}{2} \tag{i}$$

$$\frac{d}{dt}\left(\frac{d\phi}{dz}\right) = 1.46 \times 10^{-2}\,\text{hr}^{-1}\text{in.}^{-1} = 0.84°/\text{in.-hr}$$

For rapid unloading, the unloading is elastic. This will leave a residual stress in the shaft. The residual stress can be calculated by superimposing an elastic stress distribution with a moment of $-1,500$ ft-lb on the creep stress distribution. This will give a net moment of zero.

For elastic torsion

$$\sigma_{z\theta}{}^e = \frac{Mr}{I} = \frac{4Mr}{\pi R^4}$$

$$\sigma_{z\theta}{}^e = -\frac{4(18,000)r}{\pi R}$$

(j)

The residual stress is, therefore

$$\left(\sigma_{z\theta}\right)_u = 8,960 \left(\frac{r}{R}\right)^{1/8} - 22,900\left(\frac{r}{R}\right)$$

(k)

VII-7 CREEP-RESISTANT MATERIALS

VII-7-a Ferritic Steels

Plain carbon steels which are used for many structural purposes at room temperature lose strength rapidly above 800°F. Addition of up to 0.5% molybdenum provides a moderate amount of high-temperature strengthening and increases the maximum useful temperature to nearly 900°F. Above this temperature, carbon-molybdenum steels experience precipitation of the carbon into a graphitic phase which weakens the material. Graphitization can be suppressed and the oxidation resistance of the steel improved by addition of 1.5% or more of chromium. Chromium-molybdenum steels with alloy contents in the range Cr 1–10% and Mo 0.5–1.5% constitute the most creep-resistant class of ferritic steels. These steels can be used at temperatures up to 1200°F. These steels are commonly used in either a hot-worked or annealed condition for high-temperature applications.

VII-7-b Stainless Steels

Stainless steels which contain more than 11% chromium are also useful at elevated temperatures. These steels are classified into three groups based on the other alloying elements which are present. These groups are the 200, 300 and 400 series of the AISI* classification system. The 200 and 300 series steels are austenitic, meaning that they have the face-centered cubic (f.c.c.) crystal structure. Since the

*AISI stands for American Iron and Steel Institute.

f.c.c. structure is close packed, diffusion rates in austenite are lower than in body-centered cubic ferrite. Therefore, the austenitic stainless steels have somewhat higher creep resistance than the ferritic steels. The higher chromium content also gives the stainless steels improved high-temperature oxidation and corrosion resistance. The 400 series stainless steels are ferritic, so that they do not benefit from the creep resistance of the f.c.c. phase. However, the ferritic stainless steels have better room temperature strength than the austenitic steels, and still have good oxidation resistance. The major disadvantage of stainless steels for high-temperature use relative to the ferritic low-alloy steels is their higher cost.

VII-7-c Iron Base Superalloys

The iron base superalloys are basically modifications of the 300 series stainless steels. The superalloys were developed for applications requiring creep resistance above 1200°F. Most of these alloys are proprietary and therefore, produced by only one company. The naming and numbering systems for superalloys have not been systematized like those for the alloy and stainless steels. In addition to the usual constituents of a 300 series stainless steel, the superalloys may contain some of the following additional elements: Co, Mo, W, V, Nb, Ti, Al, Zr, B, and Cu. The strengthening of these alloys is accomplished by either strain hardening or precipitation hardening. In the strain hardening alloys, advantage is taken of the fact that the high alloy content has raised the recrystallization temperature to above 1700°F. Therefore, the hardening effects of straining at a warm working temperature between 1200°F and 1650°F are retained at the service temperature, which is usually in the range of 1200 to 1400°F. The precipitation hardened alloys in this class are generally hardened by precipitates of carbides, nitrides, or borides which are stable at high temperature. Intermetallic compounds may contribute to the precipitation hardening in some of the iron base superalloys. The usual service temperatures for the precipitation hardened alloys are also in the range 1200 to 1400°F.

VII-7-d Cobalt and Nickel Base Superalloys   *or Nickel*

The cobalt base superalloys are similar to the iron base superalloys with much of the iron replaced by cobalt. Hardening in these alloys is usually by precipitation of refractory metal carbides and by strain hardening. The strain hardening effects are again retained because of the high recrystallization temperatures of the alloys. The nickel base superalloys are probably the most widely used metallic materials for very severe high-temperature applications such as gas turbine engines. These alloys are strengthened by precipitates of an ordered phase of Ni_3Al or Ni_3Ti (or a mixture of the two in which the titanium and aluminum atoms act interchangeably).* This ordered phase is called the gamma prime phase. When a single dislocation moves through the ordered phase, it disrupts the ordered structure by moving some of the aluminum atoms to nickel atom sites, and vice versa. The particles therefore represent strong obstacles to the motion of dislocations. The high-temperature stability of the hardening mechanism results from the fact that the particles make up a large fraction of the total volume of the material and are as large as 0.5μ ($1\mu = 10^{-3}$ mm) in diameter, which is quite large in comparison to the particles in low-temperature precipitation hardening systems. This means that there is no space for new particles to form, and a given particle can grow only at the expense of its neighbors, which decrease in size. This process is called coarsening and takes place much more slowly than the overaging process in low-temperature alloys. Cobalt and nickel base superalloys are used at temperatures up to $1700°F$.

VII-7-e Dispersion-Hardened Alloys

Dispersion-hardened materials, of which TD Nickel is the only one currently available in greater than experimental quantities, are metals hardened by a dispersion of ultrafine nonmetallic particles. In the case of TD Nickel, the particles are thoria (ThO_2). Since these ceramic particles are stable to very high temperatures, this system offers a high-temperature stability not found in the superalloys which eventually experience overaging by particle coarsening and other structural changes. The disadvantage of the dispersion-hardened material is its high cost and the difficulty of forming it.

VII-7-f Refractory Metals

The refractory metals, of which molybdenum, tungsten, niobium, and tantalum are the most important commercially, have very high

---

*An ordered phase is one in which the two types of atoms have a fixed relationship. For example Ni_3Al has an f.c.c. structure with Al atoms at the cube corners and Ni atoms at the face centered sites. In a normal solid solution, the solute atoms are randomly distributed.

melting temperatures. Since creep rates become appreciable only above one-third of the melting temperature, these metals have an inherent advantage for elevated-temperature service. Unfortunately, these materials have poor oxidation resistance and cannot be used at high temperature in air unless they are protected by a coating which is impervious to oxygen. Although a great deal of work has been done to develop suitable coatings, a fully successful system has yet to be discovered. The primary reason for this is that even a small hole in the coating can have catastrophic results, and guaranteeing 100% integrity of the coating is presently impossible. In contrast, the nickel base superalloys, which are also used with coatings in some applications, have self-healing coatings which will close up small holes during service. Another problem with the refractory metals is their low room-temperature ductility which makes them susceptible to damage at ambient temperatures. Because of these problems, use of the refractory metals is limited to applications where a protective environment can be provided. The best known example is the tungsten filaments of incandescent lamps.

VII-7-g   Ceramics

Ceramic materials in general have very high melting temperatures and good resistance to plastic deformation and creep up to very high temperatures. Since the densities of ceramics are generally less than those of metals, they experience lower inertial loads at high speeds. Balanced against these advantages are the disadvantages of brittleness and low thermal conductivity which often causes large thermal stresses. Until recently, ceramics have been used only for relatively low-stress applications such as furnace linings and crucibles. More recently, there has been an increase in interest in using ceramics for blades in gas turbines and similar applications where the temperature requirements are particularly severe.

Cermets which are made of ceramic powder bound together by a metallic matrix, allow many of the useful properties of ceramics to be realized in a material with greatly reduced brittleness. The ceramic components of these systems are generally carbides, such as tungsten carbide (WC) or titanium carbide (TiC). The binder is usually cobalt or nickel. The most common use of these materials is for metal cutting tools. There are efforts being made to use some sintered carbide cermets, such as silicon carbide, as materials for parts in gas turbines. These materials possess very high strength and high creep resistance, but they are not presently being used because of their poor oxidation resistance.

VII-7-h Directionally Solidified Materials and Eutectic Composites

The properties of a material can be greatly affected by its structure as well as its composition. The highly alloyed systems such as the superalloys, have very complicated multiphase structures which produce strong and weak directions within the individual crystal grains. In addition, the grain boundaries are a source of structural weakness. The grain boundaries slide at high temperature and crack to produce the final fracture. When alloying elements are added to stabilize the grain boundaries at high temperature, they usually decrease low-temperature ductility by embrittling the boundaries. Because of these problems, improved behavior can be achieved by eliminating grain boundaries which are correctly oriented for sliding and by controlling the directions of the crystal grains in the material so that their strong directions are aligned with the stress direction. These structural characteristics can be achieved by controlling the direction of solidification in a casting. By causing the solidification to take place from one end of the casting to the other, the grain growth is such as to produce long columnar grains which are aligned with the growth direction. This means that most of the grain boundaries are parallel to the growth direction. If the stress direction is parallel to the growth direction, the grain boundaries are not suitably oriented for sliding and since they have no tensile stress across them they are unlikely to fracture. Furthermore, since the crystals have preferred crystallographic growth directions, most of the columnar grains will have a specific crystallographic orientation. This process has been used by Pratt and Whitney Aircraft to produce parts for the hot section of gas turbine engines. The ultimate example of this type of structural control is the production of a single crystal, which has also been done on an experimental basis by Pratt and Whitney.

In normal directional solidification, the microstructure in the columnar grains is similar to that in normal castings. By controlling the rate of solidification, the nature of the phase distribution can also be controlled. At suitable rates of solidification, a structure can be produced where the hard phase is in the form of smooth rod-shaped particles lined up parallel to the growth direction. This gives a structure similar to a fiber composite. This procedure shows promise for producing materials with outstanding high-temperature strength, but work with these directionally solidified eutectic composites is still only in the experimental stage.

REFERENCES

1. American Society for Testing and Materials: Reports of the ASTM–ASME Joint Committee on the Effect of Temperature on the Properties of Metals, *American Society for Testing and Materials Special Technical Publications Nos. 124, 151, 160, 180, 181, 187, 199, 228, 248, 291*, Philadelphia.
2. Dorn, J. E., (ed.): "Mechanical Behavior of Materials at Elevated Temperatures," McGraw-Hill, New York, 1961.
3. Finnie, I., and W. R. Heller: "Creep of Engineering Materials," McGraw-Hill, New York, 1959.
4. Grant, N. J., and A. W. Mullendore: "Deformation and Fracture at Elevated Temperatures," MIT Press, Cambridge, Mass., 1965.
5. Hult, J. A. H.: "Creep in Engineering Structures," Blaisdell, Waltham, Mass., 1966.
6. Kennedy, A. J.: "Processes of Creep and Fatigue in Metals," Oliver and Boyd, Edinburgh, 1962.
7. McClintock, F. A., and A. S. Argon: "Mechanical Behavior of Materials," Addison-Wesley, Reading, Mass., 1966.
8. Rabotnov, Y. N.: "Creep Problems in Structural Members," Wiley-Interscience, New York, 1969.
9. Clauss, F. J.: "Engineer's Guide to High Temperature Materials," Addison-Wesley, Reading, Mass., 1969.

PROBLEMS

VII.1   A high-temperature pressure vessel is cylindrical in shape with hemispherical end caps. It is to be designed so that the rates of increase in diameter of the cylinder and the caps due to creep are the same. (i.e., dR/dt for both parts is the same). This is to be done by making the wall thickness of the sphere, t_s, less than the wall thickness of the cylinder, t_c. Determine the proper ratio t_s/t_c to meet this condition. The vessel operates at 900°F and is made from 2 1/4% Cr, 1% Mo steel for which data is given in the table on the following page.

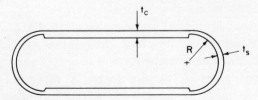

Creep Data for 2 1/4% Cr, 1% Mo Steel

850°F Stress (ksi)	Min creep rate (%/hr)	950°F Stress (ksi)	Min. creep rate (%/hr)
64	2.58	55	6.60
63	1.17	50	.205
61	.118	45	.121
58	.106	42	.0083
55	.0282	40	.0047
51	.0052		
48	.0036		

VII.2   Assume that a jet engine turbine blade must be replaced if it strains by creep more than 1%. A certain engine is designed for a life of 25,000 hr between major overhauls when all hot-section components are replaced.

a) What is the maximum allowable stress for a turbine blade made from Alloy 713C operating at 1700°F? Creep rupture data for Alloy 713C is given in the figure below.

b) The engine is to be inspected every 5,000 hr for creep damage. If it is possible that the engine may accidentally be operated above design temperature, what must be the absolute limit on the operating temperature to ensure a life of at least 5,000 hr at the stress calculated above?

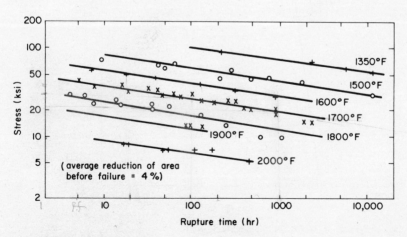

Alloy 713C stress-rupture data.

Alloy 713C is a Ni-Co base vacuum melted superalloy used for precision cast gas turbine parts. In addition to Ni and Co, it contains Cr, Mo, Nb, Ta, Al, Ti, B, and Zr.

VII.3   The bolts securing the distribution header of a tube boiler operate at a temperature of 1100°F. The bolts must be kept at a tension greater than 3,000 psi in order to keep the boiler from leaking. Since the bolts creep during service, they must be pretightened to a greater tension and periodically retightened as creep strain accumulates. Since the other

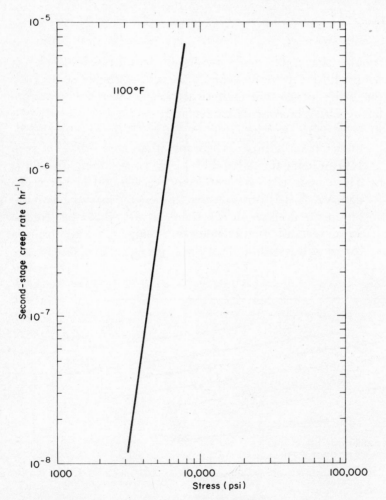

Secondary creep rate of 2% Cr, 1% Mo steel at 1100°F.

components of the header are massive, they can be considered to be rigid so that the only straining parts of the system are the bolts themselves. Therefore, the total strain in a bolt, $\epsilon_{11}{}^T$, which is a combination of the creep strain, $\epsilon_{11}{}^c$, and the elastic strain $\epsilon_{11}{}^e$, must be a constant.

$$\epsilon_{11}{}^T = \epsilon_{11}{}^e + \epsilon_{11}{}^c = \text{Constant}$$

The bolts are made from 2% Cr, 1% Mo steel for which creep data at $1100°F$ are given in the figure. The modulus of the steel at $1100°F$ is 22×10^6 psi. When the bolts were pretightened to give an initial stress at $1100°F$ of 6,000 psi they dropped below 3,000 psi in one-half of the desired time.

a) Engineer A suggests that the bolts be pretensioned to a greater stress so as to increase the service time between retensioning by the necessary factor of 2. Is this suggestion practical? How much must the tension be increased?

b) Engineer B suggests insulating the bolts and nuts from the header to lower their temperature. If the activation energy for creep is H, how much must the temperature be decreased to increase the service time by a factor of 2? Does this seem like a practical suggestion? $H = 5 \times 10^{-12}$ erg.

VII.4   A component of an experimental turbine engine is made from Waspaloy nickel based superalloy. During tests at $T = T^I$ it is found that the parts undergo excessive distortion during operation for 2,000 hr. The nominal stress in the part is σ^I. Note that the data given on the following page is plotted as log strain rate vs. linear stress.

Using an equation for creep rate which is appropriate for the data given in the figure, write an expression for the stress σ^{II} which will give a tenfold increase in life for operation at the same temperature T^I.

Give an expression for the temperature T^{II} which will give a tenfold increase in life at the original stress σ^I.

The original temperature of operation was $1400°F$. The initial stress was not known accurately, but the distortion at 2,000 hr corresponded to approximately 20% strain. Use the data given in the figures to determine σ^{II} and T^{II} from the expressions determined above.

$$[°R = °F + 460]$$

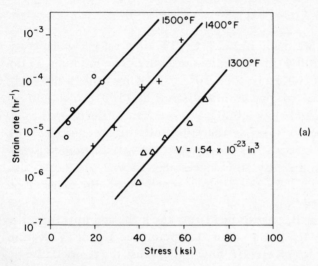

(a)

Minimum creep-rate data for Waspaloy.

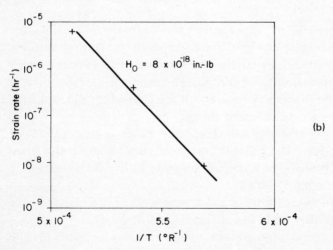

(b)

Strain rate vs. reciprocal temperature for zero stress from intercepts above.

VII.5 Give a reason why you would expect single-crystal turbine blades to have better creep resistance than polycrystalline blades.

VII.6 Creep data for precipitation-hardened aluminum alloy 2024 is shown below. Since the slope of this curve is proportional to the activation energy, it would appear that the activation energy changes as T increases. Is this a correct interpretation? Why or why not? Would you expect similar behavior for pure aluminum?

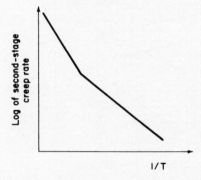

VII.7 A thin-walled hypodermic tube of $t = 0.01$-in. wall thickness and average radius $r = 0.1$ in. is wound into a helical compression spring of $R = 1$ in. mean radius having $n = 10$ turns and a free length of $L = 3$ in. The tube is made of type 309 stainless steel ($G = 10^7$ psi). At an operating temperature of $1250°F$, the spring is compressed by $\delta = 0.25$ in. between two surfaces which then remain fixed in relative spacing.
a) What is the initial shear stress in the tube material?
b) How much does the stress relax in the tube in 100 hr? Note that the creep rate decreases with decreasing stress.
c) What is the free length of the spring after 1,000 hr if the compressive load is released?
   The spring constant K of a helical spring made from thin-walled tubing is

$$K = \frac{Gtr^3}{R^3 n}$$

Creep data for type 309 stainless steel is shown below.

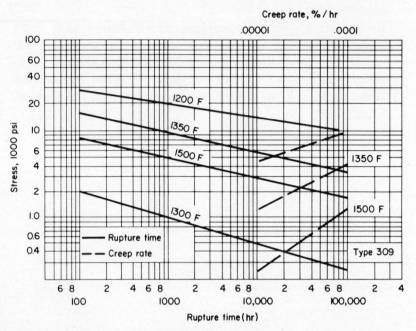

Stress vs. rupture time and creep-rate curves for annealed type 309 stainless steel, based on average data. (*Data from W. F. Simmons and H. C. Cross, The Elevated-Temperature Properties of Stainless Steels*, ASTM Special Technical Publication No. 124, *1952*.)

VII.8 During a fire in an enclosed building, the temperature may get high enough to cause some creep in the structural members. Estimate the deformation in a beam by considering a cantilever beam with rectangular cross section carrying a uniform load w per unit length at a temperature of $1000°$F. Assume that sufficient creep has occurred so that the creep strain is large compared to the elastic strain.

a) What is the bending moment M in the beam as a function of the position x_1?

b) What are the stress and strain rate distributions on the cross section at x_1 in terms of M?

c) What is the rate of change of curvature of the beam at x_1?

d) Integrate the equation for the rate of change of curvature to find the rate of sag, $\dot{\delta}$, of the end of the beam.

Dimensions of the beam and creep rupture data for low-carbon steel are given in the figures. The load per unit length on the beam is 100 lb/ft.

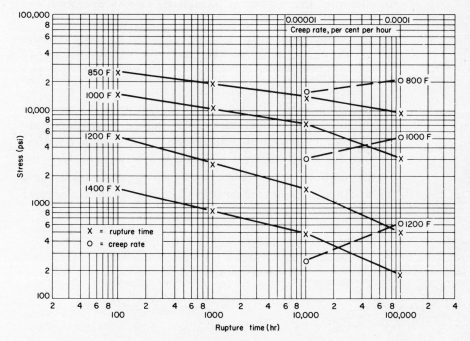

Stress vs. rupture time and creep-rate curves based on average data for killed carbon steel (0.12–0.17 C). (*Data from W. F. Simmons and H. C. Cross, Elevated-Temperature Properties of Carbon Steels,* ASTM Special Technical Publication No. 180, *p. 5, 1955.*)

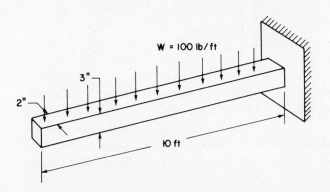

VII.9   To prevent lateral sliding of blades with fir-tree shaped roots in a small gas turbine wheel, a sheet metal "keeper" is used as shown in parts A and B of the sketch below. The medium-carbon steel keeper (with creep properties given in Fig. VII.3 of the text) is plastically bent at room temperature to the final shape given in sketch B. During operation the turbine speed is 30,000 rpm.

a) If the operation of the turbine is limited by creep fracture of the tabs, what is the maximum allowable steady operating temperature for a tab life of 1,000 hr? The radius of the turbine wheel at the blade root is 3 in.

b) If during 5% of the 100-hr life of the turbine it needs to operate at a temperature 50°F higher than the steady operating temperature what is the new allowable temperature for steady operation? State briefly but concisely the assumptions which have led you to your result, and whether or not the answer is conservative to prevent failure.

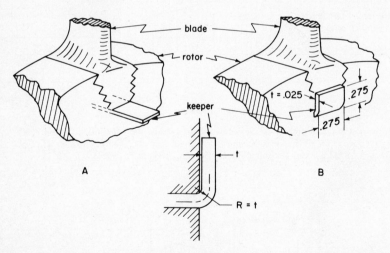

VII.10 E. R. & Company is thinking of manufacturing an indicator which shows when the lubricating oil for automobiles has to be replaced. The idea is to measure the creep of a stressed specimen submerged in the oil, so that it undergoes the same temperature history as the lubricant. Once this creep specimen reaches a critical length, it would close an electric circuit connected to an indicator light!

The activation energy for the formation of sludge in the lubricating oil was determined to be 25 kcal/mole. Suggest a material for the creep rod and justify your suggestion. The activation energy of various materials are as given below:

Medium-carbon steel: 7×10^{-18} in. lb
304 stainless steel: 5.85×10^{-18} in. lb
Waspaloy: 7.9×10^{-18} in. lb
Polyethylene (in the
   temp. range of interest):   28 kcal/mole
Polystyrene (in the
   temp. range of interest):   25 kcal/mole

Assume that the plastics behave in a linear viscoelastic manner. If you use polyethylene, how could you correct for the difference in the activation energy?

VII.11 A thin-walled tube made of Zircaloy 2 is pressurized to 300 psi for 60 days at 700°F. It is then subjected to an axial load of 100 lb for 30 days, maintaining the original internal pressure of 300 psi. The initial wall thickness and the outer diameter of the tube were 1/8 in. and 3 in. respectively. Determine the final dimension of the tube. The creep data, in terms of initial stress vs. time for various amounts of total

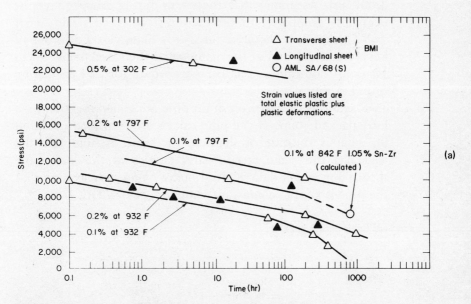

Initial stress vs. time for various amounts of total deformation for annealed Zircaloy 2.

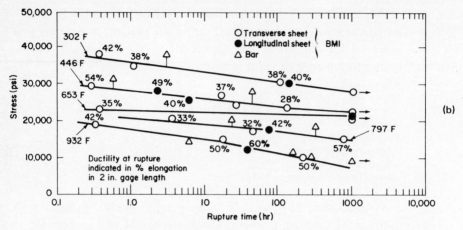

Stress rupture curves for annealed Zircaloy 2. (*Zircaloy 2 data from W. K. Anderson, C. J. Beck, A. R. Kephart, and J. S. Theilacker, Reactor Structural Materials as Affected by Nuclear Reactor Service, ASTM Special Technical Publication No. 314, p. 72, 1962.*)

deformation for annealed Zircaloy 2 and also in terms of rupture time at various initial stresses and temperatures, are given in the accompanying graphs.

VII.12 A thin-walled spherical pressure vessel made from 304 stainless steel operates at 1400°F with an internal pressure of 3,000 psi. Assuming that the strain during primary creep is negligible compared to the second-stage creep strain, what wall thickness is required for a 3-ft diam vessel in order to keep the expansion of the diameter of the vessel below 5% in 10 yr of continuous service? Creep data for 304 SS can be found in Fig. VII.6.

VII.13 Two sections of cast iron drain pipe are sealed together by a lead packed joint. If the pipe carries an axial load of 4,000 lb, what is the rate of shear of the joint? Assume that the lead has the same creep characteristics as the lead cable sheath for which data is given below. You may assume that the pipe is kept at a constant temperature of 22°C.

Creep of Lead Cable Sheath at 22°C

Stress (psi)	Creep rate ($days^{-1}$)
500	2.03×10^{-4}
600	12.58×10^{-4}

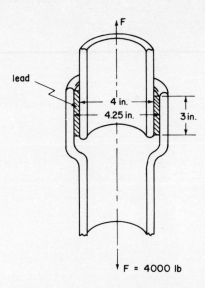

VII.14 An old lead pipe on a building in England has been in place long enough to have developed a substantial sag between its supports as a result of room-temperature creep of the lead. The pipe is approximately 1 1/4 in. ID x 1 in. ID and is loaded primarily by its own weight. The maximum sag in a section of pipe 2 ft long which can be assumed to be simply supported at its ends is 2 in. Assuming that the pipe was straight when installed, estimate the age of the pipe. Creep rate measurements on a similar lead alloy give $\dot{\epsilon} = 5.3 \times 10^{-6}$/hr at 300 psi and 1.8×10^{-3}/hr at 1,000 psi.

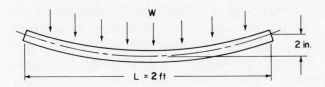

CHAPTER EIGHT

Fracture

VIII-1 INTRODUCTION

The eventual result of nearly any deformation experiment when carried far enough is the separation of the test sample into two or more pieces. Except for the case of *rupture*, where ultimate failure occurs when the cross-sectional area is reduced to zero, the ultimate failure of the sample is the result of some type of *fracture*. This broad heading of fracture encompasses a large number of phenomena. As discussed in Chapter I, when the fracture occurs at the end of the elastic extension range without extensive preceding plastic deformation, the process is called brittle fracture. Inorganic glasses, glassy polymers at low temperatures, and ceramics are subject to brittle fracture. When substantial plastic deformation precedes fracture, the process is called ductile fracture. Most metals at room temperature will undergo plastic deformation and ultimate failure by ductile fracture.

While plastic deformation occurs primarily in response to shear stress as discussed in Chapter IV, fracture, which involves pulling the atoms of a solid apart to create new surfaces, occurs primarily in response to tensile stress. Because of this difference, the state of stress can have a great effect on whether a material deforms plastically or fractures. For example, plastic deformation cannot occur under a state of hydrostatic tension because the shear stress is zero, but fracture can occur under this state of stress. Thus, a normally ductile material could presumably be made to fail in a brittle manner by subjecting it to a large hydrostatic tension. Conversely, a normally brittle metal can be made to undergo large plastic strain if a hydrostatic compression is superimposed on a shear stress so as to make all of the principal stresses compressive.

VIII-2 BRITTLE FRACTURE IN TENSION

The simplest case of fracture is the brittle fracture of glassy materials. In this case, the analysis is simplified by the fact that the prior deformation is only elastic, and also by the fact that glassy materials are more uniform on a finer scale than crystalline materials. It therefore seems that it should be a fairly easy matter to calculate the strength of such a material. The method of calculation for the cohesive strength of a *perfect solid* is very similar to the calculation of the shear strength of perfect crystals described in Chapter V. For reasonable assumptions about the nature of atomic binding forces, the theoretical tensile strength of a material, σ_T, can be related to Young's modulus, E, of the material. To within the uncertainty related to the lack of knowledge of the exact interatomic potential, the theoretical tensile strength is estimated to be

$$\sigma_T = \frac{E}{10} \text{ to } \frac{E}{20} \qquad \text{(VIII.1)}$$

For typical glasses, the tensile strength is on the order of 10^6 psi, which is much greater than the observed values of the fracture strength of commercial glasses, which fall in the range of 5×10^3 to 10^5 psi.

This large discrepancy was explained by the proposal of Griffith that glassy materials contained cracklike defects which act as stress

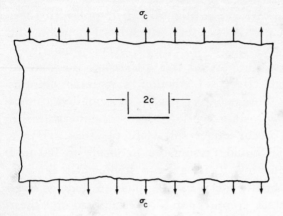

Fig, VIII.1 Sharp crack in an infinite body subjected to a critical stress normal to the crack.

raisers. Griffith argued that for the case of uniaxial tensile loading of a material containing a crack in the plane perpendicular to the tensile axis, the crack would begin to grow and cause ultimate failure at stresses below the theoretical strength. Griffith's criterion says that if the rate of increase in surface energy resulting from the creation of a free surface by the extension of the crack is equal to the sum of the rate of the decrease in elastic strain energy around the crack and the rate of the work done by the applied constant loads, fracture can occur without requiring further increase in the external loads. From this development the breaking strength in tension, σ_c, can be related to the specific surface energy of the material, α, and the half-length of the preexisting cracks, c (see Fig. VIII.1), by the expression

$$\sigma_c = \sqrt{\frac{2\alpha E}{\pi c}} \qquad\qquad \text{(VIII.2)}$$

Griffith turned to the energy analysis to develop the fracture criterion in order to avoid the necessity of being too specific about the details of the crack shape. Notice that the result is independent of the sharpness of the crack.

In an alternative development, Orowan argued that in the absence of plastic deformation the radius of curvature at the tip of the crack must be nearly equal to the atomic radius a. If this value is assumed, one can calculate the local stress at the tip of the crack by using the theory of stress concentration factors. Then the microscopic fracture stress σ_c is the stress required to make the local stress at the crack tip equal to the theoretical strength σ_T given in Eq. (VIII.1). This can be expressed as

$$\sigma_c = \frac{\sigma_T}{SCF} \qquad \text{(VIII.3)}$$

For a long sharp crack, the stress concentration factor is given by

$$SCF \simeq 1 + 2\sqrt{\frac{c}{a}} \simeq 2\sqrt{\frac{c}{a}} \qquad \text{(VIII.4)}$$

If σ_T is taken to be $E/10$, the fracture criterion for tension becomes

$$\sigma_c \simeq \frac{E}{20}\sqrt{\frac{a}{c}} \qquad \text{(VIII.5)}$$

This expression predicts results which are similar to those of the original Griffith formula, Eq. (VIII.2).

Using Eq. (VIII.5) and the measured fracture strengths of glasses, one may calculate the size of cracks necessary to account for the observed reduction in strength. Since breaking strengths of the order of .01 to .1 σ_T are commonly observed, the required stress concentration factors are of the order of 10 to 100. This requires that the material contain cracks with lengths on the order of 25 to 2,500 atomic distances, or roughly 100 to 10,000 Å.

The question of whether defects of this severity can reasonably be expected to exist in glass and, if so, what is their origin, is an interesting one to consider here. This question is not fully answered but certain aspects of it are generally accepted. (See Refs. 8 and 12 for a more complete discussion.) Most severe cracks in commercial glass seem to result from mechanical damage of the surface. Such damage occurs even during careful handling as a result of the very high contact stresses which can occur when the glass touches other hard objects. Evidence of the association of the strength-impairing defects with the surface is obtained from the fact that etching the surface of glass with hydrofluoric acid increases the strength markedly. The etch, which removes or blunts the surface cracks, can increase the strength by as much as an order of magnitude.

The observed strengths of etched glasses are still well below the theoretical strength. Therefore, other defects which cannot be removed by etching must exist. The nature of these defects is not fully understood. However, it has been shown that glass contains a fine cellular structure with dimensions on the order of 30 to 300 Å. The boundaries of these cells are believed to be the origin of the defects which determine the ultimate strength of carefully prepared glass.

Fracture of brittle materials depends very sensitively on the volume of the material at a given stress level. This is caused by the fact that the probability of finding cracks of a given length increases with an increase in the size of the specimen. Because of this size effect, the tensile fracture stress of a uniformly loaded tensile specimen is less than the maximum tensile stress at fracture of a three-point bending test specimen, since in the bending specimen only a small portion of the material experiences stress nearly equal to the maximum stress at the surface. For this reason the tensile stress at fracture of brittle materials is sensitive to the *shape of the specimen and the stress distribution*. It should also be mentioned that the fracture of brittle materials is sometimes better correlated with the maximum tensile strain than with maximum tensile stress.

VIII-3   BRITTLE FRACTURE FOR OTHER STATES OF STRESS

Brittle fracture involves pulling atoms apart. A reasonable local fracture criterion is that fracture occurs when the maximum principal stress at a point in the body reaches the theoretical strength regardless of the magnitude of other principal stresses. If σ_I^L, σ_{II}^L and σ_{III}^L are the local principal stresses at a point such that $\sigma_I^L > \sigma_{II}^L > \sigma_{III}^L$, then fracture occurs when

$$\sigma_I^L = \sigma_T \simeq \frac{E}{10} \tag{VIII.6}$$

In a perfect material, this would imply that fracture could only occur when a positive tensile stress existed in the material. However, in a material which contains an array of randomly oriented microcracks, one finds that the fracture criterion given in Eq. (VIII.6) can be satisfied near the cracks even if the nominal stress is compressive.

In order to determine the brittle fracture conditions for a material containing microcracks of every possible orientation and subjected to a complex state of stress, one must first determine what crack orientation produces the most severe stress concentration under this state of stress. Then, the magnitude of the nominal principal stresses $\sigma_I^N > \sigma_{II}^N > \sigma_{III}^N$ which will cause the local stress to reach the theoretical strength can be found. The results of such a calculation depend on the assumptions made about the shape of the micro-cracks. One set of assumptions which allows the use of known solutions for the stress distribution is that the cracks have elliptical cross section, that conditions of plane strain exist near the crack, and that the cracks cannot close up under the applied stress, i.e., the surfaces of the crack are stress free even for compressive loading. Since the stress distribution around elliptical cracks in plane strain is known, this calculation can be carried out even though it is algebraically complicated. The results expressed in terms of the tensile breaking strength σ_c, defined in Eq. (VIII.5), are

If $-3\sigma_c < \sigma_{III}^N$, fracture occurs when

$$\sigma_I^N = \sigma_c$$

(VIII.7)

If $\sigma_{III}^N < -3\sigma_c$, fracture occurs when

$$(\sigma_I^N - \sigma_{III}^N)^2 + 8\sigma_c(\sigma_I^N + \sigma_{III}^N) = 0$$

The resulting fracture locus for plane stress is shown in Fig. VIII.2. (Note that Eq. (VIII.7) holds for general states of stress.)

EXAMPLE VIII.1—Brittle Fracture of a Circular
Rod under Torsional Loading

What is the maximum torque which can be applied to a brittle circular rod with tensile strength σ_c? What is the nature of the fracture surface?

Solution:
For torsion of a circular rod, the only nonzero stress component is

$$\sigma_{z\theta} = \frac{M_t r}{I_p}$$

(a)

where M_t is the twisting moment and I_p is the polar moment of inertia of the cross section. The shear stress $\sigma_{z\theta}$ is maximum at the outside radius R, where the principal stresses are

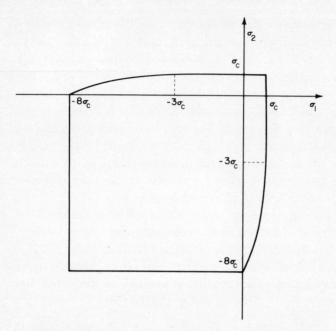

Fig. VIII.2 Brittle-fracture locus for plane stress in terms of the two in-plane principal stresses σ_1 and σ_2, when the third principal stress perpendicular to the page $\sigma_3 = 0$.

$$\sigma_I = -\sigma_{III} = \frac{M_t R}{I_p} \; ; \quad \sigma_{II} = 0 \qquad \text{(b)}$$

Since $\sigma_{III} > -3\sigma_c$, fracture occurs when

$$\sigma_I = \sigma_c = \frac{(M_t)_{max} R}{I_p} \qquad \text{(c)}$$

Solving for the maximum twisting moment

$$(M_t)_{max} = \frac{\sigma_c I_p}{R} \qquad \text{(d)}$$

Since fracture occurs as a result of the tensile principal stress, the fracture surface is expected to be perpendicular to the direction of σ_I. The surface which is everywhere perpendicular to this direction is a spiral which makes a $45°$ angle with the axis of the rod. This is the typical surface produced by torsion failure of brittle rods such as a piece of chalk.

An important result of the above derivation is that the fracture strength of glass in compression should be approximately eight times the tensile strength. This result combined with the fact that the critical defects which lead to failure are at the surface is of commercial significance, since it allows the production of tempered glass. In tempered glass, residual compressive stresses are used to strengthen the surfaces of sheets of glass. By cooling the surface of the glass while the center is still viscous, compressive residual stresses are set up on the surface when the rest of the glass cools.

EXAMPLE VIII.2–Residual Stress in Tempered Glass

By suitable thermal or chemical treatment, sheet glass can be produced which has residual biaxial compressive stresses parallel to the plane of the glass at the surface. The surface contains the most severe defects and obeys the fracture criterion given in Eq. (VIII.7). The material away from the surface has a tensile strength approximately ten times greater than the surface strength. If the glass is to be subjected to either tension or bending in service, what should be the magnitude of the residual compressive stresses in the plane of the glass at the surface so as to maximize the strength of the glass. The residual stress distribution through the thickness of the glass is shown in Fig. VIII.3(a).

Solution:

The stress distributions which result from the superposition of bending or tension on the residual stress pattern are shown in Figs. VIII.3(b) and (c). Because of the relatively low surface strength, it is clear that the bending case provides the most severe test since one must not exceed the tensile strength on the tension side of the glass ($x_3 > 0$) or the compressive strength on the compression side ($x_3 < 0$). More precisely, on the tension side the stresses are

$$\sigma_{22} = \sigma_I = \sigma_B + \sigma_r \,; \quad \sigma_{11} = \sigma_{III} = \sigma_r \,; \quad \sigma_{33} = \sigma_{II} = 0 \qquad \text{(a)}$$

It is assumed that $\sigma_{III} < -3\sigma_c$, so that fracture occurs according to Eq. (VIII.7) when

$$(\sigma_B)^2 + 8\sigma_c(\sigma_B + 2\sigma_r) = 0 \qquad \text{(b)}$$

The assumption that $\sigma_{III} < -3\sigma_c$ will have to be justified after the solution is obtained. On the compression side, the stresses are

$$\sigma_I = 0 \,; \quad \sigma_{II} = \sigma_r \,; \quad \sigma_{III} = \sigma_r - \sigma_B \qquad \text{(c)}$$

where it is again assumed that $\sigma_{III} < -3\sigma_c$ and fracture occurs when

$$(\sigma_r - \sigma_B)^2 + 8\sigma_c(\sigma_r - \sigma_B) = 0 \qquad \text{(d)}$$

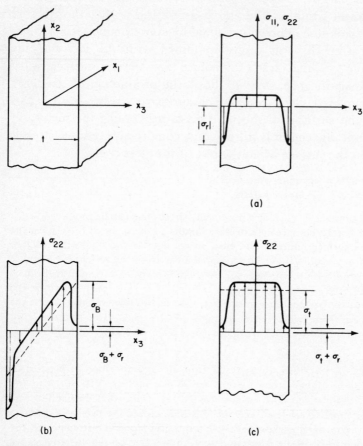

Fig. VIII.3 Schematic of the stress distributions in tempered plate glass. (a) Residual stress distribution; (b) with bending about the x_1-axis; (c) with tension along the x_2-axis.

Under ideal conditions for maximum strength, fracture will occur simultaneously on both sides of the glass in bending. From Eq. (d)

$$\sigma_r - \sigma_B = -8\sigma_c$$

$$\sigma_B = \sigma_r + 8\sigma_c \tag{e}$$

Substituting Eq. (e) into Eq. (b)

$$(\sigma_r + 8\sigma_c)^2 + 8\sigma_c(\sigma_r + 8\sigma_c + 2\sigma_r) = 0$$

$$\sigma_r^2 + 40\sigma_r\sigma_c + 128\sigma_c^2 = 0 \tag{f}$$

$$\sigma_r = -3.5\sigma_c$$

This result shows that the assumption made above, that $\sigma_{III} < -3\sigma_c$ was correct. The maximum tensile stress $\sigma_{t\ max}$ which can be applied to this glass can be calculated as follows

$$\sigma_I = \sigma_t + \sigma_r = \sigma_t - 3.5\sigma_c$$

$$\sigma_{II} = 0 \ ; \quad \sigma_{III} = -3.5\sigma_c \tag{g}$$

From Eq. (VIII.7)

$$\sigma_{t_{max}}^2 + 8\sigma_c(\sigma_{t_{max}} - 7\sigma_c) = 0 \tag{h}$$

Solving for $\sigma_{t\ max}$,

$$\sigma_{t_{max}} = 4.46\sigma_c \tag{i}$$

VIII-4 FRACTURE IN MATERIALS WITH LIMITED DUCTILITY

VIII-4-a Limitations of the Griffith Formulation

Unlike glasses and ceramics which nearly always fail in a brittle manner, at room temperature, metals are generally considered to be ductile. In fact nearly all metals can be made to deform plastically at room temperature when tested in the form of a smooth tensile bar. Some metals, particularly those with high plastic yield stresses, however, can be made to fracture without undergoing general yield when they contain notches, slots, or other severe stress concentrators. In this case, the flaws which produce the stress concentrations are not microscopic defects which are essentially inherent to the material, as are the flaws which account for the low strengths of glasses, but are macroscopic flaws. They may be introduced into the structure either purposely, as in the case of designed in notches, or by a fault in a fabrication process, such as imperfect welding, or by the previous history of the body, as in the case of fatigue cracks or cracks caused by wear of the surfaces. Since brittle behavior is observed for the material only when such macroscopic flaws are present, it is not proper to talk about brittle materials in these cases. It is more accurate to talk about brittle structures made from materials with limited ductility.

In recent years, high-strength alloys have become increasingly widely used. Because fabrication methods are often imperfect, many structures made from these materials contain cracklike defects. As a consequence of this, fracture prior to plastic yield has become an increasingly familiar mode of failure. Under these circumstances, designs based solely on resistance to plastic deformation are often inadequate when high-strength materials are used. It is, therefore, important for many applications to have a theory for metals which can either predict the breaking loads for structures containing flaws or cracks of known geometry, or predict the maximum tolerable flaw size for a given load. Being able to do the latter is extremely important in cases where nondestructive inspection methods are to be used to detect flaws, because it allows calculation of the size of dangerous flaws which the inspection must be designed to detect. The subject of relating the fracture strength of a part to the size of the flaws it contains is called *fracture mechanics*.

A starting point for developing a theory for fracture in metals would seem to be the Griffith fracture theory, which works well for glasses and other brittle materials such as ceramics. However, this theory, either as formulated by Griffith or by Orowan, is meant to apply only to materials which do not undergo plastic deformation during the fracture process. Both the Griffith and the Orowan developments include assumptions which cannot be satisfied if plastic deformation occurs around the crack. Griffith made the assumption that all of the work done during fracture goes into the creation of new free surface, which does not allow for any dissipation of energy by plastic deformation. Orowan's stress concentration analysis assumes purely elastic behavior. These requirements are reasonably satisfied in the case of glasses and ceramics where shear deformation at room temperature requires exceedingly high stresses, but they are almost certainly not satisfied by metals. It is well documented that even the most brittle fracture in a metal is accompanied by considerable plastic deformation in a small region at the tip of the crack. It is therefore, clear that the arguments which are used to justify the Griffith theory cannot be easily extended to apply to brittle fracture in metal structures.

VIII-4-b Generalization of the Griffith Criterion

In spite of the fact that the physical arguments on which the Griffith theory is based do not carry over to fracture of metals, the important functional relationships predicted by the theory are often found to work. Both the Griffith formula, Eq. (VIII.2) and the Orowan formula, Eq. (VIII.5), predict that the fracture criterion can be expressed in the form

$$\sigma_c \sqrt{c} = \text{constant} \tag{VIII.8}$$

or that the tensile fracture stress σ_c is inversely proportional to the square root of the crack length c. In experiments on metals containing macroscopic precracks, of the order of several tenths of an inch in length, it is also found that the tensile fracture strength is inversely proportional to the square root of the crack length just as it is for brittle materials. This could be simply accepted as an empirical law without further justification, but by basing the fracture criterion on a physical argument which is not limited by the assumption that the deformation is purely elastic, it is hoped that added insight into the phenomenon can be gained. In particular, it is hoped that a physical model will help to answer the question of when the formulation is valid and when it is not.

The first argument used to justify the use of a fracture criterion of the type given in Eq. (VIII.8) was a direct extension of the Griffith argument. It was suggested that the energy dissipated by plastic deformation could be added to the surface energy to result in an *effective surface energy* which could be supplied from the elastic strain energy and by the work done by the applied loads during the fracture process. If this were true, the Griffith formula could be directly applied by simply substituting the effective surface energy for the true surface energy α in the Griffith equation. The effective surface energy cannot be directly calculated from the plastic work because it would require a solution to the elastic-plastic problem of the deformation at the crack tip. However, in most cases, the plastic work will be much larger than the true surface energy, and the plastic work portion of the effective surface energy will dominate the behavior. Because of the large magnitude of the effective surface energy, the crack sizes in metals will have to be much larger than in glasses in order to cause fracture. This agrees well with experimental observations that brittle fracture in metals occurs only in the presence of macroscopic precracks. Although this argument does predict that the fracture criterion will be of the form given in Eq. (VIII.8), it is less than fully convincing since it does not answer the question of why the plastic work per unit area of fracture surface created should be a material constant which does not depend on the crack length. It also provides little insight into the question of when such a fracture criterion should be valid. Consequently, an alternative argument will be used to justify the fracture criterion in the presence of a limited amount of plastic deformation. This alternative development introduces the concept of the *stress intensity factor*.

VIII-4-c Linear Elastic Fracture Mechanics

The concept of a stress intensity factor begins with the elastic stress distribution about a sharp crack in an infinite elastic body under conditions of plane strain. The treatment given here will be concerned exclusively with cracks which are loaded by tension perpendicular to the crack, as shown in Fig. VIII.4. This is known as mode I loading. There are two other modes of loading which may cause crack growth, but will not be discussed here. These other loading modes are shear on the plane of the crack perpendicular to the crack front, mode II, and shear on the plane of the crack parallel to the crack front, mode III. In terms of the coordinates in Fig. VIII.4, the stresses for mode I, mode II, and mode III loadings are

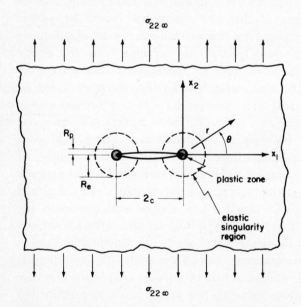

Fig. VIII.4 Schematic of the state of stress and strain around the tip of an initially sharp crack. The stress distribution inside a radius R_e is essentially given by the crack tip singularity solution. Plastic deformation has occurred in a region of radius R_p which is much smaller than R_e.

σ_{22}, σ_{21}, and σ_{23}, respectively. Mode II and III loadings are covered in a number of the references at the end of the chapter (see for example Refs. 1, 7, and 8). The reason for dealing exclusively with tensile cracking is that this is the mode which is of greatest engineering significance and for which there is the most experimental verification.

For the case of tensile loading, mode I, the elastic stress distribution near the crack tip is

$$\sigma_{11} = \frac{k_1}{\sqrt{2r}} \cos\frac{\theta}{2}\left(1 - \sin\frac{\theta}{2}\sin\frac{3\theta}{2}\right)$$

$$\sigma_{22} = \frac{k_1}{\sqrt{2r}} \cos\frac{\theta}{2}\left(1 + \sin\frac{\theta}{2}\sin\frac{3\theta}{2}\right) \qquad \text{(VIII.9)}$$

$$\sigma_{12} = \frac{k_1}{\sqrt{2r}} \sin\frac{\theta}{2}\cos\frac{\theta}{2}\cos\frac{3\theta}{2}$$

where, in the infinite body, $k_1 = \sigma_{22\infty}\sqrt{c}$,* and is called the *stress intensity factor*. The parameter $\sigma_{22\infty}$ is the tensile stress perpendicular to the crack applied far away from the crack, and c is the crack half-length. This solution is valid in a region where r is much less than the crack length c. The $1/\sqrt{r}$ singularity in the stresses is characteristic of the elastic stress distribution around a sharp crack. The stress intensity factor is, therefore, a parameter which describes the strength of the elastic crack tip singularity. The stress intensity factor is a parameter which combines information about the applied loads $\sigma_{22\infty}$ and the geometry of the body, in this case the crack length c. Since the stress distribution in the region is dominated by the $1/\sqrt{r}$ crack tip singularity, the stress intensity factor completely describes the local stress distribution at the tip of a crack in an elastic body. Any combination of applied stress σ_{22} and crack length c which gives the same value of k_1 will therefore give the same stress distribution near the crack tip, even though the stresses may be different far away from the crack tip. Since the fracture process should depend only on the local stresses at or near the tip of the crack, the fracture criterion for elastic bodies should be expressible as

---

*Some other authors use a different symbol and definition for the stress intensity factor, $K_I = k_1\sqrt{\pi}$. This definition is used for example in ASTM standards and publications.

TABLE VIII.1 Stress Intensity Factors for Various Geometries

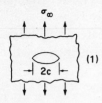

(1)

1. Crack in an infinite body*

$$k_1 = \sigma_\infty \sqrt{c}$$

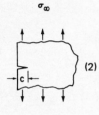

(2)

2. Edge crack in a semi-infinite body.*

$$k_1 = 1.12\sigma_\infty\sqrt{c}$$

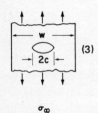

(3)

3. Center crack in a strip of finite width.*

$$k_1 = \sigma_\infty\sqrt{c}\left(\frac{w}{\pi c}\tan\frac{\pi c}{w}\right)^{1/2}$$

$$\text{for } c < \frac{w}{4}$$

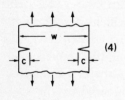

(4)

4. Symmetric edge cracks in a strip of finite width.*

$$k_1 = \sigma_\infty\sqrt{c}\left[\frac{w}{\pi c}\left(\tan\frac{\pi c}{w} + .1\sin\frac{2\pi c}{w}\right)\right]^{1/2}$$

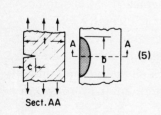

(5)

Sect. AA

5. A semi-elliptical surface crack in a plate of finite width.*

$$k_1 = \left[1 + .12\left(1 - \frac{c}{b}\right)\right]\left(\frac{\sigma_\infty\sqrt{c}}{\Phi_o}\right)$$

$$\left(\frac{2t}{\pi c}\tan\frac{\pi c}{2t}\right)^{1/2}$$

$$\Phi_o = \int_0^{\pi/2}\left[1 - \left(\frac{b^2 - c^2}{b^2}\right)\sin^2\theta\right]^{1/2}d\theta$$

(complete elliptic integral of the second kind).

For $b \gg c$ and $t \gg c$, $k_1 \approx 1.12\,\sigma_\infty\sqrt{c}$

*P. C. Paris and G. C. Sih, Stress Analysis of Cracks, in Ref. 1.

TABLE VIII.1   Stress Intensity Factors for Various Geometries (*Continued*)

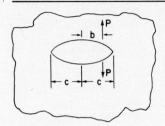

6. Crack in an infinite body with crack face loading.[†]

$$k_1 = \frac{P\sqrt{c}}{2ct}\left[\frac{2}{\pi}\left(\frac{c+b}{c-b}\right)^{1/2}\right]$$

t is the thickness.

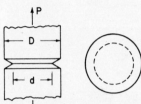

7. Circumferentially notched rod.[*]

$$k_1 \approx \frac{0.932P\sqrt{D}}{\pi d^2}$$

$$1.2 \le \frac{D}{d} \le 2.1$$

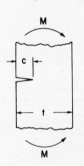

8. Edge notched plate in bending.[*]

$$k_1 = \frac{6Mg\left(\dfrac{c}{t}\right)}{\sqrt{\pi}\,(t-c)^{3/2}}$$

c/t	.05	.1	.2	.3	.4	.5	$\ge$.6
$g(c/t)$	.36	.49	.60	.66	.69	.72	.73

For $.1 \le \dfrac{c}{t} \le .6$

$$g\left(\frac{c}{t}\right) \approx .38 + 1.3\frac{c}{t} - 1.2\left(\frac{c}{t}\right)^2$$

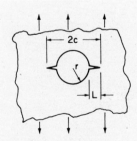

9. Cracks emanating from a hole in an infinite body.[*]

$$k_1 = \sigma_\infty\sqrt{L}\,F\left(\frac{L}{r}\right)$$

L/r	0.1	0.2	0.4	0.6	0.8	1.0
$F(L/r)$	2.73	2.41	1.96	1.71	1.58	1.45

For $\dfrac{L}{r} > 0.2;\ k_1 \approx \sigma_\infty\sqrt{L+r} = \sigma_\infty\sqrt{c}$

[*]P. C. Paris and G. C. Sih, Stress Analysis of Cracks, in Ref. 1.

[†]G. R. Irwin, Analysis of Stresses and Strains near End of a Crack, *J. Appl. Mech.*, vol. 24, pp. 361–364, 1957.

TABLE VIII.1   Stress Intensity Factors for Various Geometries (*Continued*)

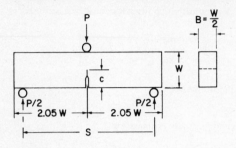

10. Proportions and loading for an ASTM bend-test specimen.‡

$$k_1 = \frac{PS}{BW^{3/2}}\left[1.6\left(\frac{c}{W}\right)^{1/2} - 2.6\left(\frac{c}{W}\right)^{3/2} + 12.3\left(\frac{c}{W}\right)^{5/2}\right.$$
$$\left. - 21.2\left(\frac{c}{W}\right)^{7/2} + 21.8\left(\frac{c}{W}\right)^{9/2}\right]$$

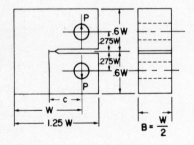

11. Proportions and loading for ASTM compact tensile test specimen.‡

$$k_1 = \frac{P}{BW^{1/2}}\left[16.7\left(\frac{c}{W}\right)^{1/2} - 104.7\left(\frac{c}{W}\right)^{3/2} + 369.9\left(\frac{c}{W}\right)^{5/2}\right.$$
$$\left. - 573.8\left(\frac{c}{W}\right)^{7/2} + 360.5\left(\frac{c}{W}\right)^{9/2}\right]$$

‡American Society for Testing and Materials Standard No. E399.

$$k_1 = k_{1c} \qquad\qquad (VIII.10)$$

at fracture where k_{1c} is the critical stress intensity factor. The critical stress intensity factor is a material constant which in theory can be derived from the microscopic details of the fracture process.

The concept of the stress intensity factor can be generalized to situations other than the crack in an infinite body, such as edge cracks in a semi-infinite body and cracks in finite bodies. Since the $1/\sqrt{r}$ singularity caused by the presence of the crack tip itself dominates the stress distribution near the crack tip, the expressions for the stress distributions in other geometric situations can also be put in the form given by Eq. (VIII.9) plus additional terms which become negligible relative to the $1/\sqrt{r}$ term, when r is sufficiently small. When the stress distributions for these other cases are put in this form, all of the terms which describe the nature of the applied loads and the geometry of the body appear as part of the parameter k_1 and yield expressions like the ones shown in Table VIII.1. Each geometric situation, therefore, has associated with it an expression for the stress intensity factor k_1 which is a function of the applied loads and the geometric parameters of the body. Once this expression is known, the stress distribution around the crack tip is given by the expressions in Eq. (VIII.9). The functional form of k_1 has been determined for a number of different geometries. Since even for geometries other than a crack in an infinite body, the stress distribution near the crack tip is uniquely determined by the stress intensity factor, the fracture criterion for these cases should also be expressible in terms of k_1 as given in Eq. (VIII.10).

VIII-4-d   Generalization of Fracture Mechanics to Elastic-Plastic Fracture

Up to this point, the discussion has been limited to cases where the deformation of the body is completely elastic. As might be expected, the fracture criterion given by Eq. (VIII.10) is completely equivalent to the Griffith fracture criterion. If this were all that could be done with the concept of the stress intensity factor, it would have little relevance to fracture in metals. Fortunately, it can be shown that in many cases, in metals, when the plastic deformation is limited to a small area near the crack tip, the argument justifying the fracture criterion is still valid.

Consider the case of a metal sample containing a sharp crack, and subjected to mode I stresses which increase slowly with time. When the stress is still very small, the deformation is elastic, and the stress distribution near the crack is given by Eq. (VIII.9) which is valid in a region of radius R_e such that $R_e \ll c$ and $\sqrt{c/r} \gg 1$ (see Fig. VIII.4). Because of the singularity in the stresses, the yield condition is quickly reached at the crack tip, and plastic flow begins in a region near the crack tip. Assume that the plastic deformation is limited to a region of radius R_p which is much smaller than R_e. The plastic deformation disturbs the stress distribution inside R_p and also in the elastic region outside R_p. However, since the plastic deformation does not change the net tractions on the boundary of the plastic region, St. Venant's principle should be applicable to this situation. Thus, the difference between the stress distribution in the real elastic-plastic problem and an idealized case where plastic flow is not allowed, should decrease as the distance from the plastic zone increases. When $R_p \ll R_e$, St. Venant's principle would indicate that a region must still exist outside of R_p where the stress distribution is still given by Eq. (VIII.9) even after plastic flow has occurred. The stress distribution in this elastic region is uniquely specified by the value of the stress intensity factor k_1.

As the applied stresses continue to increase, the size of the plastic zone, R_p, continues to increase, but as long as R_p remains small compared to R_e, the plastic region is always surrounded by an elastic region with a stress distribution given by Eq. (VIII.9). Since the tractions of the plastic region at any time are uniquely specified by the value of k_1, the stress and strain distribution inside the plastic region must also be uniquely determined by the value of k_1. Thus, two different bodies of the same material containing cracks which have been loaded from zero to the same value of k_1 will experience the same stress-strain history at the crack tips even when limited plastic deformation occurs.

As the applied stresses are increased further, the plastic strain at the crack tip increases, and if general plastic yield of the body does not occur first, the microscopic fracture condition will eventually be satisfied at the crack tip. What this microscopic fracture criterion is cannot be specifically stated, but it should be possible to express it in terms of the local stress and strain history at the crack tip, even when the strain is partially plastic. When the plastic zone size R_p remains small compared to the singularity region bounded by R_e, all the way to fracture, the stress and strain distributions in the plastic zone are at all times in the monotonic loading program uniquely determined by the current value of k_1. In this case, the microscopic conditions for crack advance should be produced at a specific value of k_1, independent of the stress distribution outside the singularity region. This would mean that a fracture criterion of the type $k_1 = k_{1c}$ at fracture should still be applicable even for elastic-plastic fracture. The critical stress intensity factor k_{1c} should be a material constant which is independent of the details of the shape of the body in question, provided that the plastic zone at fracture is sufficiently small. Table VIII.2 gives values of the critical stress intensity factors for a number of metals.

It will be noted that according to this argument, it is not necessary to specify the microscopic fracture criterion, or to know the details of the plastic deformation at the crack tip. It is only necessary to argue that since the elastic stress distribution in the region R_e outside of the plastic zone bounded by R_p is uniquely given by k_1, the stress-strain in the plastic region inside R_p must also be determined by k_1 and its history. Therefore, any microscopic fracture criterion which can be expressed as a condition on the stress, strain, and strain history at the crack tip can be equally well expressed in terms of the stress intensity k_1 and its history. For the case of monotonic loading to fracture, the history is given by the current value, and fracture should always occur at the same value of k_1 as predicted by Eq. (VIII.10), even when some plastic deformation occurs. The critical value of the stress intensity factor, k_{1c}, can be determined from any valid fracture experiment where the assumptions about the size of the plastic zone are satisfied.*

VIII-4-e Limitations on Elastic-Plastic Fracture Mechanics

Although the argument above, which justifies the validity of Eq. (VIII.10) in cases where plastic deformation occurs has avoided the need to specify the details of the plastic deformation, it does require

---

*Procedures for experimentally determining k_{1c} are covered by ASTM standard E399, "Tentative Method of Test for Plane-Strain Fracture Toughness of Metallic Materials." This standard deals extensively with the criteria which must be met in order for the measurement to be valid.

TABLE VIII.2 Values of the Plane Strain Critical Stress Intensity Factor for Several Metals

Material and condition	Yield strength σ_Y, psi	Critical stress intensity factor k_{1C}, psi in.$^{1/2}$	Reference
Steels			
4340	230,000	38,000	a
4340, 900° F temper	175,000	90,000	b
4340, 800° F temper	205,000	90,000	b
18% Ni Maraging			
200 Grade	215,000	57,000	c
250 Grade	245,000	48,000	c
300 Grade	290,000	33,500	c
17-7 PH stainless			
condition TH 1050 (age hardened at 1050° F)	165,000	22,500–28,000	b
AM 355 PH stainless			
aged at 925° F	200,000	42,000	b
aged at 850° F	205,000	22,500	b
Aluminum Alloys			
2024–T3	47,000	33,000	a
7075–T6	68,000	20,000	a
7075–T651	78,000	12,500	b
Titanium Alloys			
Ti 6Al 6V 2Sn			
aged at 900° F	185,000	11,300	b
aged at 1300° F	145,000	22,600	b
Ti 6Al 2Sn 4Zr 6Mo			
α-β forged at 1650° F	170,000	13,000	b
Ti 6Al 4V			
aged at 1000° F	170,000	22,600	b
aged at 1200° F	160,000	28,000	b

[a]F. A. McClintock and A. S. Argon, "Mechanical Behavior of Materials," p. 539, Addison-Wesley, Reading, Mass., 1966.

[b]"Aerospace Structural Metals Handbook," Mechanical Properties Data Center, Belfour Stulen, Inc., Traverse City, Mich., 1973.

[c]United States Steel Corp., typical properties data sheet.

that a number of assumptions about the size of the plastic zone, R_p, be satisfied. It is these assumptions which place limitations on the validity of the fracture criterion, and they explain why some materials are not embrittled by the presence of cracks. The argument above will not be valid unless the plastic zone is limited to a suitably small region so that the $1/\sqrt{r}$ behavior of the stresses is still valid in the surrounding region. This will only be true if the plastic zone radius R_p is small compared to the crack half-length c and small compared to the distance from the crack tip to any other boundaries of the body. If these requirements are not met, the argument that a region still exists in which the crack tip stress distribution, Eq. (VIII.9), still describes the stress distribution is no longer valid. For example, if the plastic zone gets to be nearly the same size as the crack length, the plastic zones at opposite ends of the crack will interact with each other and link up to form a plastic zone completely surrounding the crack. In this case, the $1/\sqrt{r}$ singularities in the stress will no longer exist. If the plastic zone extends too near to a free surface, the plastic flow may extend through from the crack tip to the free surface, and a plastic failure could result.

There are two additional conditions which must be satisfied in order for the development proposed to be completely valid. One is that conditions of plane strain must be present at the crack tip, and the other is that the crack tip must be sufficiently sharp. The elastic solution is based on a crack with zero radius of curvature at the tip. In practice this requirement can be relaxed when plastic deformation occurs at the crack tip. During the plastic deformation, the crack tip opens up and blunts out. If the initial crack tip radius is small compared to the displacement which occurs at the crack tip during loading, the initial shape of the crack will have only a secondary effect on the shape of the crack tip which results from the deformation. It is difficult to state this requirement quantitatively, but sufficiently sharp cracks can be produced by fatigue cycling at stress amplitudes small enough to keep the stress intensity k_1 always well below the critical value k_{1c}.* Therefore, the ASTM method for determining the critical stress intensity factor requires precracking the sample in fatigue. When the crack is not sharp, the stress intensity required to cause fracture will be if anything greater than for a sharp crack.

The plane strain requirement can be related to the size of the plastic zone. If the plastic zone radius R_p is small compared to the thickness of the part, the plastic zone has the shape of a long slender cylinder. In this case the constraints imposed by the surrounding elastic material keep the material inside the plastic zone from moving

---

*Fatigue and crack growth during fatigue cycling is discussed in Chapter IX.

parallel to the crack tip so that the strain in the thickness direction must be zero. Because of the constraint on the strains parallel to the crack tip, a tensile stress develops through the thickness of the part. If the plastic zone radius is large compared to the part thickness, the plastic zone has the shape of a thin disk, and there is no effective restraint on a contraction in the thickness direction. In this case, the stress through the thickness should be nearly zero so that conditions of plane stress would prevail at the crack tip. When the plastic zone is of an intermediate size such that R_p is nearly equal to the thickness of the part, the constraint of the surrounding material is insufficient to enforce conditions of plane strain, but the thickness is too great to give conditions of plane stress, and a more complex state of stress is present. Once a part is thick enough to give conditions of plane strain at the crack tip, a further increase in thickness will not change the state of stress or strain at the crack tip away from the free surfaces. Therefore, any parts which satisfy the plane strain condition should obey the fracture criterion given in Eq. (VIII.10). If conditions of plane stress exist, a fracture criterion of the same form should be applicable because the same argument could be repeated for plane stress. Since the states of stress at the tip of the crack are not the same in plane stress and plane strain, it should not be expected that the critical value of the stress intensity measured in plane stress will be the same as that measured in plane strain. Since the tension through the thickness in plane strain makes the stress more hydrostatic, thus suppressing plastic flow and increasing the tendency to fracture, it is expected that the measured value of the critical stress intensity will be greater for plane stress than for plane strain. This is observed experimentally. Very thin samples fracture at relatively high values of the stress intensity, k_1. In the intermediate region where neither plane stress nor plane strain are present the stress intensity at fracture decreases with increasing thickness. Finally when plane strain conditions are achieved, the critical stress intensity factor reaches its plane strain value, k_{1c}, and remains constant for further increases in thickness. This behavior is shown in Fig. VIII.5.

In order to determine whether the size restrictions stated above are satisfied in any given situation, it is necessary to estimate the size of the plastic zone, R_p. Although an exact value for the plastic zone size can be determined only by solving the elastic-plastic problem, an estimated value can be determined from the undisturbed elastic stress

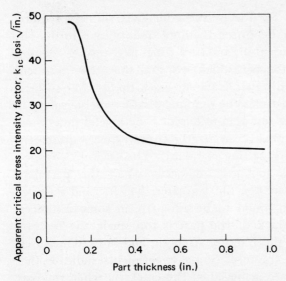

Fig. VIII.5 Effect of the thickness of the part on the apparent critical stress intensity factor. (Schematic representation based on data for 7075-T6 aluminum.)

distribution given in Eq. (VIII.9). From that stress distribution it is determined that the equivalent stress along the maximum stress directions has a magnitude of approximately $k_1 \sqrt{2r}$. This implies that the material is at a stress above yield, according to the elastic solution, out to a radius R_p given by

$$R_p \approx \frac{1}{2}\left(\frac{k_1^2}{\sigma_Y^2}\right) \qquad \text{(VIII.11)}$$

where σ_Y is the tensile yield strength of the material. The maximum value of k_1 is k_{1c} which occurs at fracture. Therefore, the maximum plastic zone size R_c is the one which is present just before the fracture condition is reached.

$$R_c \approx \frac{1}{2}\left(\frac{k_{1c}^2}{\sigma_Y^2}\right) \qquad \text{(VIII.12)}$$

Using these estimates for the plastic zone size, the size requirements for the part can be put into more quantitative form. The ASTM standard on plane strain fracture toughness testing requires that the crack length, the part thickness, and the other pertinent dimensions, such as the distance from the crack tip to a free surface, be approximately fifteen times the nominal plastic zone size given in Eq. (VIII.12) before the test is considered to be a valid determination of k_{1c}. This is probably an overly conservative requirement since it is concerned with setting standards. The fracture criterion gives reasonable results in most materials when the pertinent dimensions of the part are ten times greater than R_c, and ratios as small as three have been found to be adequate in some instances. When the size of the crack or the part is too small, the fracture criterion underestimates the fracture strength, and in extreme cases, plastic failure precedes fracture. Some experiments indicate that plane strain conditions are achieved at the crack tip when the part thickness is greater than three times the nominal plastic zone radius, R_c, and that plane stress exists if the thickness is less than one-third of R_c. Other investigators would prefer to put the plane strain limit at a larger thickness and the plane stress limit at a smaller one. When the result is very important, experiments which measure the apparent critical stress intensity as a function of thickness, such as the case shown in Fig. VIII.5, must be performed to determine the extent of the region where the strength is a function of thickness.

EXAMPLE VIII.3—Fracture Toughness of a Welded Steel Pressure Vessel

A spherical pressure vessel is to be constructed by welding formed steel plates together. The welds are to be inspected by ultrasonics or x-rays using techniques which are virtually certain to detect any crack greater than 0.1-in. long. The three grades of maraging steel with properties given in Table VIII.2 are being considered. Assuming the worst possible undetected cracks to be present, which steel will give the maximum pressure capability? The wall thickness of the vessel is 1 in. and the diameter is 10 ft.

Solution:

Assume that either yield or fracture constitutes failure. For a thin-walled spherical pressure vessel the stresses are

$$\sigma_{\theta\theta} = \sigma_{\phi\phi} = \frac{pr}{2t} ; \quad \sigma_{rr} \text{ is negligible} \tag{a}$$

where p is the pressure, r the radius of the sphere, and t is the wall thickness. The equivalent stress is

$$\bar{\sigma} = \sigma_{\theta\theta} = \frac{pr}{2t} \qquad (b)$$

Using the Mises yield criterion and solving for the pressure at yield, p_Y (σ_Y is the yield stress),

$$p_Y = \frac{2\sigma_Y t}{r} \qquad (c)$$

For fracture, only the stress component perpendicular to the crack matters. Cracks less than 0.1 in. long can be treated as a crack in an infinite body with half-length $c = 0.05$ in. (See Table VIII.1, case 1.)

$$k_1 = \sigma_\infty \sqrt{c} = \frac{pr}{2t} \sqrt{c} \qquad (d)$$

Solving for the pressure at fracture, p_f,

$$p_f = \frac{2tk_{1c}}{r\sqrt{c}} \qquad (e)$$

Using Eq. (VIII.11) one can calculate the critical plastic zone radius to check the validity of the analysis. The results are shown in Table VIII.3

The results indicate that grade 200 fails by yield at about the same pressure that causes fracture in grade 250. However the result for grade 250 is questionable because the plastic zone radius is 38% of the crack half-length c.

TABLE VIII.3   Failure Characteristics for 18% Ni Maraging Steels

Grade	$p_Y r/t$ (ksi)	$p_f r/t$ (ksi)	R_c (in.)
200	430	~~510~~	.035
250	490	430	.019
300	~~530~~	300	.007

$\lesssim 80$

Grade 250 would probably be somewhat better than grade 200. It is clear that nothing is gained by using grade 300.

EXAMPLE VIII.4—Fracture of a Milling Cutter

Determine the maximum cutting force F_{max} which may be applied to the 5-in.-diam 1/10-in.-thick cutter shown in Fig. VIII.6(a). It is to be assumed that the center hole and the sharp cornered keyway have the same effect as a central crack of length 1.1 in. loaded 0.1 in. from the end as shown in Fig. VIII.6(b). This corresponds to case 6 of Table VIII.1. A test sample cut from

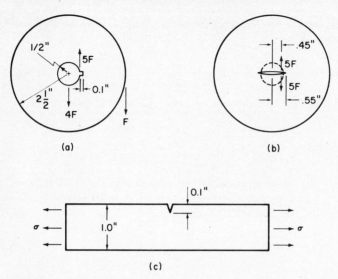

Fig. VIII.6 Geometry of fractured milling cutter and sample for determination of the critical stress intensity factor for the cutter material. (a) Actual cutter geometry and loading; (b) idealization of cutter geometry and loading; (c) test sample cut from fractured cutter.

an identical cutter and having the shape shown in Fig. VIII.6(c) broke at a stress of 61,000 psi. The tensile strength of the cutter is approximately 250,000 psi.

Solution:

If the milling cutter actually contained only the hole and square cornered keyway, fracture mechanics would not be applicable. However, during service, the cutter is subjected to intermittent loading, and fatigue can occur. Since the square corner of the keyway is a point of stress concentration, a fatigue crack may form at the corner. Brittle failure could then occur if the stress intensity at the tip of the fatigue crack exceeds the critical value. In this case, fracture mechanics would be applicable because a sharp crack exists, but the geometry is quite complicated. A rough estimate of the stress intensity can be obtained using the approximation suggested in the problem statement, where the crack length is considered to span the entire center hole. Before any fatigue crack forms in the cutter, it will be much stronger than indicated by the fracture mechanics calculation presented here. This calculation is, therefore, a conservative estimate of the strength of the cutter.

The test sample configuration is approximately that of case 2 of Table VIII.1 with $c = 0.1$ in. The table gives

$$k_1 = 1.1\sigma_\infty c^{1/2}$$

$$k_{1c} = 1.1(61 \text{ ksi})(0.1 \text{ in.})^{1/2} = 21 \text{ ksi in.}^{1/2}$$

(a)

Since the yield strength of the steel is not given, R_c will be estimated from

$$R_c \approx \frac{1}{2}\left(\frac{k_{1c}}{\sigma_u}\right)^2 = 0.004 \text{ in.} \tag{b}$$

where σ_u is the tensile strength of the steel. Because of the low ductility of tool steels, the tensile strength is not much greater than the yield strength, so this should be a good approximation.

The dimension R_c is sufficiently small to justify the analysis and to give conditions of plane strain. From case 6 of Table VIII.1, k_1 is found to be

$$k_1 = \frac{5F_{\max}}{2ct} c^{1/2}\left[\frac{2}{\pi}\left(\frac{c+b}{c-b}\right)^{1/2}\right] \tag{c}$$

where $c = 0.55$ in. and $b = 0.45$ in. Solving for $F_{\max}$ using the value of k_{1c} from Eq. (a)

$$F_{\max} = \frac{\pi(21 \text{ ksi in.}^{1/2})(0.55 \text{ in.})^{1/2}(0.1 \text{ in.})}{5\sqrt{10}} \tag{d}$$

$$F_{\max} = 310 \text{ lb}$$

This value is within about 20% of the best estimate of the force which was exerted when the cutter actually failed.

VIII-5 DUCTILE FRACTURE

The processes of ductile fracture are extremely complex. The phenomenon depends not only on the current value of the stress and equivalent strain as does the process of plastic deformation, but on all of the details of the deformation history. Because of the complexity of the process it is not yet possible to formulate a predictive theory of ductile fracture. The discussion in this section will therefore be limited to a qualitative description of the features which are usually observed in ductile fracture. Since it is virtually impossible to give a complete treatment of a subject which is still the object of active research, this discussion will be brief and will leave out many special cases and exceptions. The reader with a special interest in this area is urged to consult other works which treat the subject of fracture in more detail.

Most ductile fracture in tension appears to be associated with the development of porosity in the material caused by the growth of holes from initially small defects. The eventual separation of the specimen results from the linking up of the holes to form the fracture surface. On the fracture surface of some materials the ridges which separate holes are very sharp, indicating that the internal strain between holes proceeded almost to rupture. The discussion here will be separated into a discussion of the origin of the holes and a discussion of their growth to produce the final failure.

In commercially pure materials and other nominally single-phase alloys, the origin of the holes most frequently appears to be impurity inclusions, most often oxides. This association is graphically clear on the fracture surfaces of many materials where the impurity particles are found at the bottom of many dimples on the fracture surface after the failure. The origin of the hole in this case can be either a separation of the particle from the metal matrix or the cracking of the brittle particle itself. In two-phase systems the origin of the holes may be the separation of the phase boundaries or cracking of the more brittle phase. In mild steel for instance, the source of the holes may be cracks in the pearlite. In some materials, especially at low temperatures, holes may result from cleavage cracks in favorably oriented grains.*

The initiation of the holes is not the critical step in the ductile fracture process, since the plasticity of such materials is sufficient to quickly blunt the cracks and reduce any associated stress concentration. The development of the initial holes to the larger voids which are present just prior to fracture will depend on the subsequent plastic strain and on the state of stress. If the strain in the sample were to be strictly uniaxial, the holes would tend to elongate in the tension direction and become smaller in the transverse direction. This would result in long stringlike holes which would have little detrimental effect on the strength of the body. However, in a tensile specimen which is deforming in a neck (i.e., which has reached the tensile instability) there are large transverse tensile stresses caused by the constraint imposed by the neck shoulders. It is the action of these stresses which tend to open up the holes into more or less spherical voids which substantially reduce the actual load-carrying area of the sample much more rapidly than the external cross-sectional area decreases.

---

*A cleavage crack is a brittle crack in a single crystal. The crack follows specific crystallographic planes in the material and therefore gives a very flat surface after fracture.

Because of the fact that the transverse stresses are greatest in the center of the specimen, the ductile fracture usually begins at the center and grows outward. This is the origin of the typical cup and cone fracture. This type of fracture is composed of a central portion perpendicular to the tensile axis formed by the growth of holes in response to the transverse stresses. This is surrounded by a conical ring formed by shear deformation after the wall thickness became too small to support the transverse stress. The experiments of Bridgman in which a hydrostatic pressure is superimposed on the tensile stress also show the importance of the transverse stresses in the neck of a tensile specimen for the development of ductile fracture. As predicted by the Mises yield criterion, the hydrostatic pressure has little effect on the yield stress of the material. However, the increased pressure does greatly increase the reduction in cross-sectional area prior to fracture. From a calculation of the stress distribution in the neck of a tensile specimen, Bridgman was in fact able to show that the fracture does not occur until the tensile transverse stresses from the constraint effect become greater than the compressive transverse stresses from the hydrostatic pressure.

In the discussion above it was assumed that the deformation was simple tension. Clearly the structure developed from the initial defects will depend strongly on the nature of the strain history subsequent to the formation of the defect. In particular the shear experienced in torsion will tend to elongate the holes along spiral surfaces around the axis. For this reason the character of the fracture surface in a tensile test can be markedly changed by twisting the specimen in torsion to some fairly large strain prior to the tensile test. The fracture surfaces resulting from this type of experiment are shown in Fig. VIII.7(a). Since most of the defect development occurs during the twisting, the resulting fracture reflects the nature of the prior strain. If the torsional strain is sufficiently large, the fracture in subsequent tension will follow the spiral surface on which the defects have been elongated. This experiment clearly demonstrates that the process which leads eventually to ductile fracture begins quite early in the deformation process. It is this sensitive dependence on previous history which makes the study of ductile fracture so difficult.

If the entire strain to fracture is pure shear, holes cannot open in the material in the way that they do in tension. In such cases, the nature of the final failure is strongly dependent on the

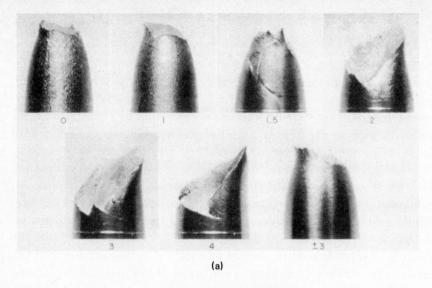

(a)

(b)

Fig. VIII.7 Fracture surfaces for a variety of materials. (a) Tensile fracture in pre-twisted specimen. Numbers below the photographs indicate the surface shear strain before tensile testing. Sample labeled ±3 was twisted to a surface shear strain of 3 and completely untwisted before tensile testing. (b) Typical tensile fractures: A—woody fracture of wrought iron; B—cup and cone fracture of alloy steel; C—rosette fracture of alloy steel; D—apparent shear fracture of duralumin; E—granular fracture of gray cast iron. (*[a] from W. A. Backofen, A. J. Shaler, and B. B. Hundy, ASM Trans., vol. 46, p. 655, 1954; [b] from I. H. Cowdrey and R. G. Adams, "Materials Testing," John Wiley, New York, 1935.*)

microstructure of the material. In a normal commercial material, inclusions and other second-phase particles create perturbations in the stress and strain which cause tension to occur locally around the particles. Cracks and holes can be formed near the particles much as they are formed in tension even when the average strain in the body is pure shear. Deformation causes the holes to elongate along the shear strain directions, and they can eventually link up to form the fracture surface. However, in shear, the holes do not interact strongly with each other as they do in tension, so that much more strain is required to produce the fracture in shear. In nearly homogeneous materials, fracture is often preceded by localization of the strain on very narrow bands so that the local strain in the band becomes very large and accelerates the process of spreading the fracture. In shear fracture the fracture surface follows the surface of maximum shear strain.

Finally, fracture surface appearances, such as those in Fig. VIII.7, will be discussed briefly. In a typical fine-grained ductile material, the typical fracture surface in tension is the cup and cone fracture. However, the sensitivity of the fracture process to the defect structure means that the fracture surface appearance can be markedly changed if any anisotropy or inhomogeneity has been produced by the processing of the material. The example of prior torsion discussed above is an example of this type of effect which is not likely to arise from normal material processing. Other effects which do arise from processing are: 1) Apparent shear fracture where the fracture occurs on a plane which makes an angle with the tensile axis. This occurs most often in rolled products which have a texture or anisotropy from the rolling. 2) In drawn or extruded products the impurity inclusions are drawn out along the axis. This often causes the fracture to spread out on planes parallel to the specimen axis. In extreme cases the material almost seems fibrous so the fracture is given the name woody fracture. Less extreme cases are called rosette fracture. 3) In relatively inhomogeneous materials with low ductility such as cast iron, the fracture follows the weaknesses in the material. This usually gives rise to a planar fracture perpendicular to the specimen axis without the shear lips which characterize the cup and cone fracture.

VIII-6 THE DUCTILE–BRITTLE FRACTURE
TRANSITION

In a sufficiently ductile material, the notches and cracks, which are built into the structure by design or due to fabrication errors as well as cracks inherent in the material, do not have a significant effect on the load-bearing capability of the structure. From the analysis of Sec. VIII.4 it is clear that when the stress intensity factor for the notches in the structure do not exceed the critical stress intensity before the nominal stress reaches the yield stress, brittle fracture will not result. This is normally the case with face-centered cubic metals such as aluminum and copper. Some body-centered cubic materials, notably low-carbon steels, are found to behave in a ductile manner at elevated temperature and/or low strain rates, exhibiting little notch sensitivity; but at low temperatures or high strain rates they behave in a brittle manner, failing with no general plastic strain. Since the temperature for which this transition occurs can be within the range of ambient temperature fluctuations in the colder parts of the world, this transition behavior has considerable practical importance. For example, ships have

Fig. VIII.8 A T-2 tanker that fractured at pier. (*From E. R. Parker, "Brittle Behavior of Engineering Structure," John Wiley, New York, 1957.*)

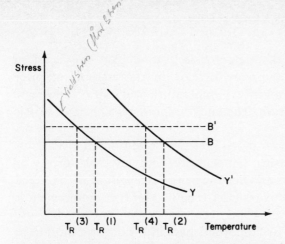

Fig. VIII.9 Davidenkov diagram for the ductile-brittle transition. The yield stress curve Y and the brittle fracture strength line B intersect at the transition temperature $T_R^{(1)}$. Increasing the yield stress to curve Y' increases the transition temperature to $T_R^{(2)}$. Increasing the brittle fracture strength to B', without changing the yield strength, gives a lower transition temperature, $T_R^{(3)}$.

broken in two by brittle fracture in the winter both at sea and in the harbor (see Fig. VIII.8), steel pipelines have cracked over miles of length in a matter of a few seconds, and because of the embrittlement of steel parts, vehicles and machinery were nearly useless to early explorers of the Arctic and Antarctic.

An approach to understanding this behavior is to use a diagram proposed by Davidenkov like the one shown in Fig. VIII.9. In the diagram, the flow stress is plotted as a function of temperature on the line labelled Y. The critical stress for brittle fracture, which is generally assumed to be independent of temperature, is plotted as line B in the Fig. VIII.9. For materials which exhibit a ductile-brittle transition, the flow stress rises sufficiently as the temperature decreases so that it intersects the brittle fracture stress. The temperature of the intersection will be the transition temperature, because below this temperature the yield stress cannot be reached before fracture occurs. The curve Y is a curve for yield stress vs. temperature assuming that all other variables remain constant. The yield stress also depends on the strain rates, prior hardening by cold work, precipitates, impurities, or radiation. A new line such as Y' must be drawn for any change in any of these other variables. Thus any change which increases the yield stress, such as an increase in strain rate, impurity or secondary-phase particle concentration, or amount of cold work, will cause a rise in the transition temperature. Conversely, any change which increases the brittle strength, such as a decrease in the defect size (the defect size is often related to the grain size) or an increase in the amount of prior cold work above the transition temperature will cause a decrease in the transition temperature. The line B' represents a changed brittle fracture stress.

Transition behavior therefore requires two conditions; a yield stress which rises with decreasing temperature, and internal defects or external notches which can lower the brittle strength of the material substantially below the theoretical strength. The first requirement is generally satisfied by the body-centered cubic metals because thermal activation plays an important part in dislocation motion in these materials. The second condition is satisfied for most body-centered cubic metals by the existence of relatively weak cleavage planes in the crystal. Because of various different interactions of the dislocations with the material structure such as grain boundaries, dislocation walls, and impurities, it is not uncommon for cleavage fractures to occur across single grains at a very early stage of the deformation process. Above the transition temperature these small cracks will not proceed into adjacent grains, but below the transition temperature they can be of critical size below the yield stress leading

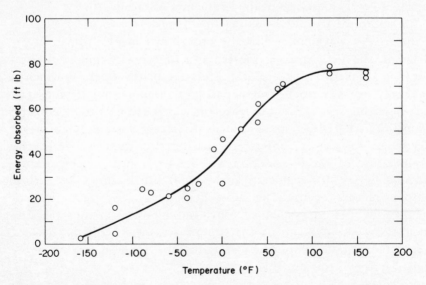

Fig. VIII.10 Charpy V-notch impact tests for A533-B Class 2 steel plate, showing the ductile-brittle transition. (*From J. R. Hawthorne and T. R. Mager, Relationship between Charpy V and Fracture Mechanics K_{1C} Assessments of A533-B Class 2 Pressure Vessel Steel*, Fracture Toughness, ASTM Special Technical Publication No. 514, p. 157, 1972.)

to brittle fracture. In the presence of external notches either built in or caused by corrosion, internal cleavage cracks may not be necessary for the nucleation of brittle fracture.

Typical tests for the transition temperature measure the energy required to break a standard specimen as a function of the temperature of the test. Typical results for steel are shown in Fig. VIII.10. As the temperature passes through the transition temperature, the energy absorbed drops markedly. The energy absorbed well below the transition temperature is only 5 to 10% of that absorbed well above the transition temperature. Figure VIII.11 shows the configurations for the three standard impact tests. All of the three standard fracture toughness tests involve the use of a massive pendulum dropped from a standard height in order to get reproducible conditions of strain rate. In all of the tests the energy absorbed is measured by comparing the height that the pendulum reaches on the follow-through with the initial drop height. The principal difference between the three tests is in the configuration of the standard test specimen. In the Izod test a notched bar is loaded as a cantilever beam. In the Charpy test the bar is simply supported at the ends and struck behind the notch at the center so that it is loaded in three-point bending. Both of these tests use a 1 x 1 x 5 cm bar usually notched at the center. In the tensile impact test a circumferentially notched tensile specimen is loaded in tension by the pendulum.

Since the transition temperature is a function of so many variables, it is unlikely that the transition temperature in service will be exactly the same as that measured in a standard impact test. Instead the conditions of the test with high strain rates and relatively sharp notches are designed to provide a severe test which, if anything, will overestimate the transition temperature. Experience is then generally utilized to choose an energy absorption value in the standard test which represents adequate behavior for a particular application. For example, a Charpy Impact value of 15 ft-lb at the lowest anticipated service temperature is considered adequate for steel plate for ships. One should be cautioned that under extreme conditions such as sharp cracks built into heavy sections by welding and subjected to triaxial tension from the residual stress of the weld and the applied loads, the transition temperature in service may be higher than in the standard tests. In some cases the transition temperature is determined from the fracture surface appearance on standard specimens. As the temperature increases the percentage of the area of the fracture surface which failed by cleavage decreases, and the portion

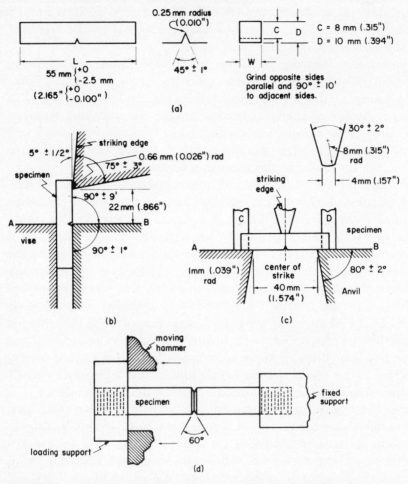

Fig. VIII.11 Impact test geometries. (a) Charpy and Izod V-notch specimen; (b) Izod cantilever beam test; (c) Charpy three-point bending test; (d) notched tensile impact test. (*[a]*, *[b]*, *and [c] from* ASTM Standard E23.)

which failed in a ductile manner increases. A plot of the percentage ductile fracture vs. temperature gives the fracture surface appearance transition temperature, which may be as much as 100°C above the energy transition temperature.

VIII-7 FRACTURE OF AMORPHOUS POLYMERS

The most significant difference between fracture of amorphous (glassy) polymers and fracture of any other material, in which only localized deformation precedes fracture, is the phenomenon of crazing.* A craze is a region of material, usually in the form of a very thin plate perpendicular to the tensile direction, which has undergone a dilatational type of permanent deformation. During this deformation the density of the material can decrease by fifty percent or more. The deformation appears to take the form of formation of micropores surrounded by material in which the molecules are drawn out into the direction of the tensile axis. Crazes apparently result because the stress necessary to break the intermolecular bonds can be reached much before the stress is sufficient to break the intramolecular bonds. The craze is formed when molecules are pulled apart from each other but the material continues to be held together by the molecular chains threading into the craze. Because of the gross rearrangement of the molecules and the decrease in density which occur during crazing, the craze material has different optical and mechanical properties than the bulk material. In particular, the modulus of the craze is usually lower than that of the bulk material but, because of molecular orientation the tensile flow strength of the craze may be greater than that of the bulk.

When fracture occurs in these amorphous polymers, the crack is always preceded by a craze. Since the craze will transmit stress, the presence of the craze serves to reduce the stress concentration effect at the tip of the craze and the crack. Flow in the craze material blunts the crack. In addition, the formation and subsequent deformation of the craze material requires substantial amounts of work thus raising the work done during fracture well above that required to simply produce new surfaces. This is in much the same way that localized plastic flow in metals increases the fracture work. Many of the details of the formation of crazes and subsequent cracking of the crazed material are not completely understood. However, in an amorphous polymer, nearly all of the important features of the fracture process seem to be related to the behavior of crazes.

Toughened polymers such as ABS and impact-grade polystyrene are made to resist fracture due to impact. They are produced by mixing a finely divided rubbery phase with the parent polymer. The toughening effect of these particles is related to their ability to nucleate crazes. Fracture in the pure polymer takes place along a single surface with only one craze preceding the crack, or at most

*Crazing is also discussed briefly in Chapter VI.

only a limited number of crazes. In the toughened polymers many fine crazes form at the rubbery particles near the crack. These crazes reduce the stress in the vicinity of the crack and lead to fragmentation of the crack front. This makes it more difficult for the crack to propagate and increases the resistance to fracture on impact.

VIII-8   ANALYSIS OF FRACTURE SURFACE MARKINGS

When metal, glass or a polymer fractures, the markings left on the surfaces of the fracture can be analyzed to determine a great deal about the conditions under which the fracture occurred. These fracture surface markings occur on several size scales, from those visible to the naked eye down to those which can only be observed with the large magnifications that can be achieved with the electron microscope. These fracture markings can be used to trace the crack to its origin for the purpose of identifying the crack nucleus, to judge the speed of the crack, and to judge whether the crack propagated as a ductile or brittle failure.

VIII-8-a   Macroscopic Fracture Appearance

The general appearance of the fracture surface, as observed without magnification, in terms of its roughness and reflectivity often identify the type of fracture mode and indicate the speed of the crack. A metal fracture surface which sparkles when tilted in light is usually found, by examination under high magnification, to be composed of cleavage facets formed when individual grains failed in a brittle manner. A metal fracture surface which is dull and does not sparkle or change brightness when tilted back and forth in light is usually composed of rounded dimples characteristic of ductile fracture. For fractures of glass and amorphous polymers, a smooth mirrorlike surface indicates slow crack growth and a rough surface indicates rapid crack growth. Smooth surfaces usually indicate the location of the fracture origin where the crack was still moving relatively slowly as shown in Fig. VIII.12(a). When a crack grows in a metal by fatigue (fatigue is discussed in Chapter IX), one often finds a smooth or polished region near the fracture origin. This surface results from the surfaces rubbing together on the compression portion of the cyclic loading.

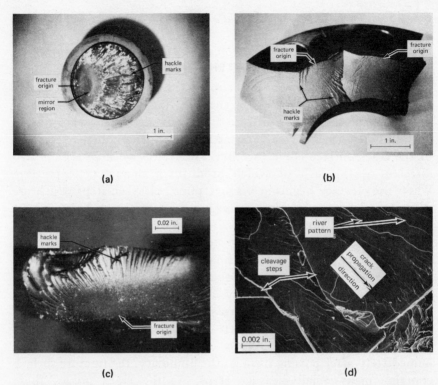

(a)

(b)

(c)

(d)

Fig. VIII.12 Fracture surface markings. (a) Fractured glass rod. Note the smooth mirror region near the fracture origin created by a slowly moving crack, and the rough region, with ridge lines radiating away from the fracture origin, created after the crack had accelerated. (b) Fractured steel flange. Note the hackle marks radiating from the fracture origin. (c) Fractured part of molded nylon. Note hackle mark pattern pointing to the fracture origin. (d) Scanning electron microscope photograph of the cleavage surface of a grain of Fe-3%Si alloy fractured at 77°K. Note the cleavage steps forming a river pattern which converges in the direction of crack propagation.

The most obvious fracture surface features which can be observed with the naked eye are hackle marks or chevron markings, as shown in Figs. VIII.12(a), (b), and (c). They are everywhere perpendicular to the crack front or parallel to the direction of propagation. These ridges originate where two regions of the crack front which are growing on slightly different surfaces come together and join by forming a step. Markings of this type are quite common. They occur on fracture surfaces of glass, semibrittle metals and on some polymers. These markings are often visible to the naked eye or with only a small amount of magnification. Since they are parallel to the direction of crack propagation, they radiate out from the crack origin. They can be used to identify the origin and to determine local directions of propagation.

VIII-8-b   Microscopic Fracture Surface Markings

Cleavage steps are roughly the microscopic equivalent to hackle marks. When a grain cracks by cleavage, the crack follows specific crystallographic planes. If different sections of the crack get started on different planes, a cleavage step is formed where the two sections of the crack surface come together. These cleavage steps arise in two different ways, each of which produces a characteristic pattern. When the cleavage begins at a single point, cleavage steps which are started at the fracture origin propagate with the crack front and get farther apart as the crack grows. This produces a fan pattern of diverging cleavage steps. Cleavage steps can also originate along a line when the fracture encounters a small-angle boundary in a grain.* If the angle across the boundary is such that the cleavage planes do not match perfectly at the boundary, a fracture on a single plane will spread onto several planes as it crosses the boundary. Each section of the fracture surface is connected to adjacent sections by a small cleavage step. When this process occurs, it is usually found that immediately after passing through the boundary, the fracture surface contains many small cleavage steps. As the crack grows, these cleavage steps converge together to form fewer but larger steps. This produces a converging pattern of cleavage steps called a river pattern because of its appearance of many fine tributaries converging into larger rivers (see Fig. VIII.12). The differences in the two patterns are that in the fan pattern the cleavage steps begin at a point and diverge in the direction of crack growth, while in the river pattern, the cleavage steps originate along a line of small angle boundary and converge in the direction of crack growth.

The characteristic microscopic markings on ductile fracture surfaces are the dimples formed when microvoids grow together to produce the fracture. These round bottom pits or dimples often contain impurity particles which nucleated the void. This dimpled appearance is found in nearly all cases of ductile tensile fracture of metals, and clearly differentiates this type of fracture from brittle fracture, cleavage, or fatigue.

---

*A small-angle grain boundary is a network of dislocations in a grain which produces a small change in the orientation of the crystal lattice across the boundary. The angle change associated with a small-angle grain boundary is less than five degrees.

Many fatigue fracture surfaces exhibit striations or beach markings. These markings which indicate successive positions of the crack result from the irreversible opening and closing of the crack tip during the cyclic loading. These striations clearly indicate the local direction of crack propagation. Since one striation is formed per cycle, the striation spacing indicates the rate of crack advance per cycle. A more complete description of the markings on fatigue fracture surfaces is included in Chapter IX.

REFERENCES

1. American Society for Testing and Materials, "Fracture Toughness Testing and Its Applications," *American Society for Testing and Materials Special Technical Publication No. 381*, Philadelphia, 1965.
2. Andrews, E. H.: "Fracture in Polymers," Oliver and Boyd, Edinburgh, 1968.
3. Averbach, B. L., D. K. Felbeck, G. T. Hahn, and D. A. Thomas (eds.): "Fracture," MIT Press and John Wiley, New York, 1959.
4. Brown, W. F., and J. E. Srawley: "Plane Strain Crack Toughness Testing of High-Strength Metallic Materials," *American Society for Testing and Materials Special Technical Publication No. 410*, Philadelphia, 1966.
5. Brown, W. F.: "Review of Developments in Plane Strain Fracture Toughness Testing," *American Society for Testing and Materials Special Technical Publication No. 463*, Philadelphia, 1970.
6. Dobson, M. O. (ed.): "Practical Fractural Mechanics for Structural Steel," U. K. Atomic Energy Authority, Risley, Warrington, 1969.
7. Liebowitz, H.: "Fracture," Academic Press, New York, 1968.
8. McClintock, F. A., and A. S. Argon: "Mechanical Behavior of Materials," Addison-Wesley, Reading, Mass., 1966.
9. Parker, E. R.: "Brittle Behavior of Engineering Structures," John Wiley, New York, 1957.
10. Polakowski, N. H., and E. J. Ripling: "Strength and Structure of Engineering Materials," Prentice-Hall, Englewood Cliffs, N. J., 1966.
11. Rosen, B., (ed.): "Fracture Processes in Polymeric Solids," Wiley-Interscience, New York, 1964.
12. Stanworth, J. E.: "Physical Properties of Glass," Oxford University Press, London, 1950.
13. Tetelman, A. S., and A. J. McEvily: "Fracture of Structural Materials," John Wiley, New York, 1967.

PROBLEMS

VIII.1   The tensile strength of granite rock is measured to be 2,500 psi. If it is estimated that the shear stress necessary to cause plastic deformation is 300,000 psi, determine the confining hydrostatic pressure which must be superimposed on this shear stress to allow plastic deformation without fracture.

Estimate the depth in the crust of the earth at which plastic deformation of this rock can occur.

VIII.2   How deep can a hole be drilled through basalt which has tensile strength $\sigma_c = 4,000$ psi before the sides of the hole begin to crumble? The state of stress in the rock before the hole is drilled can be assumed to be hydrostatic compression which increases linearly with depth. The state of stress near the hole can be determined by considering it as a superposition of a uniaxial compression parallel to the hole which is unaffected by the hole and a compression in two directions in the plane perpendicular to the hole with a circular stress concentration.

[The specific gravity of granite is approximately 2.7 and basalt is nearly the same.]

VIII.3   A truss system in a multistory building is designed so that in the event of a major earthquake, the horizontal chords of the truss will yield and increase the damping of the

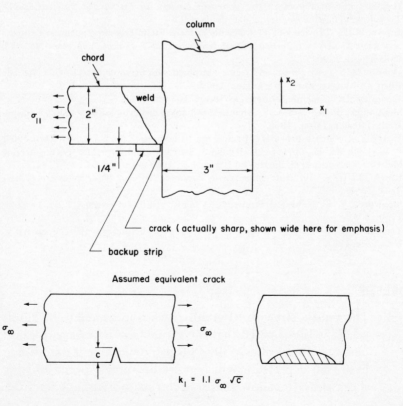

crack (actually sharp, shown wide here for emphasis)

backup strip

Assumed equivalent crack

$$k_1 = 1.1\, \sigma_\infty \sqrt{c}$$

structure. A view of one of the joints of the truss is shown
in the figures. Note that the backup strip which was placed
to facilitate the welding, creates a 1/4-in. crack at the
bottom of the beam. After construction it is discovered that
the weld metal was improperly cooled, so that it is more
brittle than desired. Assuming that k_1 for this crack in a
corner is the same as for a crack on the outside of a straight
bar such as shown in the figures, is the truss safe in its
present condition? The properties of the weld metal are:
yield strength, 70 ksi; tensile strength, 80 ksi; k_{1c}, 50 ksi
in.$^{1/2}$.

   a) Assume that σ_{11} is the only stress.

   b) Assume that thermal contraction after welding has
      created residual stress $\sigma_{33} = 2Y/3$, $\sigma_{22} = Y/3$ in addition
      to σ_{11}. Y is the yield strength.

VIII.4   Rectangle Tool Company in North Carolina manufactured
5,000 special-purpose diagonal cutters for National Business
Machine Corporation in New York City. After the tools
were delivered to the customer, it was found that many of
the pliers broke near the jaw (at A) when 56 lb of load was
applied at B. The wire to be cut is normally placed at C.
When the broken pliers were inspected carefully by the
engineers at RTC, it was found that a sharp notch (0.010-in.
deep) was produced inadvertently by the grinding wheel
which ground the cutting edge.

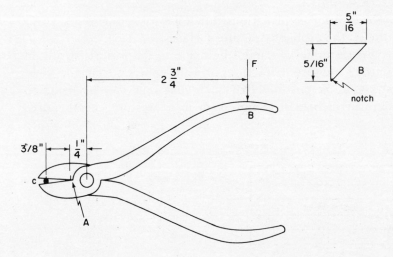

The cross section of the jaw is shown on page 451. Estimate k_{1c} for the plier material. How would you increase the strength of the pliers?

VIII.5   A crack initiated at a point on the free surface of an isotropic, brittle material when the strains at that point were $\epsilon_{11} = .0005$, $\epsilon_{22} = -.0003$, and $2\epsilon_{12} = -.0006$. The crack ran perpendicular to the maximum principal stress at the point.

a) Sketch the relative orientation of the crack and the x_1-x_2-axes at that point.

b) What stress would cause fracture of this material in a tensile test if $E = 10^7$ psi, $\nu = 0.2$?

VIII.6   Plot the probable stress-strain curves of the following metals under the conditions given, showing the relative magnitude and shape of each on the same graph:

a) .18% C steel, annealed, grain size = 0.3 mm, $\dot{\epsilon} = 10^{-3}$ sec^{-1}, $T = 25°C$.

b) .18% C, cold rolled steel; testing conditions and grain size the same as a).

c) .30% C steel, annealed, grain size = 0.07 mm, $\dot{\epsilon} = 10^2$ sec^{-1}, $T = 0°C$.

d) Commercially pure aluminum, grain size = 0.3 mm, $\dot{\epsilon} = 10^{-5}$ sec^{-1}, $T = 100°C$.

e) The same as (d) except $\dot{\epsilon} = 10^{-7}$ sec^{-1}, $T = 400°C$.

f) Beryllium copper, grain size = 0.07 mm, $\dot{\epsilon} = 10^{-6}$ sec^{-1}, $T = 200°C$.

Give brief explanations for each of your plots. Also discuss the probable fracture surfaces.

VIII.7   A brittle material cannot be made to deform plastically in tension because it fractures at a lower stress than the yield stress. It is possible that such materials can be made to deform in biaxial compression in a pressure chamber like

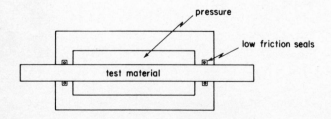

the one shown. A rod of the test material passes through low-friction seals at the ends of a pressure chamber, so that it is subjected to compression in two directions and is relatively stress free in the axial direction. If the tensile fracture stress is σ_c, and the tensile yield stress is σ_y, how large can the ratio σ_y/σ_c be without fracture preceding yield? What is the pressure required?

VIII.8   Consider a ring of circular cross section which is loaded by a force F, as shown below (similar to Problem III.7). The material is the same as that of the milling cutter discussed in Example VIII.4, and a sharp notch of 0.030 depth is located on the inside of the ring at $\theta = 0$. What is the maximum force F possible before fracture? How does this change if the notch occurs at $\theta = \pi/4$? at $\pi/2$?

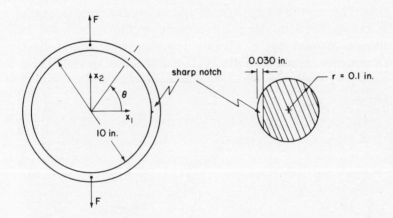

VIII.9   A support bracket made from AM355 precipitation-hardened stainless steel failed by fracture when the applied load $F = 2,000$ lb. Subsequent examination revealed that a fatigue crack had been created at the base of the bracket as indicated. The material was age hardened at $850°$F. Estimate the length of the crack which caused the failure. A larger crack could have been tolerated if the material had been aged at $925°$F. What would be its size? Note that aging

at the higher temperature reduces the yield strength by only 2.5%. (See Table VIII.2 for data.)

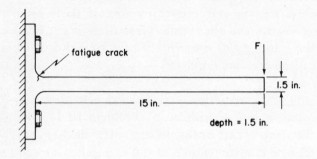

VIII.10   A skin panel of an experimental aircraft is to be made from 1/8 in. thick titanium alloy. The panel is attached by a row of rivets along one edge and is loaded in tension. The design criteria state that the maximum applied stress must be less than 75% of the yield strength and below the fracture stress if cracks 1/16 in. long are present at the rivet holes as shown. Consider the various titanium alloys and heat treatments shown in Table VIII.2 and select the alloy and heat treatment which will allow the maximum applied stress.

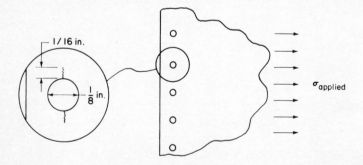

VIII.11   It is found that drawn glass ribbon made by pulling molten glass through a slit shaped die has highly anisotropic strength. The tensile strength of such ribbon in the direction along the axis is quite good for glass, being on the order of 20,000–30,000 psi, but the strength in the transverse

direction may be only 1–10% of this value. Give a reason for strength anisotropy.

VIII.12   A ·semiconductor strain gage is a silicon semiconductor resistor. The silicon has a coefficient of thermal expansion of $3 \times 10^{-6}/°C$. The gage is bonded to steel which has a coefficient of thermal expansion of $12 \times 10^{-6}/°C$. The bonded gage is stress free at the curing temperature of the adhesive which is 100°C. If the tensile breaking strength of the silicon is 10,000 psi and Young's modulus and Poisson's ratio are 16×10^6 psi and 0.25, respectively, what are the maximum and minimum temperatures to which the gage can be taken without fracturing it?

VIII.13   If the strain gage in Problem VIII.12 is to be used to measure a uniaxial stress in the steel which may have a value in the range ±3,000 psi, what are the maximum and minimum useful temperatures for the gage?

VIII.14   A silicon semiconductor device in part consists of a silicon substrate 3μ thick with an oxidized insulating layer of SiO_2 1μ thick. The oxide layer may contain cracklike defects of a length comparable to the thickness of the layer. Estimate the tensile breaking stress of the oxide layer, assuming Young's modulus for SiO_2 to be 10^7 psi. Calculate the stresses induced by differential thermal expansion including bending effects. The coefficients of thermal expansion are

$$\alpha_{Si} = 3 \times 10^{-6}/°C; \qquad \alpha_{SiO_2} = 2.2 \times 10^{-6}/°C$$

and Young's modulus for silicon is 1.6×10^7 psi. What temperature change is required to break the oxide layer?

VIII.15   In Chapter IV it was shown that in the plastic expansion of a thin-walled tube, or sphere, the biaxial state of stress caused the instability condition for necking to be reached at a smaller equivalent strain than that required for necking in a tensile specimen. This means that the average strain to fracture for the part is smaller of the tube or sphere than for the tensile specimen. Consider now the *actual true strain at fracture at the failure site*. Do you expect this to be greater than, less than, or equal to the true strain at the fracture surface of a tensile specimen?

CHAPTER NINE
Fatigue

IX-1 INTRODUCTION

Fatigue fracture resulting from cyclic deformation through a few cycles of large plastic strain amplitude is familiar to most people, and must have been known since the earliest attempts at metal working several thousand years ago. This phenomenon is called *low-cycle fatigue*, since fracture occurs after only a few strain cycles, ranging from less than ten to several hundred, depending on the amount of strain in each cycle. The fact that fracture could also result from many thousand cycles of stress below the elastic limit was not discovered until comparatively recently, in the middle of the nineteenth century. The obvious reason for the timing of the discovery of this phenomenon, which is called *high-cycle fatigue*, is that it was not until this time that machines were built which required their parts to withstand such repeated loading for so many cycles. Since fatigue occurs at stresses below those required for plastic failure, it is often the limiting consideration for design of machinery. Fatigue was first recognized as a problem in the failure of

456

railroad car axles, and its importance has increased as mechanization has increased.

Because it is so important as a mode of failure, the phenomenon of fatigue has been extensively studied for more than one hundred years. Most of the early research on fatigue was confined to obtaining data for correlating the expected lifetime of a sample with the stress amplitude. The empirical relationships for fatigue lifetimes which have emerged from this research are of great engineering value. They provide the basis for the design criteria for parts subjected to cyclic load that are described in subsequent sections of this chapter.

More recently, the emphasis in fatigue research has shifted to probing the microscopic mechanisms of the fatigue process. It is recognized that the total process of fatigue fracture consists of two parts. First, a crack must be formed in the material where no crack originally existed, and then it must grow until it is of sufficient size to cause the body to fail. The first portion is called crack initiation and the second is called crack propagation. Much of the research on the crack initiation process has been made possible by the improved understanding of the mechanisms of plastic deformation which has followed the introduction of dislocation theory. It has also benefitted greatly from recent advances in experimental techniques such as electron microscopy of metal foils for studying dislocation structures and improved transducers which allow very precise measurement of strain. From this recent work, a picture of the process of damage accumulation leading to crack initiation is emerging. This picture is still predominantly qualitative and gaps still exist in understanding the sequence of events which leads to the formation of the crack, but research in this area is continuing and additions to the understanding of the mechanisms involved are still being made. Once a crack is formed, the process of interest becomes the growth of the crack. In this area, the greatest aid in understanding the phenomenon has come from the theory of fracture mechanics and the concept of the stress intensity factor presented in Chapter VIII. Although much work remains to be done on the process of fatigue crack propagation, fracture mechanics has provided a reasonable qualitative description of the process and has established a physically reasonable framework for the correlation of experimental results.

IX-2 CRACK INITIATION

When a metal is subjected to a strain cycle of sufficient amplitude to cause some plastic strain on each cycle, the irreversible nature of plastic deformation would lead one to expect that a gradual accumulation of damage should be occurring in the material. For this reason, the fact that eventual failure by fracture should occur is perhaps not very surprising. When one does such a cyclic-straining experiment at constant strain amplitude, it is found that initially the material work hardens quite rapidly, as does also the stress required to maintain the strain amplitude of the cycle. However, as the number of cycles increases, the work hardening rate decreases markedly and in many cases, stops completely, so that the stress amplitude required to maintain a constant strain amplitude becomes constant. The plastic strain amplitude in such experiments is of the same order of magnitude as the Bauschinger strain discussed in Chapter V. For this reason, the concept that work hardening is related to the integrated equivalent plastic strain is not expected to be valid.

The fact that the work hardening rate eventually decreases to zero would seem to indicate that the structure of the material has reached steady state. Transmission electron microscopy studies of the dislocation structures which result from cyclic straining tend to confirm this idea. It is found that the effect of cyclic straining is to create a cellular dislocation structure, where the cell walls are complex networks and tangles of dislocations, and the cell interiors are relatively dislocation free. When the stress amplitude reaches a steady-state value in a cyclic experiment, it is found that the dislocation cell structure has reached a steady-state size. It is believed that the bulk of the plastic strain beyond this point is accounted for by dislocations crossing the cells from one cell wall to another. However, since fracture eventually occurs, the steady state achieved in fatigue must be only a pseudosteady state, and the accumulation of damage which will lead to the formation of a crack must be occurring somewhere in the sample. The exact nature of the process by which the crack is formed is not known, but is presumed that strain concentrations caused by grain boundaries, second-phase particles, or surface flaws must play a role. Such irregularities would cause damage accumulation to continue at some sites even after the majority of the material has reached its steady-state condition.

Under high-cycle fatigue conditions, the stress amplitude is below the yield strength of the material, so that the strain is nominally elastic. If the strain were literally purely elastic, fatigue could not result because elastic straining is, by definition, a reversible process. However, this difficulty is associated with the oversimplification introduced by the concept of a yield strength and the assumption of purely elastic deformation below this yield strength. As discussed in Chapters III, IV, and V, nearly all metals undergo a minor amount of plastic strain even at low stresses. This is called microstrain, because at stresses well below the yield strength the magnitude of the plastic strain is small compared to the elastic strains.

Microscopic examination of the surfaces of samples which have been subjected to cyclic loading reveals that the microstrain occurs inhomogeneously in the sample, with the entire strain seemingly concentrated in a relatively few isolated slip bands. These slip bands form during the first few thousand cycles and remain active until after a crack is formed. Because the straining in these bands continues after the bulk of the material has stopped undergoing strain, they are called *persistent slip bands*. Since the strain is so inhomogeneous, the plastic strain amplitude in the persistent slip bands is quite large compared to the average strain amplitude (averaged over the entire sample). Thus, damage accumulation leading to crack formation can continue in the persistent slip bands at very low average plastic strain amplitudes. The nature of the damage which leads to crack formation in high-cycle fatigue seems to be related to the formation of intrusions and extrusions within the slip bands as shown in Fig. IX.1. In this phenomenon, material is pushed out of the surface at one point in the band and material is drawn in to form deep valleys at other points in the band. In some cases, the slip bands appear to act together, so that one entire band becomes an intrusion and the material is extruded from an intersecting band. The exact mechanism which accounts for this behavior is at present unknown. Transmission electron microscopy studies have shown that the persistent slip bands which are seen on the surface, penetrate into the material parallel to the dislocation slip plane, as shown in Fig. IX.1. The persistent slip band is relatively free of dislocations, but is periodically divided up by dislocation walls which cross the thickness of the band.

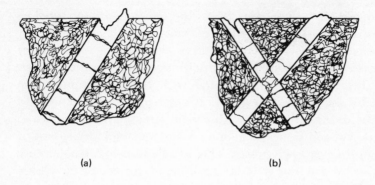

(a) (b)

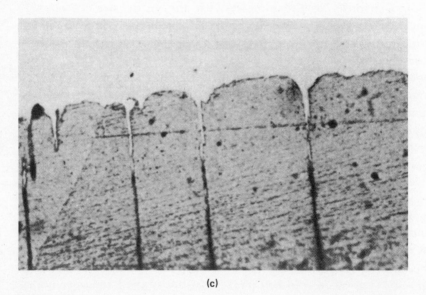

(c)

Fig. IX.1 Persistent slip bands forming intrusions and extrusions. (a) Schematic of the structure of a persistent slip band showing that it is relatively free of dislocations except for the periodic walls of dislocations across the band. (b) Two intersecting persistent slip bands forming an intrusion where one intersects the surface and an extrusion where the other intersects the surface. (c) Tapered section through the surface of copper showing slip bands developing into notches (intrusions). Cutting the sample at a small angle to the surface magnifies the apparent depth of the crack by a factor of 20. Approximate magnification 1200. (*From D. McLean, "Mechanical Properties of Metals," p. 351, John Wiley, New York, 1962.*)

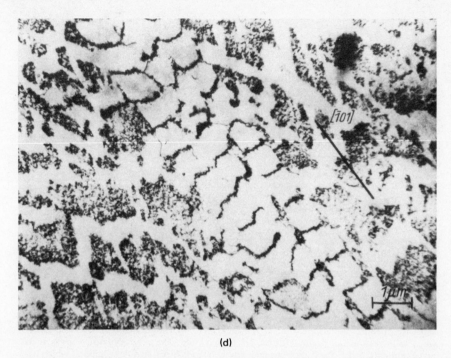

(d)

Fig. IX.1 (*Continued*) Persistent slip bands forming intrusions and extrusions. (d) Transmission electron microscope picture of a persistent slip band in a fatigue copper single crystal. (*From P. Lukáš, M. Klesnil, and J. Krejčí*, Phys. Stat. Sol., *vol. 27, p. 549, 1968.*)

IX-3 CRACK PROPAGATION

Once a true crack has formed in a material, the presence of the crack itself dominates the stress and strain behavior in its vicinity. The development of the theory of fracture mechanics to describe the behavior of bodies which contain cracks has been quite useful in reaching an understanding of the process of crack propagation in fatigue. A schematic representation of the sequence of events which occurs at the tip of the crack during a fatigue cycle is shown in Fig. IX.2. As the crack is pulled in tension, plastic flow occurs at the tip of the crack, creating the plastic zone indicated by the shaded area in the figure. The crack tip opens up and blunts out. The crack tip opening displacement (CTOD) at the maximum tension is indicated in Fig. IX.2a. As the stress decreases, the crack closes. Initially, the strain during closing is elastic, but because of the residual stresses created during tension, the crack tip closing cannot be completely elastic. At some point during the unloading or compressive portion of the cycle, the material at the crack tip begins to flow plastically in

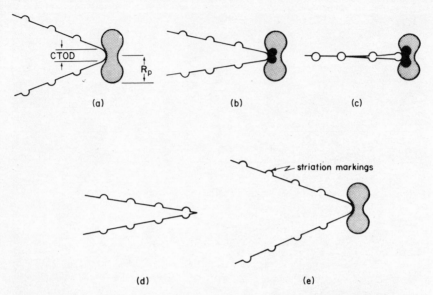

Fig. IX.2 Schematic illustration of the deformation at the crack tip during a fatigue cycle. (a) Maximum tension. The plastic zone is indicated by the shaded area and the plastic crack tip opening displacement (CTOD) is indicated. (b) Partially closed. A reverse plastic zone, solid region, has begun to form. (c) Fully closed. The reverse plastic zone is at its maximum extent. (d) Partially reloaded. The crack advances into new material which has been strained plastically on the previous cycle. (e) Fully opened. The cycle begins to repeat.

the opposite direction from the deformation in tension. The reversible plastic strain is limited to a region smaller than the forward plastic zone. The reverse plastic zone is indicated by the solid region in Fig. IX.2b. In the fully closed condition at maximum compression, the reverse plastic zone has grown to its maximum extent, but it is still much smaller than the forward plastic zone. This is because in compression, the flanks of the crack come together and transmit the compressive load. As the load is reapplied, the fracture proceeds into the region in front of the old crack tip where the material has been processed by the plastic strain from the previous cycle. (Actually, the advance of the crack can occur at any stage of the cycle. The details of the process are not sufficiently well known to state when the crack actually moves forward.) When maximum tension is reached, the process begins again as the cycle repeats itself. Thus, on each cycle, the crack will move forward by a small amount which is determined by the details of the deformation history of the crack tip.

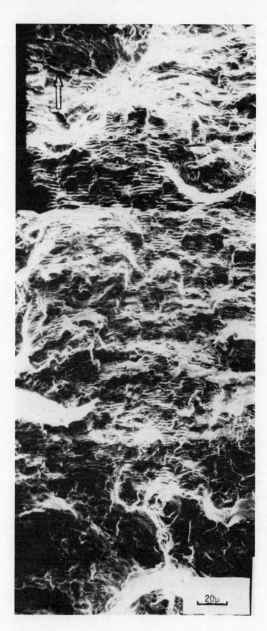

Fig. IX.3 Striation markings on a fatigue fracture surface. Annealed maraging steel with stress intensity factor amplitude $\Delta k_1 = 28$ ksi in.$^{1/2}$. (*From C. Bathias and R. M. Pelloux*, Met. Trans., *vol. 4, p. 1269, 1973.*)

The stepwise advance of the crack gives rise to the characteristic striation marks (or beach marks) which are often observed on fatigue fracture surfaces. Because of the plastic deformation which occurs in the vicinity of the crack tip, the surface of the fatigue crack is often grooved or rippled, as indicated in the schematic diagram of Fig. IX.2 and depicted in Fig. IX.3. These markings are called *fatigue striations*, and their spacing is equal to the distance which the crack advanced on each cycle. The fatigue striations on a fatigue fracture surface, therefore, represent a record of the local crack propagation rate which can be examined after the failure to determine the conditions under which the fatigue occurred.

Since the stress-strain history of the crack tip is uniquely specified by the history of the stress intensity factor when the plastic deformation is limited to a suitably small region at the crack tip, it is expected that the rate of crack growth per cycle, $\Delta c/\Delta N$ or dc/dN, should be a function of the amplitude of the change Δk_1 of the stress intensity factor in the cycle. Because the crack closes during a compression, it is assumed that k_1 is zero when the applied stress is less than or equal to zero. Therefore, only the positive stress portion of the fatigue cycle contributes to the crack growth. Because the stress intensity factor amplitude is the important parameter in the crack propagation process outlined above, one would expect that crack growth rates measured for a variety of applied stress amplitudes with varying crack lengths could be correlated with the parameter Δk_1 for the cycle. An example of such a correlation when the crack growth rate is plotted as a function of Δk_1 is shown in Fig. IX.4. Although this type of correlation definitely indicates that crack growth rates can be expressed as functions of the stress intensity factor amplitude, the exact nature of the functional relationship between the two quantities is still subject to controversy.

The qualitative argument illustrated in Fig. IX.2 would suggest that the crack growth per cycle should be proportional to the crack tip opening displacement, CTOD, or the plastic zone size, R_p. These two quantities are proportional to each other and to the square of the stress intensity factor, so that it might be expected that the crack growth rate, dc/dN, can be expressed as

$$\frac{dc}{dN} = A \left(\frac{\Delta k_1}{\sigma_Y}\right)^n \qquad (IX.1)$$

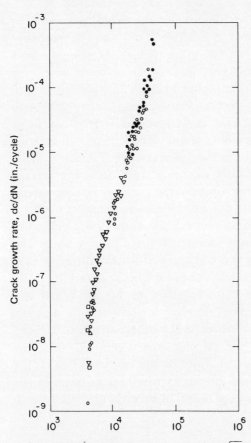

Fig. IX.4 Correlation of crack growth rates with the stress intensity factor amplitude of the load cycle for A533-B steel. Different symbols are for different tests under different conditions. (*From P. C. Paris, R. J. Bucci, W. G. Wessel, W. G. Clark, and T. R. Mager, Extensive Study of Low Fatigue Crack Growth Rates in A533 and A508 Steels,* Stress Analysis and Growth of Cracks, ASTM Special Technical Publication No. 513, *p. 143, 1972.)*

with the exponent n equal to 2. The expression σ_Y is the tensile yield strength. Many experimental measurements however indicate that the crack growth rate is proportional to a higher power of the stress intensity factor amplitude. Experimental values of n can be as high as five. A model which predicts the higher exponent has not been firmly established. However, there is some experimental evidence that the crack growth rates determined from measurements of the striation spacing on the fracture surfaces do not agree with those determined from external measurements of the crack length during the fatigue test.

The crack growth rate can be determined either from the derivative of the crack length vs. number of cycles curve, or from the striation spacing at various points on the fracture surface. When the results of these two methods are compared, it is often found that the exponent n of the crack growth rate law given in Eq. (IX.1), determined from the striation spacing measurements, is close to 2 even when the macroscopic measurements give a higher value. This discrepancy can be explained if either of two events occur during the crack growth process. If the crack growth is periodically arrested by an obstacle, or if the crack periodically makes a large jump when it meets an existing crack, defect, or brittle region, the microscopic fracture surface measurements of striation spacing will indicate a different average crack growth rate than the macroscopic measurements. The arrests (or jumps) are averaged into the macroscopic measurements made by measuring the crack length as a function of the number of cycles, but are not accounted for by the measurements of striation spacing. The discrepancy in the measurements could account for the differences in the exponent n and explain why the exponent determined by external measurements of the crack length does not agree with the prediction of the crack-opening model for crack propagation. Alternative explanations are possible. As for other aspects of fatigue, the environmental conditions, in particular, have a great effect on the observed crack growth behavior. In spite of the uncertainties of the model, experimental correlations of crack growth rates, such as the one shown in Fig. IX.4, can be very useful in predicting crack growth rates and service lifetimes.

IX-4 HIGH–CYCLE FATIGUE UNDER FULLY REVERSED TENSION–COMPRESSION LOADING

By far the majority of experimental work on fatigue has been concerned with establishing the relationship between the stress amplitude and the number of cycles-to-failure. This work has predominantly used tension-compression or bending for loading. Since bending produces a state of stress which is locally tension or compression, these methods are equivalent. In the usual fully reversed test, the magnitude of the compression is equal to the magnitude of the tension, so that the mean stress is equal to zero. Results of such experiments are generally presented on S-N diagrams, where the stress amplitude or its logarithm is plotted as a function of the logarithm of the number of cycles-to-failure. Examples of

S-N curves are shown in Fig. IX.5. Certain characteristics of such plots are found to be common to most metals. Lifetimes below approximately 1,000 cycles belong to the region of low-cycle fatigue. In this region, the behavior can be more readily correlated with strain amplitude. The stress amplitude is nearly equal to the tensile strength in low-cycle fatigue. (Low-cycle fatigue will be discussed in more

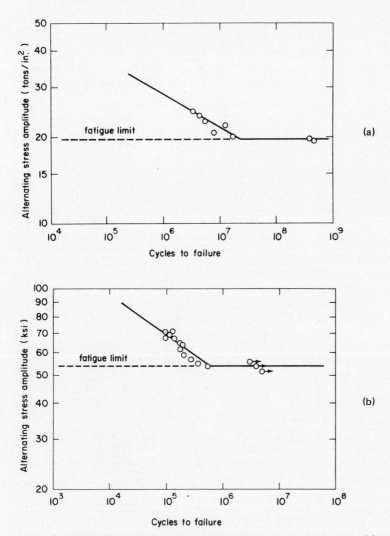

Fig. IX.5 Fatigue *S-N* curves for several common structural metals. (a) Cold-rolled mild steel; (b) 4340 steel in the annealed condition.

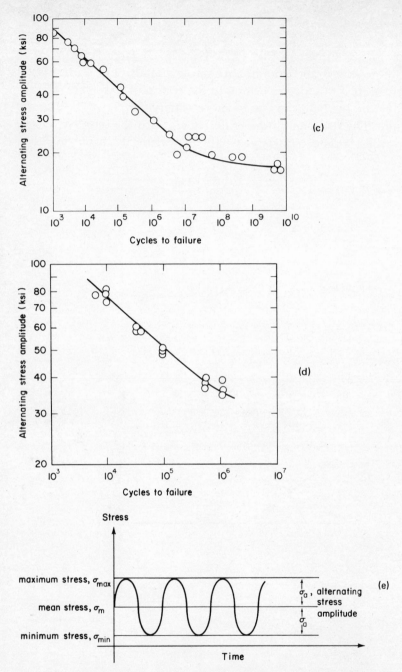

Fig. IX.5 (*Continued*) Fatigue S-N curves for several common structural metals. (c) 7075 T-6 aluminum; (d) 2024 T-3 aluminum; (e) schematic representation of a fatigue cycle, showing definitions of the alternating stress amplitude, maximum stress, minimum stress, and mean stress. ([d] *from K. Walker, Effect of Environment and Complex Load History on Fatigue*, ASTM Special Technical Publication No. 462, *1970*.)

detail later.) The region where the life is in excess of 1,000 cycles is high-cycle fatigue. In this region, the stress amplitude is below that required to produce general plastic flow on each cycle. The lifetime expressed in terms of the number of cycles-to-failure increases rapidly as the stress amplitude decreases. In the region of life-times from 10^3 to 10^6 cycles, the behavior of the S-N curve relating to the logarithm of the stress amplitude to the logarithm of the number of cycles is nearly a straight line. This suggests that a functional relationship of the following form should describe fatigue life:

$$\sigma_a{}^m N = C \text{ (constant)} \tag{IX.2}$$

where N is the number of cycles-to-failure, σ_a is the alternating stress amplitude, and C is a constant. The values found for m in metals range from 8 to 15. The values of m observed for steels are much less widely scattered, with a characteristic value of 10.

The most pronounced difference in fatigue behavior among different metals is the behavior below about half of the tensile strength, for lifetimes above 10^6 to 10^7 cycles. In steel the fatigue process seems to stop at stress amplitudes below this value, so that the S-N curve turns nearly horizontal at approximately 10^6 cycles. Therefore, ferrous materials are said to exhibit a definite *fatigue limit* or *endurance limit*. At stress amplitudes below this limit stress, failure never occurs. Titanium alloys have also been found to have a fatigue limit. In most other metals, such as aluminum, for example, the fatigue process seems to continue to occur beyond 10^7 cycles. The sloping behavior of the S-N curve has been found to extend well beyond 10^8 cycles in such materials. When no fatigue limit exists, the stress which will give a life of 10^8 cycles is often reported as the *fatigue strength* of the material. Such fatigue strengths are generally much less than half of the tensile stength, often being as low as 25% of the tensile strength. In designing very-high-speed equipment which may accumulate more than 10^8 cycles during its design life, one should be conscious of the distinction between the fatigue strength of nonferrous materials and the fatigue limit of steels. (For example, at a rate of 10,000 cycles/min, only seven days are needed to accumulate 10^8 cycles.)

*EXAMPLE IX.1—Fatigue from Misalignment
  of a Shaft*

The drive shaft which connects an electric motor with the machine it drives is misaligned. The nature of the misalignment is such that the section of the shaft between the motor bearings and the machine bearings is in pure bending. The bearings are 2 in. apart and the shaft has a 1/4-in. diameter. The shaft is made from cold-rolled mild steel, for which the *S-N* curve appears in Fig. IX.5(a). The speed of the motor is 1,800 rpm.

(a) What is the lifetime of the shaft, if the angular misalignment is $1.5°$?

(b) What is the maximum angle of misalignment which can be tolerated and still have unlimited life for the shaft?

Assume that the stress from the torque is negligible compared with the bending stress.

Solution:

In pure bending, the shaft is the arc of a circle. The angular difference between the ends is determined from

$$\theta = \frac{L}{\rho} \qquad \theta = 1.5° = \frac{\pi}{180}(1.5) \tag{a}$$

where L is the shaft length, ρ is the radius of curvature and θ is the angle between the ends. From the formula for deflection of elastic beams

$$\frac{1}{\rho} = \frac{M}{EI} \tag{b}$$

where E is Young's modulus, and I is the moment of inertia of the cross section; $I = \pi r^4/4$ for a circle. Combining Eqs. (a) and (b) and solving for the maximum stress σ_{max},

$$\theta = \frac{ML}{EI}; \qquad \sigma_{max} = \frac{Mr}{I} = \frac{\theta Er}{L} \tag{c}$$

Substituting the values from above

$$\sigma_{max} = \frac{1.5\pi}{180}\frac{(3 \times 10^7)(1/8)}{2} = 49,000 \text{ psi} \tag{d}$$

From the *S-N* curve of Fig. IX.5(a), $N = 4 \times 10^5$ so the system will last for 2.2×10^2 min.

The fatigue limit for this steel is 39,000 psi, so that a stress below the fatigue limit requires a misalignment of less than $1.2°$.

IX-5 FATIGUE WITH NONZERO MEAN STRESS

Although most testing is done with fully reversed stress cycles, fatigue often occurs in service with an alternating stress superimposed upon a static mean stress. Such data as exist (Fig. IX.6) indicate that it is neither the maximum tensile stress nor the alternating stress amplitude which determines fatigue life, but a combination of the two. Experimental data for fatigue with nonzero mean stress is reported in several ways. It can be reported in terms of the alternating stress amplitude and the mean stress. They are either given as curves of alternating stress amplitude vs. mean stress for a constant life (Fig. IX.6a) or as S-N curves of cycles-to-failure vs. alternating stress amplitude for a constant mean stress (Fig. IX.6b). Alternatively, the fatigue behavior can be presented in terms of the maximum stress and the stress ratio $R = \sigma_{min}/\sigma_{max}$. When these variables are used, the data are usually presented as S-N curves of number of cycles-to-failure vs. maximum stress at constant stress ratio R (see Fig. IX.6c). In this type of presentation, $R = -1$ is fully reversed tension-compression.

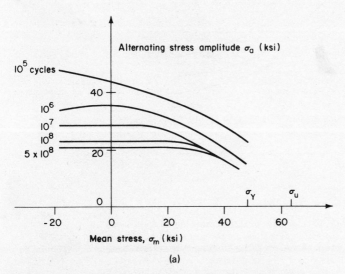

(a)

Fig. IX.6 Various ways of representing the effect of mean stress on fatigue life. (a) Alternating stress amplitude vs. mean stress for constant fatigue life curves for 2024 T-4 aluminum. (*From F. M. Howell and J. L. Miller, Axial Stress Fatigue Strengths of Several Structural Aluminum Alloys, ASTM, Proc., No. 55, 1955.*)

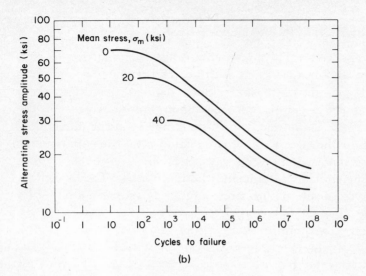

(b)

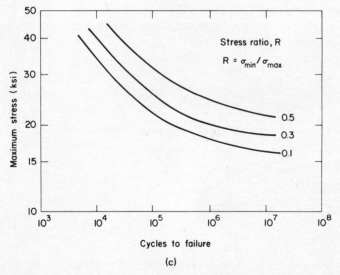

(c)

Fig. IX.6 (*Continued*) Various ways of representing the effect of mean stress on fatigue life. (b) Alternating stress amplitude vs. cycles-to-failure for constant mean stress curves for 2014-T5 aluminum. (*From E. C. Hartmann and F. M. Howell, Laboratory Fatigue Testing of Materials, in G. Sines and J. L. Waisman, "Metal Fatigue," p. 100, McGraw-Hill Book Company, New York, 1959.*) (c) Maximum stress vs. cycles-to-failure at constant stress ratio curves for notched 7075-T6 aluminum sheet. Stress concentration factor K_t = 3.0. (*From M. Field et al., "Machining of High-Strength Steels with Emphasis on Surface Integrity," p. 9, Air Force Machinability Data Center, Cincinnati, Ohio, 1970.*)

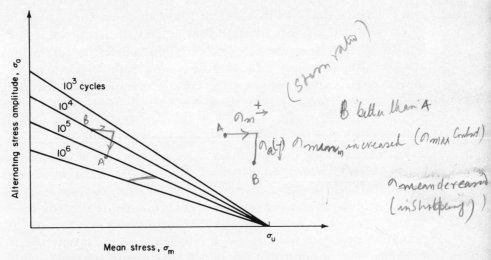

Fig. IX.7 Goodman diagram for the effect of mean stress. Each line represents constant life. Alternating stress amplitude is measured on the vertical axis and the mean stress on the horizontal axis.

The mean stress is most detrimental when it is positive. Therefore, the initial discussion will be limited to this condition. One approach to the effect of mean stress is that proposed independently by Goodman and Soderberg, and illustrated in Fig. IX.7. The allowable alternating stress for a given life must certainly go to zero as the mean stress approaches the tensile strength. Goodman and Soderberg proposed that an approximate curve of alternating stress amplitude vs. mean stress, for a fixed life, could be constructed by using a straight-line relationship which passes through the known zero mean stress point and through zero alternating stress amplitude for a mean stress equal to the tensile strength. (Actually in Soderberg's formulation, the line passes through zero stress amplitude when the mean stress equals the yield strength instead of the ultimate tensile strength.) *If σ_{a0} is the alternating stress amplitude which gives a life N at zero mean stress,* then the Goodman diagram predicts that *the alternating stress amplitude, σ_a, which gives a life N with a mean stress σ_m, is given by*

$$\frac{\sigma_a}{\sigma_{a0}} + \frac{\sigma_m}{\sigma_u} = 1 \qquad\qquad (IX.3)$$

where σ_u is the tensile strength of the material.

In the actual data of Fig. IX.6, the stress amplitude for a given life falls off <u>less</u> rapidly with increasing mean stress than is predicted by Eq. (IX.3). Therefore, the Goodman relation underestimates the life of a part. The Goodman relation gives a conservative estimate of the allowable stress and is used in design criteria, even though it does not give an accurate prediction for the lifetime.

Fatigue behavior with negative mean stress has been less extensively studied than that with positive mean stress. The data presented in Fig. IX.6 indicate that compressive mean stress generally improves fatigue life as might be expected. This is the basis for using operations such as shot peening and rolling to produce residual compressive stresses at the surfaces of parts in order to improve their fatigue lives. No generally accepted relationship is known to quantitatively relate the increase in fatigue life to the magnitude of the compressive mean stress. The simple extension of the straight line on the Goodman diagram to the left of the zero mean stress axis greatly overestimates the fatigue life with negative mean stress. Best practice is probably to assume that the compressive mean stress does not affect the fatigue life, since this is usually a conservative estimate. It should also be pointed out that if the minimum stress on a cycle can be increased without at the same time increasing the maximum stress, this will improve the fatigue life. When this is done, the mean stress is increased, but the alternating stress amplitude is reduced and the net effect is beneficial. This is perhaps clearer in terms of the stress ratio—maximum stress formulation. Increasing the minimum stress for a given maximum stress increases the stress ratio R without changing the maximum stress, and therefore increases the fatigue life. In many machines, it is possible to increase the minimum stress on a part, independently of the maximum stress, by, for example, limiting the travel of a section of the machine. This type of change in the cycle cannot be achieved through residual stresses, however, because the principle of superposition implies that the residual stress will not affect the cycle amplitude, but will instead increase both the maximum and minimum stresses by the same amount. This will increase the mean stress only and will therefore reduce the fatigue life.

EXAMPLE IX.2—Fatigue of a Leaf Spring

A leaf spring is made from a steel bar with dimensions 2 × 0.5 in. The distance between the supports is 3 ft. The bar is simply supported at the ends, and the load is applied at the center. It is found that the fatigue limit load of the spring expressed for a zero mean load test is 1,000 lb. The spring is to be used to carry a static load of 1,000 lb and may experience occasional vibrations superimposed on the mean load. (a) What is the maximum vibrational stress amplitude which will give a life of at least 50,000 cycles, with the 1,000-lb mean load? (b) What is the maximum vibrational stress amplitude which will be below the fatigue limit? No additional information is available about the steel other than the fatigue limit given above, but it is known to be a common alloy steel.

Solution:

Since the exact properties of the steel are not known, it is necessary to approximate the *S-N* curve. The fatigue limit for nearly all steels falls between 0.45 and 0.5 σ_u, where σ_u is the tensile strength. The fatigue limit can be assumed to occur at 10^6 cycles. The *S-N* curve can also be assumed to pass through the tensile strength at 10^3 cycles and to obey Eq. (IX.2). Therefore, the two points determine the *S-N* curve. One should note that the error in this approximate *S-N* curve is usually small compared to the error which will be introduced by using the Goodman-Soderberg relation to evaluate the mean stress effect.

Working in terms of the loads, which are proportional to the stresses, the fatigue limit (given as 1,000 lb) will be taken to be 0.5 σ_u. Then, the load at the tensile strength is 2,000 lb. The parameters of Eq. (IX.2) are given by

$$\left(\tfrac{1}{2}\sigma_u\right)^m 10^6 = (\sigma_u)^m 10^3 \tag{a}$$

where the two sides of the equation are the two points on the fatigue curve assumed above. Solving Eq. (a) for *m*

$$10^3 = (2)^m ; \quad m = 10 \tag{b}$$

The fatigue law, Eq. (IX.2), can be written in terms of load amplitude F_{ao} since the stress amplitude for elastic deformation is proportional to the load amplitude

$$F_{ao}^{10}N = (2,000)^{10} 10^3 \tag{c}$$

The Goodman-Soderberg law can also be written in terms of loads.

$$\frac{F_a}{F_{ao}} + \frac{F_m}{F(\sigma_u)} = 1 \tag{d}$$

where F_a is the oscillating load amplitude, F_{ao} is the equivalent load amplitude for zero mean load, F_m is the mean load (F_m = 1,000 lb), and $F(\sigma_u)$ was assumed to be 2,000 lb. Using these values

$$\frac{F_a}{F_{ao}} = \frac{1}{2} \tag{e}$$

(a) For N = 50,000, Eq. (c) can be solved for F_{a0}:

$$F_{ao}{}^{10}(50,000) = (2,000)^{10} 10^3$$

$$F_{ao} = 2,000 \left(\frac{1}{50}\right)^{1/10} = \frac{2,000}{1.48} = 1,350 \text{ lb} \tag{f}$$

With a mean load of 1,000 lb, the maximum amplitude of F_a is 675 lb from Eq. (e).

(b) For unlimited life

$$F_{ao} = 1,000 \text{ lb} \quad \text{and} \quad F_a = 500 \text{ lb} \tag{g}$$

IX-6 FATIGUE FOR COMPLEX STATES OF STRESS

In the absence of a complete understanding of the mechanism of fatigue, no theoretical basis exists for using tension-compression fatigue results to predict fatigue under other states of stress. The effects of various states of stress should not be expected to be the same in both the initiation and propagation phases of fatigue. Crack initiation, which at least in its early stages depends on plastic deformation, might be expected to correlate with the maximum shear stress or the equivalent stress. Crack propagation, which requires opening up the crack, may depend more on the nominal maximum principal stress.

If the fatigue limit occurs when the local plastic strain amplitude falls below some critical value, it should depend on the plastic processes alone. Then the fatigue limit should depend on the shear stress or equivalent stress alone. In work done by Ransom and Mehl* using combined bending and torsion this was found to be the case. They found that fatigue did not occur when

---

*J. T. Ransom and R. R. Mehl, "The Anisotropy of the Fatigue Properties of SAE 4340 Steel Forgings," *Proc. ASTM*, vol. 62, 1952.

$$\left(\frac{\sigma_B}{2}\right)^2 + \sigma_T{}^2 \leq \left(\frac{\sigma_l}{2}\right)^2 \qquad\qquad \text{(IX.4)}$$

where σ_B is the bending stress amplitude, σ_T is the torsional stress amplitude, and σ_l is the tensile fatigue limit stress. Since the left side of Eq. (IX.4) is the square of the radius of Mohr's circle, this is a maximum shear stress criterion. One should note that this equation is only applicable for determining the boundary between fatigue failure and no failure. It cannot necessarily be used to calculate lifetimes for conditions where there are a finite number of cycles-to-failure.

IX-7 LOW–CYCLE FATIGUE

When the stress amplitude of a cycle becomes great enough, plastic deformation occurs on each cycle. A typical stress-strain cycle for such a test is shown in Fig. IX.8. Under these conditions, the life can best be correlated with the alternating plastic strain amplitude ϵ_a. When the logarithm of the alternating plastic strain amplitude is plotted as a function of the logarithm of the number of cycles-to-failure, the result is almost always found to be a straight line following the relationship

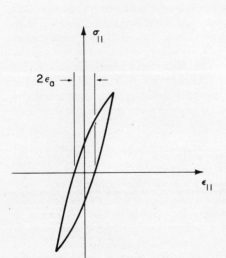

Fig. IX.8 Typical hysteresis curve for a low-cycle fatigue test. The width of the loop at zero stress is twice the plastic strain amplitude.

$$\epsilon_a N^{1/2} = C' \qquad \text{(IX.5)}$$

where N is the number of cycles-to-failure and C' is a constant. The plastic strain amplitude, ϵ_a, can either be measured directly during the experiment, as indicated in Fig. IX.8, or it can be estimated from the total strain amplitude, ϵ_t, using $\epsilon_a \approx \epsilon_t - \sigma_Y/E$. In many cases, the plastic strain amplitude is much larger than the elastic strain amplitude, and the plastic strain amplitude is nearly equal to the total strain amplitude. The constant C' can be related to the tensile fracture strain. If the tensile test is considered to be the extreme case of fatigue in one quarter cycle, and if the relationship in Eq. (IX.5) can be extended to this low value of N, then the constant in the equation is given by

$$\epsilon_f \left(\frac{1}{4}\right)^{1/2} = C' = \frac{\epsilon_f}{2} \qquad \text{(IX.6)}$$

where ϵ_f is the true plastic strain at failure in a tensile test. Using this value for C', quite excellent correlations with experimental results have been obtained.

EXAMPLE IX.3—Thermal Fatigue of Turbine Vanes

One of the most common types of low-cycle fatigue is thermal fatigue brought on by rapid heating of a part. Since the outside surface heats up faster than the interior, the surface will be put in compression. If the temperature graident is very steep, this compression layer is very thin and the stresses and strains in the interior are negligible. The elastic and plastic strains in the surface layer must compensate for the thermal expansion difference between the surface and the interior because the underlying material constrains the total strain to be zero. The surface should expand more than the interior only if large shears took place parallel to the surface. Therefore,

$$\epsilon^e + \epsilon^p + (\alpha T_S - \alpha T_I) = 0 \qquad \text{(a)}$$

where T_S is the surface temperature, and T_I is the interior temperature.

This type of failure is common in the stationary vanes of jet engines which experience thermal cycles each time that the power level of the engine is changed quickly, such as on takeoff. A turbine vane in a jet engine is made from Udimet 500. The blade is suddenly heated from 1400 to 1800°F when the power is increased from idle to takeoff power. The vane experiences the reverse cycle when the engine is shut down. If all other cycles are much less

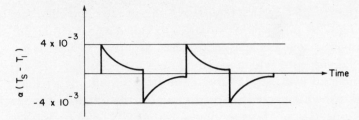

Fig. IX.9 Strain-time history for thermally shocked jet engine turbine vane.

severe, how many flights can the engine make before failure. The important properties of Udimet 500 in this temperature range are: yield strength, 40 ksi; tensile strength, 50 ksi; fracture strain (reduction of area), 25%; coefficient of thermal expansion, $10^{-5}/°F$; modulus, 20×10^6 psi.

Solution:

The cycle of $\alpha(T_S - T_I)$ is sketched above. The total strain amplitude ϵ_T is equal to the amplitude of the cycle $\alpha(T_S - T_I)$ (see Fig. IX.9). The elastic strain amplitude is approximately equal to $\epsilon^e \simeq \sigma_Y/E = 2 \times 10^{-3}$ where σ_Y is the tensile yield strength. Subtracting the elastic strain amplitude from the total strain amplitude to find the plastic strain amplitude gives

$$\epsilon_a = \alpha(T_S - T_I) - \frac{\sigma_Y}{E} = 2 \times 10^{-3} \tag{b}$$

From Eq. (IX.6)

$$N^{1/2} = \frac{\epsilon_f}{2\epsilon_a} \tag{c}$$

Solving for N

$$N = \frac{1}{4}\left(\frac{\epsilon_f}{\epsilon_a}\right)^2 = \frac{1}{4}\left(\frac{0.25}{2 \times 10^{-3}}\right)^2 = 3{,}900 \text{ cycles} \tag{d}$$

Although this value for N is somewhat high for low-cycle fatigue, Eq. (IX.6) usually gives good results for thermal fatigue in this range.

IX-8 FATIGUE WITH VARYING STRESS AMPLITUDE (CUMULATIVE DAMAGE)

Although tests are generally done with a constant stress amplitude so that each cycle of the test is identical to the others, service conditions often involve cycles of varying amplitude. The simplest rule for dealing with this type of situation is called Miner's rule. In Miner's treatment, each cycle is considered to have used a fraction of the fatigue life, which is equal to the reciprocal of the expected number of cycles-to-failure at the stress amplitude of that cycle. Failure is then expected when the sum of the fractions for all cycles equals one. If the fatigue obeys Eq. (IX.2) and n_i is the number of cycles which occur at a stress amplitude σ_i, Miner's rule states that failure occurs when

$$\sum_{i=1}^{k} \frac{n_i}{N_i} = \sum_{i=1}^{k} \frac{n_i}{C} \sigma_i^{m} = 1 \qquad \text{(IX.7)}$$

where C is the constant of Eq. (IX.2), N_i is the expected number of cycles-to-failure at the stress amplitude σ_i, and k is the total number of different amplitudes which occur. For a continuous spectrum of that any cycle will have a stress amplitude in the interval between σ and $\sigma + d\sigma$ is $p(\sigma)d\sigma$. Then, if the total number of cycles-to-failure is N_f, the number of cycles, dn, with amplitude in the interval $d\sigma$ is $dn = N_f p(\sigma)d\sigma$. Replacing the summation in Eq. (IX.7) by an integral and substituting dn for n_i gives the formula for Miner's rule for continuous distributions of the stress amplitude

$$N_f \int_{\sigma_l}^{\sigma_{\max}} \frac{p(\sigma)}{C} \sigma^m \, d\sigma = 1 \qquad \text{(IX.8)}$$

where σ_l is the fatigue limit and $\sigma_{\max}$ is the maximum stress amplitude with a greater-than-zero probability. (If the material has no fatigue limit, σ_l is replaced by zero.) If, for a certain vibration, the stress amplitude probability distribution is known, the expected lifetime can be calculated.

The phenomenon called coaxing and the effect of occasional overloads during the fatigue life give results which are contrary to Miner's rule. In coaxing, a part is cycled for a large number of cycles just below the fatigue limit. Miner's rule predicts no effect from cycles below the fatigue limit. For materials which show the coaxing effect, the subsequent fatigue behavior is improved. It is believed that the coaxing effect is associated with the phenomenon of strain aging. Strain aging is a process by which the action of the stress accelerates the rate of precipitation of a hardening phase, or the precipitation of impurities onto dislocations. The application of the stress cycles therefore increases the strength of the material so that it can better resist fatigue damage when the stress amplitude is raised.

Miner's rule predicts that occasionally overloading a part which is subjected to fatigue will use up part of its fatigue life and therefore shorten the life. In many cases it is found that occasionally overloading a part for a single half-cycle in tension actually improves the fatigue life. The probable explanation for this phenomenon is that the tensile overload causes plastic flow at the fatigue cracks which have been nucleated. This blunts them and slows their growth. As might be expected from this explanation, overloading the part in compression reduces the fatigue life.

EXAMPLE IX.4—Fatigue of Drill Pipe in Drilling
Operations from a Floating Vessel

The drill pipe used in drilling operations from a floating ship is bent by the ship's roll. Since the pipe is being rotated during the drilling, it experiences a fatigue cycle for each rotation. In addition to the bending stress the pipe also carries a static tension from the weight of the pipe below. (The torsional stress is usually much less than the tension and bending stresses.) Since the roll of the ship depends on the wave spectrum of the sea, the bending stress amplitude is statistical in nature. Consider a case where the water depth is 5,000 ft. The maximum stress occurs just below the ship; therefore, calculations will be made for that section of the pipe. The bending stress (σ_B) amplitude distribution is calculated to be

$$p(\sigma_B) = A \exp(-A\sigma_B) \quad \text{for} \quad \sigma_B \leq \sigma_{\max} = 150,000 \text{ psi} \qquad \text{(a)}$$

where $A = 10/\sigma_{\max}$ and $\exp(-A\sigma_{\max}) \ll 1$.* The cutoff at high stress results

---

*The probability distribution for the stress depends on the ship and the pipe characteristics, and the wave height and frequency. The function chosen here to describe the stress amplitude probability distribution is chosen mainly for computational simplicity and because it correctly predicts that the probability drops rapidly as the stress amplitude increases. A better, although more unwieldy, description of the probability function in the low-stress region would be obtained with a Poisson distribution, which predicts maximum probability for some value greater than zero. However, since most of the fatigue damage results from the higher stress cycles, the approximation used here should be reasonably accurate.

from the fact that drilling operations will be stopped if the sea gets too rough. The pipe is type 135 drill pipe for which data for fatigue in seawater is given in Fig. IX.13 (see Sec. IX-9e). What is the expected life of the section of pipe just below the ship? The tensile strength of the pipe, σ_u, is 160,000 psi.

Solution:

The mean stress σ_m in the pipe due to the weight of the pipe below is

$$\sigma_m = L(\rho_{\text{steel}} - \rho_{\text{water}}) = 14,700 \text{ psi} \tag{b}$$

Using the Goodman relationship, the equivalent zero-mean-stress amplitude σ_{a0} is given by

$$\frac{\sigma_B}{\sigma_{a0}} + \frac{\sigma_m}{\sigma_u} = 1 \tag{c}$$

$$\sigma_{a0} = \frac{\sigma_B}{1 - \sigma_m/\sigma_u} = \frac{\sigma_B}{1 - 14,700/160,000} \tag{d}$$

$$\sigma_{a0} = 1.1\sigma_B \tag{e}$$

Therefore, by substituting for σ_B in Eq. (a) one obtains the probability distribution for σ_{a0}.

$$p(\sigma_{a0}) = A \exp\left(-\frac{A\sigma_{a0}}{1.1}\right) \tag{f}$$

From Eq. (IX.8)

$$N_f \int_{\sigma_l}^{(\sigma_{a0})_{\text{max}}} \frac{p(\sigma_{a0})}{C} \sigma_{a0}{}^m \, d\sigma_{a0} = 1 \tag{g}$$

Substituting Eq. (f), we have

$$N_f = \frac{1}{\displaystyle\int_{\sigma_l}^{1.1\sigma_{\text{max}}} \frac{A\exp(-A\sigma_{a0}/1.1)}{C} \sigma_{a0}{}^m \, d\sigma_{a0}} \tag{h}$$

In seawater the fatigue limit disappears. Thus, it may be assumed that $\sigma_l = 0$. The integral becomes

$$\frac{A}{C} \int_0^{1.1\sigma_{\text{max}}} \sigma_{a0}{}^m \exp\left(\frac{A}{1.1}\sigma_{a0}\right) d\sigma_{a0} \tag{i}$$

Before evaluating the integral, m must be known. Evaluating m and C from Fig. IX.13, one finds

$$m = 10 \qquad C = (10^5\,\text{psi})^{10}\,10^{5.7} \tag{j}$$

When m is an integer, the integral in Eq. (g) can be reduced to a sum of a finite number of terms. When this is evaluated numerically, the result is

$$N_f = \frac{1}{(A/C)(1.1\sigma_{max})^{11}1.46 \times 10^{-5}} \tag{k}$$

where A is $10/\sigma_{max}$. Evaluating Eq. (k) gives

$$N_f = \frac{1}{3.72 \times 10^{-8}} = 2.7 \times 10^7 \text{ cycles} \tag{l}$$

IX-9 OTHER FACTORS WHICH AFFECT FATIGUE IN METALS

IX-9-a Stress Concentrations

Since fatigue fracture is a local process, it is expected that it will be sensitive to any stress concentrations. This is indeed found to be the case and many fatigue failures are found to originate from stress concentrations such as from the keyways of a shaft, shoulders, and holes. Because of the fact that fatigue begins at discrete sites on the sample where microscopic structural conditions are favorable, stress concentrations are not fully effective. For example, in the case of a very sharp notch, it is unlikely that the metallurgical conditions will be suitable for crack initiation at the point where the stress is greatest. It is more likely that a fatigue crack will begin at some point near the notch where the stress is still above the nominal but is not as great as at the maximum. As the notch becomes more blunt, the probability of having suitable conditions for fatigue where the stress is near the maximum increases as the volume of the material subjected to the maximum stress increases. For this reason there is a size effect for stress concentrations in fatigue. For stress concentrators of small physical size such as sharp notches in small parts, the effective stress concentration evaluated by the effect on the fatigue life is less than the theoretical elastic stress concentration. As the size

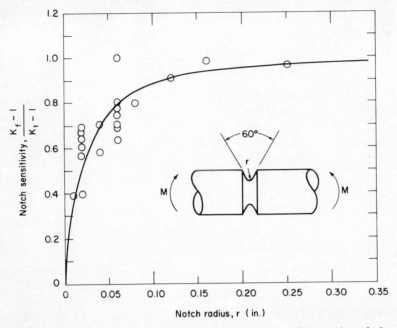

Fig. IX.10 Size effects on notch senstitivity in fatigue. The severity of the effective stress concentration in fatigue increases with increasing notch size. Data for normalized 1020 steel tested in rotating bending with circumferential grooves. (*From R. E. Peterson, Notch Sensitivity, in G. Sines and J. L. Waisman, "Metal Fatigue," p. 293, McGraw-Hill Book Company, New York, 1959.*)

of the stress concentrator increases, the effective stress concentration approaches its theoretical value. Figure IX.10 shows an example of this effect. In these figures, K_f is the apparent stress concentration factor evaluated from the reduction of fatigue life, and K_t is the theoretical elastic stress concentration. The factor K_f is evaluated from the fatigue law and the nominal applied stress amplitude. If the sample fails after N cycles, and the material has an S-N curve described by $\sigma_a{}^m N = C$, the effective stress amplitude is $\sigma_a = (C/N)^{1/m}$. If the nominal stress amplitude is $\sigma_{a_{\text{nom}}}$, then $K_f = \sigma_a/\sigma_{a_{\text{nom}}}$.

IX-9-b Surface Quality

The surface quality of machined parts affects the fatigue resistance of metals. The problem of controlling the surface quality has become important in the aerospace industry as structural metals* with higher and higher strengths are employed. The two most important surface qualities are *surface roughness* and *surface integrity*. Surface roughness is a measure of the surface topography, while surface integrity denotes the relative freedom from any metallurgical and mechanical alterations in the surface layer. Machined parts with poor surface quality can have as much as 40% lower fatigue life than parts which are properly machined.

A rough machined surface can provide stress concentration points for localized crack nucleation. Similarly, built-up edge (BUE)† adhering on the machined surface can also lower fatigue life by providing stress concentration points. However, simply making certain that the machined surface is smooth cannot guarantee good fatigue resistance, because the machining process can damage the surface layer. Therefore, the integrity of the surface layer must also be considered.

Surface alterations are introduced during materials processing in the form of

1) Mechanical damage such as microcracks and macrocracks in the surface layer due to the plastic deformation of the layer, and also tears and laps associated with BUE and burrs. Cracks are most likely to form at inclusions.

2) Metallurgical changes in the surface layer due to excessive heating and uncontrolled quenching of the layer.

These surface alterations can be introduced when dull cutting tools are used or when grinding is done abusively at excessively high speeds and heavy feeds. Metallurgical changes are also introduced during electric discharge machining (EDM) because of melting and recasting of the molten metal on the surface.

One of the metallurgical changes which may occur is the formation of hard and brittle untempered martensite from the excessive heating and sudden quenching of the surface layer by the base material or by

*Steels commonly used for aerospace applications, such as in landing gears and wing assemblies, are AISI 4340, 4130, 4330 (modified), D6AC, and 18% Ni maraging steels. These steels, when heat treated have yield strengths ranging from 150,000 to 400,000 psi. This gives very high strength-to-weight ratios, equal to those of titanium alloys and better than those of aluminum, magnesium, and nickel base alloys.

†BUE is highly deformed work material which attaches itself to the rake face of the cutting tool and acts as the cutting edge during machining. The BUE will sometimes detach from the tool and adhere to the surface of the work piece. BUE is normally a problem at intermediate cutting speeds.

the cutting fluid. The layer of untempered martensite is shown in Fig. IX.11(b) and is called "white layer." Cracks often form in these hard untempered martensite surfaces because of the low ductility of the martensite and the high residual stresses caused by the transformation. Another phase change which may occur during machining is the formation of an overtempered martensite layer beneath the untempered layer. This overtempered layer is softer than the base metal due to the additional tempering which takes place when it is heated during a cutting or grinding process (see Fig. IX.11). A soft layer can also form at the surface of a maraging steel by the process of "austenite reversion" when the material returns to its austenitic state as a result of the high temperature generated locally by the machining process. Electric discharge machining (EDM) can also produce a damaged surface, because the process involves the local

(a)

Fig. IX.11 Surface characteristics of AISI 4340 (quenched and tempered, 50 R_c) produced by grinding. (a) Gentle conditions—no visible surface alterations in microstructure were detected. 1000X.

(b)

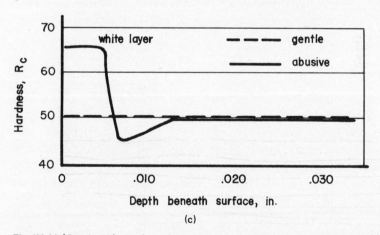

(c)

Fig. IX.11 (*Continued*) Surface characteristics of AISI 4340 (quenched and untempered, 50 R_C) produced by grinding. (b) Abusive conditions—a total heat-affected zone of 0.013 in. was produced. A white layer or rehardened martensite at 65 R_C up to 0.005 in. thick was followed by an overtempered zone with a minimum hardness of 46 R_C. 1000X. (c) Hardness variation beneath the surface. (*From M. Field and W. P. Koster, Surface Integrity in Conventional Machining-Chip Removal Processes, ASTME Paper No. EM68-516, Dearborn, Mich., 1968.*)

melting and recasting of the material at the surface. The recast layer is often porous and cracked.

In addition to the formation of cracks and metallurgical changes, machining processes also impart residual stress to the surface. These residual stresses will also change the fatigue behavior of the material. When the surface is in compression, the fatigue life improves, whereas the fatigue life deteriorates when the surface is in tension. Excessive heating and sudden quenching of the surface layer during abusive grinding and EDM leave a surface layer with a residual tensile stress. In these machining processes, the metal at the surface is raised very nearly to or to the melting temperature of the metal. When the layer cools down, tensile stresses are set up due to the thermal contraction. Milling and turning operations as well as gentle grinding processes leave a surface layer with compressive stresses. Electro-chemical machining (ECM) and electropolishing (ELP) leave a stress-free surface; also, in some cases, they preferentially etch fine cracks.

The combined effects of various grinding processes and ELP on the fatigue life of 4340 steel is shown in Fig. IX.12(a). The corresponding residual stress patterns are shown in Fig. IX.12(b).

Surface quality can be significantly improved by employing gentle machining operations, eliminating BUE, surface scratches and laps, and by shot peening the surface after machining. Shot peening leaves the surface in compression.

IX-9-c Temperature

The effect of temperature on fatigue life roughly parallels the effect of temperature on the yield strength or the tensile strength. At very high temperatures the effect of creep must be included and atmospheric effects are much more important. The fatigue limit in iron alloys disappears at temperatures in the hot creep stage. For structural materials around room temperature, the temperature effect is small and usually ignored.

IX-9-d Speed of Testing

The effect of strain rate on fatigue behavior is similar to the strain rate dependence of the flow stress. At room temperature this is negligibly small. At very high cycle rates the plastic work can cause a temperature rise in the sample. The strain rate is therefore coupled with the temperature effect. For reasonable temperature rises, the combined effect is small in metals. This result has considerable practical importance in that it allows accelerated-rate laboratory testing.

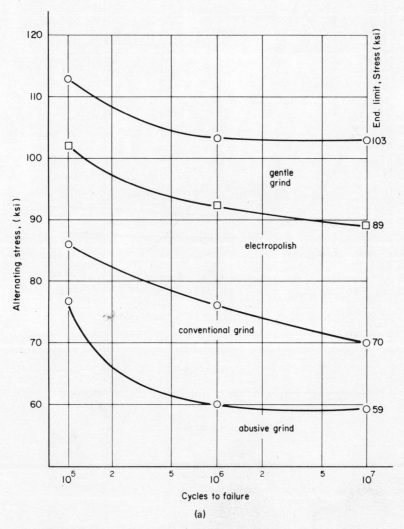

Fig. IX.12 (a) Effect of various machining processes on the fatigue life of 4340 steel quenched and tempered to give a hardness of Rockwell C 50.

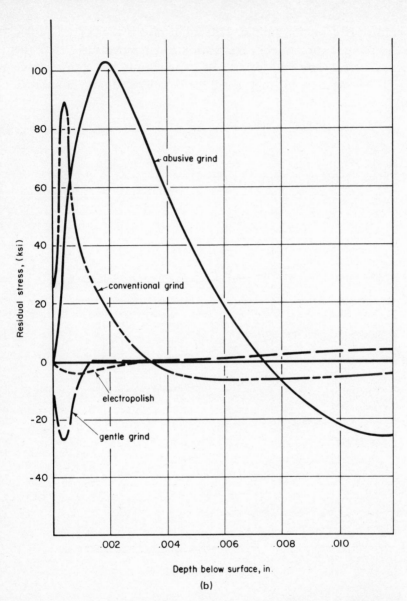

Depth below surface, in.

(b)

Fig. IX.12 (*Continued*) (b) Residual stresses in the surface layer 4340 steel produced by various machining processes. (*From M. Field et al., "Machining of High-Strength Steels with Emphasis on Surface Integrity," Air Force Machinability Data Center, Cincinnati, Ohio, 1970.*)

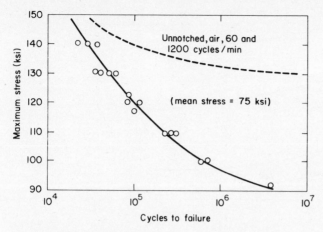

Fig. IX.13 Fatigue behavior of unnotched specimens of type 135 drill pipe tested in deaerated seawater at 60 cpm (solid line). (*From D. E. Pettit et al., Effect of Environment and Complex Load History on Fatigue,* ASTM Special Technical Publication No. 462, *1970.*)

IX-9-e Atmospheric Effects

When fatigue takes place in a corrosive medium such as salt water, a coupling takes place between corrosion and fatigue. The two processes together cause failure to occur much faster than with either process separately. The understanding of the process of corrosion fatigue is very limited. Very few principles seem to have general applicability. Therefore, design for fatigue in the presence of a corrosive medium must be based on experimental results for the specific material and environment expected in service (see Fig. IX.13). It should be emphasized that even a mildly corrosive medium can have a substantial effect on the fatigue behavior of some materials. The presence of a mildly corrosive substance can, in some cases, reduce the fatigue life by several orders of magnitude.

IX-10 FATIGUE IN NONMETALS

The process of fatigue seems to be a general problem common to all materials which can deform plastically. It is therefore observed in polymers and nonmetallic crystalline materials which can deform at elevated temperatures. It does not seem to occur in glasses or brittle

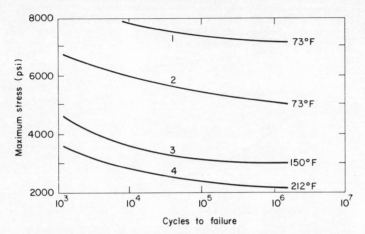

Fig. IX.14 Fatigue behavior of Delrin (polyacetal). (1) Tensile stress only (0 to maximum stress); (2), (3) and (4) completely reversed tension-compression. (Maximum stress equals alternating stress amplitude). (*From C. C. Osgood, "Fatigue Design," Wiley-Interscience, New York, 1970.*)

ceramics.* Fatigue in polymers is quite similar to that in metals. Polymer *S-N* curves show many of the same general features as those for metals, especially the rapid increase in life with decreasing stress

*This is, in spite of the terminology, used to describe *static fatigue* in glass. Static fatigue is the phenomenon in which glass, carrying a tensile load for a period of time, eventually cracks without an increase in the load. This effect is believed to be the result of the stress-enhanced corrosion of the glass by water vapor in the air. It is not phenomenologically similar to the fatigue of metals.

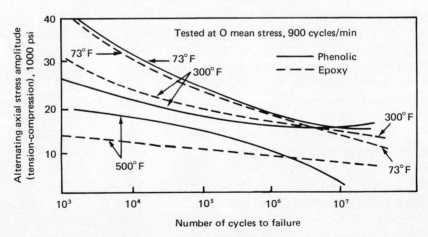

Fig. IX.15 Fatigue behavior of glass-fiber reinforced phenolic and epoxy. (*From J. W. Davis, J. W. McCarthy, and J. N. Schurb, The Fatigue Resistance of Reinforced Plastics, Materials Des. Eng., vol. 60, no. 7, p. 91, 1964.*)

amplitude. Because of the nature of deformation in polymers, fatigue in these materials is much more dependent on temperature and speed of testing than in metals. The experimental difficulties introduced by the speed and temperature dependences, along with the relatively short history of the use of polymers as structural materials, account for the limited extent of available data on fatigue of polymers. Figures IX.14 and IX.15 show typical fatigue curves for a solid polymer and a glass-reinforced polymer matrix composite material.

REFERENCES

1. American Society for Testing and Materials, Fatigue Crack Propagation, *American Society for Testing and Materials Special Technical Publication No. 415*, Philadelphia, 1967.
2. Averback, B. L., D. K. Felbeck, G. T. Hahn, and D. A. Thomas (eds.): "Fracture," MIT Press and John Wiley, New York, 1959.
3. Freudenthal, A. M. (ed.): "Fatigue in Aircraft Structures," Academic Press, New York, 1956.
4. Kennedy, A. J.: "Processes of Creep and Fatigue in Metals," Oliver and Boyd, Edinburgh, 1962.
5. Madayag, A. F.: "Metal Fatigue: Theory and Design," John Wiley, New York, 1969.
6. McClintock, F. A., and A. S. Argon: "Mechanical Behavior of Materials," Addison-Wesley, Reading, Mass., 1966.
7. Osgood, C. C.: "Fatigue Design," Wiley-Interscience, New York, 1970.
8. Polakowski, N. H. and E. J. Ripling: "Strength and Structure of Engineering Materials," Prentice-Hall, Englewood Cliffs, N. J., 1966.
9. Sines, G., and J. L. Waisman: "Metal Fatigue," McGraw-Hill Book Company, New York, 1959.

PROBLEMS

IX.1   A support bracket for a piece of machinery, with dimensions as shown in the accompanying figure, is made from a high-strength steel. The bracket is subjected to tension-release loading of unknown but reproducible amplitude. The bracket fails in fatigue after a period of use. Examination of the fracture surface reveals that the fatigue crack had grown only a short way into the part before the final failure occurred. The dimensions of the fatigue crack are shown in the figure.
The material has the following properties:

Tensile strength	190,000 psi
Yield strength	160,000 psi
Fatigue limit	90,000 psi
Critical stress intensity factor k_{1c}	85,000 psi-in.$^{1/2}$

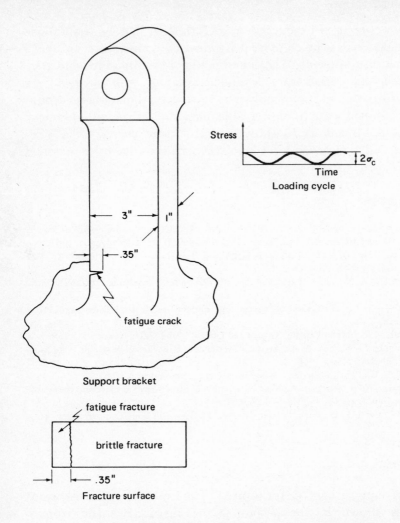

Support bracket

Fracture surface

a) From the critical crack length, determine the magnitude of the stress amplitude. (Specify the model you use for k_1.)

b) Construct an approximate S-N curve from the data above and estimate the number of cycles which occurred before failure.

c) How must the stress amplitude be reduced in order to eliminate the fatigue problem?

IX.2 In an ideal bolted joint where the pieces joined are essentially rigid relative to the bolt itself, the force in the bolt is constant until the applied load F_a exceeds the initial preload in the bolt. (See the figures for a more complete description of the forces in the bolted joint.)

A bolted joint in which the bolts have been preloaded by tightening them to a torque of 50 ft-lb is subjected to an oscillating applied load of 18,000 lb per bolt. The bolts in the joint are found to fail after only 100,000 cycles. The bolts are made from 1040 steel, which has a tensile strength of 100,000 psi and a fatigue limit of 50,000 psi.

a) Construct an approximate *S-N* curve (or give an approximate fatigue life equation for the steel) and determine the *equivalent* stress amplitude in the bolt.

b) Estimate the actual preload stress in the bolt and sketch the actual fatigue cycle.

c) If the preload stress in the bolts is proportional to the tightening torque, to what torque must the bolts be tightened to eliminate the fatigue problem?

The dimensions of the bolt are given in the figure. The relief of the diameter of the shank of the bolt is sufficient to cause failure in the straight section away from the threads.

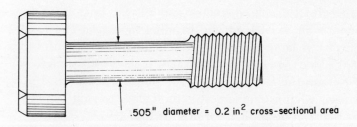

.505" diameter = 0.2 in.^{2} cross-sectional area

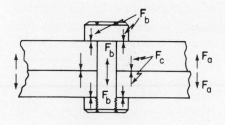

In a bolted joint, the force F_b in the bolt is always equal to the sum of the applied loads F_a and the contact load F_c between the plates. If the plates are rigid, the length of the bolt cannot change unless the two plates come out of contact (i.e., after F_c goes to zero). Therefore, the strain and stress in the bolt are constant for F_a less than F_p, the preload in the bolt. The contact force F_c between the plates is equal to the preload F_p when the applied force is zero. The sketch below shows the behavior of F_c and F_b as the applied load is increased.

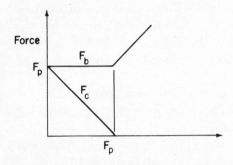

Applied force, F_a

IX.3   A steel shaft 2 in. in diameter and 30 in. long is to be subjected to an axial load and bending moment. The axial load may be as large as the critical buckling load when both ends of the shaft are simply supported. If the yield strength is 50,000 psi and the fatigue limit is 38,000 psi when there is no mean stress, what is the largest completely reversed bending moment that can be applied for infinite life?

IX.4   A drive shaft made of 4340 steel, supporting a rotating flywheel in an erosion apparatus, failed first by fatigue and then catastrophically by fracture. It was assumed that the bearings used were not long enough to prevent dynamic whirl in the mode shown. The bearings are approximated as pinned joints for the dynamic analysis given.

Dynamic considerations indicate that the shaft, while rotating at the velocity Ω (shaft speed), will whirl in the shown mode at the velocity ω_{whirl}. A reaction moment M_B is required to cause the angular momentum vector to change in the assumed manner. If $\omega_{\text{whirl}} \neq \Omega_{\text{shaft}}$, the shaft "sees" a

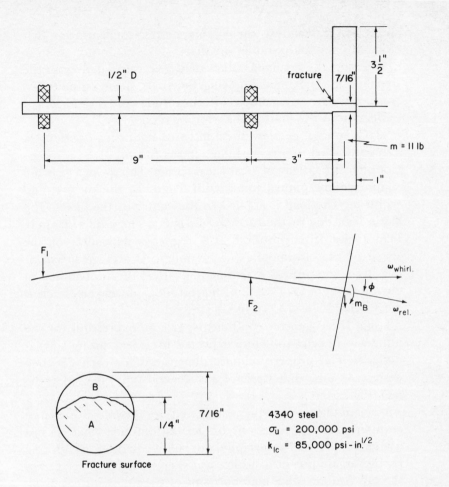

1/2" D

fracture   7/16"

$3\frac{1}{2}$"

m = 11 lb

9"

3"

1"

F_1

$\omega_{whirl.}$

F_2

ϕ

m_B

$\omega_{rel.}$

B

A

7/16"

1/4"

4340 steel

σ_u = 200,000 psi

k_{Ic} = 85,000 psi - in.$^{1/2}$

Fracture surface

rotating (cycling) bending moment of magnitude M_B. It is this moment that causes the fatigue failure.

The failure occurred at the necked-down region of the shaft where the flywheel is attached. Region B of the fracture surface was smooth and shiny, and Region A was rough (see the attached figures).

1) Estimate the magnitude of M_B which caused fracture after fatigue.

2) Making assumptions about the fatigue behavior of 4340 steel, and assuming the moment M_B to be constant during crack propagation, was M_B sufficient to cause fatigue of the

shaft? If not, what was the *minimum* stress concentration that existed at the sharp step in the shaft?

3) Using the assumed values of ω_{whirl} calculated from the dynamic analysis, and knowing the shaft speed to be 9,000 rpm, estimate the order of magnitude of the angle of deflection of the shaft in the whirling mode.

4) Comment on a few possibilities of stabilizing or strengthening the shaft to prevent failure in the future.

IX.5 Consider the fatigue of a car's leaf spring. The spring is a beam supported by pinned joints with length L, thickness t, and width w. The load is applied to the center of the beam. The static load due to the car's weight is F_0. The road will apply load cycles of magnitude F at a frequency depending on the speed of the automobile (i.e., N/mile). At average speeds, it has been determined experimentally that the probability of encountering a load cycle of magnitude F during any cycle is $p(F) = Ae^{-BF}$.

Using these general coefficients, give an expression for the number of cycles-to-failure expected for a leaf spring so that designers can properly choose dimensions for each car they design. Assume that steel (of any general type) is used as the material.

IX.6 An accumulator is being considered to provide the energy required for acceleration of an automobile that has a hydraulic drive. The accumulator allows braking by storing energy as air pressure, which can then be used to accelerate the car. Among other materials, one of the following maraging steels is being considered:

	Grade	
	200	300
Yield strength (0.2% offset), ksi	215	290
Tensile strength, ksi	225	290
Elongation (% in 2 in.)	12	8
Reduction of area (rect. cross sect.), %	40	30
Rockwell C hardness	46–50	52–56
Critical stress intensity k_{Ic}, ksi-in.$^{1/2}$	100	60

It is proposed that the accumulator be a cylinder with an 8-in. ID and a length of 12 in. between hemispherical ends.

The working pressure is to be 4,000 psi. Conventionally, the working stress is half the tensile strength, based on thin-walled-cylinder equations. Check this procedure by carrying out the following calculations.

a) What is the required thickness?

b) The cylinder is inspected periodically by a method that detects cracks whose depth is of the order of 10% of the thickness. Under repeated cyclic loading from 50 to 100 percent of the rated working stress, how long will it take such a crack to grow to one-third of the thickness? Assume that the cracks are wide enough compared to their depth to be in plane strain and that the crack growth per cycle is the crack tip opening displacement, which is estimated to be 0.01 R_p.

IX.7 An aluminum alloy has an endurace stress of σ_e at a life of 10^8 cycles. In service it is subjected to a nonuniform alternating stress. A spectral analysis of the stress amplitudes of the service load discloses that 80% of the load cycles are at the endurance stress σ_e, while 10% each take place at $1.1\sigma_e$ and $1.2\sigma_e$. If σ_e is 30% of the tensile strength, calculate the number of cycles the part will survive during service.

IX.8 In order to eliminate side-to-side vibrations during fatigue testing in tension using an SF-2U fatigue testing machine, it is proposed that the top of the push rod be constrained by a spring, as shown in the figure below. The spring is in the form of a beam built in at both ends working in bending. Since the fatigue machine will accumulate many million cycles, it is

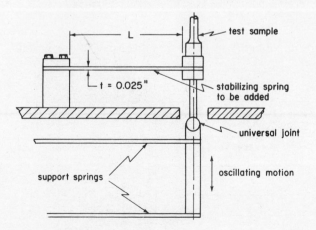

necessary that the spring be stressed below the fatigue limit. The spring is to be made of spring steel with a tensile strength equal to 200,000 psi. The thickness of the beam is 0.025 in. The maximum amplitude for which the machine is designed is 0.25 in. Estimate the value of L which will give the required life. During the tuning operation required to set up the machine, the amplitude sometimes exceeds 0.25 in. This amplitude is also exceeded for a few cycles when the sample breaks. Assume that the machine runs for 1,000 cycles at an amplitude of 0.32 in. during tuning and for 10 cycles at an amplitude of 0.50 in. when the sample breaks for each test. How many tests can be run before the spring fails assuming that the spring was designed to be working at its fatigue limit for an amplitude of 0.25 in.?

IX.9   In a hydraulic turbine pump the actuating levers which control the position of the wicket gates are provided with notched shear pins for overload protection. During normal operation the shear pins are subjected to fluctuating shear loads which can produce fatigue crack growth out of the notches and eventual failure.

The dimensions of the tubular shear pins are shown in the sketch below. The pins are made of a deep-hardening 4340 steel which has, in the heat-treated condition, a critical fracture toughness $k_{1c} = 38,000$ lb-in.$^{-3/2}$ and a fatigue crack propagation law of

$$\frac{dc}{dn} = A(k_a)^4$$

in mode I loading, where k_a is the stress intensity factor based

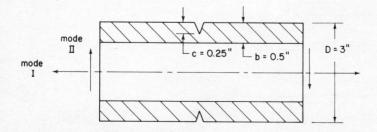

on the amplitude of the tensile stress, and $A = 1.7 \times 10^{-23}$ in.^{7} lb^{-4} cycle^{-1} is the scale factor of the crack propagation rate.

Based on this available information calculate the nominal stress amplitude for a 10^5-cycle fatigue life in the shear pin, assuming that the shear case can be approximated by mode I loading in tension.

IX.10 An underground high-voltage power transmission line consists of a central conductor in the shape of a 4-in.-OD thick-wall tube made of type 1100 aluminum (in the fully annealed condition), coaxially located inside a large-diameter gas-filled aluminum pipe. The current conducting tube is positioned in the center of the pipe by a number of equally spaced, "frictionless," insulated sleeve bearings. The annular gap between rod and pipe is filled with a noncorrosive, electrically insulating gas. The general design of the transmission line is shown in Fig. 1 on page 502. When the transmission line is assembled, buried underground, and reaches thermal equilibrium with the ground at 5°C, it is completely unstressed if it carries no current. Under peak-power operation, the temperature of the center conductor could reach 60°C, while under slack condition, the temperature is not much above the 5°C of the ground temperature. The outer pipe is assumed always to remain at the ground temperature of 5°C. It is anticipated that there would be one power cycle per day.

To prevent buckling of the center conductor when it is hot, expansion joints of the type seen in Fig. 2 are to be introduced at regular intervals along the shaft. These joints consist of a series of split washers which are dip brazed to the two half-moon shaped extensions of the conductors. As shown in Fig. 2, when the conductors expand, the ring washers undergo double flexure. Consider the dip-brazed ends of the split washers at points B as being built-in at their two ends into two parallel walls which can undergo relative translation without change of separation, as shown in the last sketch.

For the given dimensions of the expansion joint, calculate the maximum allowable distance between joints for a minimum fatigue life of 50 years. The coefficient of expansion of

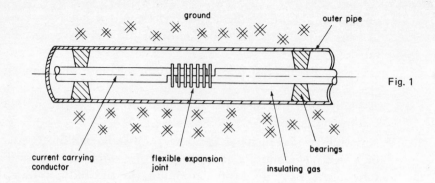

Fig. 1

ground

outer pipe

current carrying
conductor

flexible expansion
joint

insulating gas

bearings

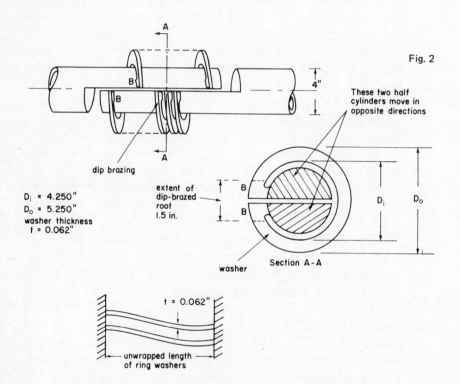

Fig. 2

A

B

B

4"

These two half
cylinders move in
opposite directions

A

dip brazing

$D_i = 4.250"$
$D_o = 5.250"$
washer thickness
$t = 0.062"$

extent of
dip-brazed
root
1.5 in.

B

B

D_i

D_o

washer

Section A-A

$t = 0.062"$

unwrapped length
of ring washers

aluminum is $\alpha = 24 \times 10^{-6}\,^{\circ}\mathrm{C}^{-1}$. For all other materials data,
consult your text or handbooks.

Discuss whether or not your answer is conservative and
what other problem areas you would want to consider to

make certain that your answer will not lead to premature failure.

IX.11 A cylindrical pressure vessel of diameter 3 in. and wall thickness 0.3 in. is fabricated by forming curved sections of A533-B steel and welding them together with longitudinal welds and capping the cylinder with hemispherical ends. The vessel is subjected to a pressure which cycles between zero and 3,000 psi. Because of imperfections in the weld, it is expected that the pressure vessel will fatigue and that the cracks will start in the weld and propagate along the axis of the cylinder perpendicular to the maximum-stress direction. To ensure safety, the cylinder is to be proof tested to a higher pressure every 1,000 cycles. Data for crack growth rates in A533-B steel are given in Fig. IX.4. In order to ensure that failure does not occur in the subsequent 1,000 cycles, what should the test pressure have to be? You may assume that failure occurs when a crack reaches critical size for fracture at 3,000-psi pressure and that the test cycle does not affect the subsequent crack growth rate. (Is this last assumption conservative or nonconservative?) The factor k_{1c} for A533-B steel is $45,000 \text{ psi} = \text{in.}^{1/2}$.

IX.12 To allow for an internal drive shaft, the axle of a riding lawn mower was made from tubular steel. The design and dimensions are shown on the sketch below. Due to the rotation of the axle, a cyclic stress is applied to it. What is the maximum

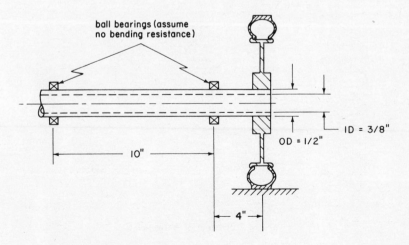

ball bearings (assume no bending resistance)

OD = 1/2"

ID = 3/8"

10"

4"

load F that can be carried by the wheel, such that the fatigue life of the axle is infinite. The factor $K_f = 2$ and $\sigma_u = 70 \times 10^3$ psi.

IX.13 Tests on the steering knuckle of an automobile traveling at a speed of 50 mph on a moderately rough cobblestone test track show that it is subjected to a loading cycle which gives 10^5 cycles per mile, with the probability density distribution for the load amplitudes of the cycle given by

$$p(F) = \frac{F}{s^2} \, e^{-F^2/25^2}$$

where $p(F)$ is the probability distribution that a cycle will have an amplitude of force F, and s is the variance of the distribution with a value of 350 lb. The steering knuckle is to be made from 1040 steel with a tensile strength of 85,000 psi and a fatigue limit of 40,000 psi. The steering knuckle is designed so that a force of 1 lb gives a maximum stress of 40 psi. How many miles can the car be driven over roads similar to the test track before the steering knuckle breaks?

IX.14 Metal bellows are often used for seals for shafts which transmit a small amplitude reciprocating motion, as, for

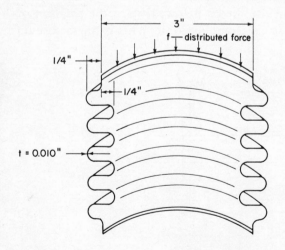

Spring constant:

k = 30 lb/in. of deflection

example, seals for valve stems in vacuum systems. The bellows shown above is made from 304 stainless steel. Its maximum deflection is 1 in. and its spring constant is 30 lb/in. It is operated so that it cycles between its free length L_f and a compressed length $L_f - 1$ in. What is its expected fatigue life?

IX.15 In the paper punch described in Prob. III.11, the punching fingers which are blocked by a gate, where the holes are not desired, are forced to buckle to accommodate the motion of the punch head. The displacement which must be accommodated by the buckling is 2 mm. The punch fingers are made from steel with a tensile strength of 1.7×10^9 N/m². Estimate the fatigue life of a punch finger.

IX.16 The tungsten filament of a vacuum tube operates at a temperature of 1600°C. The bending stress in the filament due to its own weight is approximately 3,000 psi. Because of the steep temperature gradient and geometrical constraints at the welds to the filament support, there is a total change in strain of approximately $1/2(\alpha\Delta T)$ [i.e., $\Delta\epsilon^e + \Delta\epsilon^p = 1/2(\alpha\Delta T)$]. The filament will eventually fail, either from fatigue caused by the off-on cycles, or because creep causes excessive sag in the filament. (You can assume from spring theory that 2% strain constitutes failure.) Give expressions for the lifetime of the tube before failure by each of the two mechanisms. If the instrument which contains the tube is used

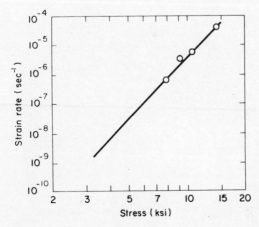

Fig. 1 Second-stage creep strain rate for tungsten at 1600°C.

for a period of 8 hr each day, is it better to leave the instrument on when not in use, or to turn it off? The properties of tungsten are:

$$E = 5 \times 10^7 \text{ psi}$$
$$\alpha = 4.5 \times 10^{-6}/°C$$
$$\epsilon_f = 0.2$$
$$\left. \begin{array}{l} \sigma_Y = 24 \text{ ksi} \\ \sigma_u = 36 \text{ ksi} \end{array} \right\} \quad \text{(at temperature of interest)}$$

See Fig. 1 for creep data for tungsten.

IX.17 Consider a steel ring of rectangular cross section which is located by two internal rollers. These rollers simultaneously rotate the ring with angular velocity ω and load it with a force F, as shown. Determine the maximum force that can be applied for infinite fatigue life. The material properties are given in Prob. IX.1.

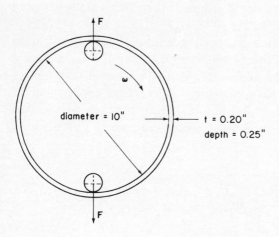

CHAPTER TEN

Surface Phenomena

X-1 INTRODUCTION

X-1-a Engineering Importance of Surface Phenomena

The subject matter of the preceding nine chapters was concerned with the bulk behavior of materials. An understanding of bulk behavior is important in analyzing the deformation of materials under a given set of external loads and in determining the fracture load of structural parts. Attention will now shift to another aspect of the mechanical behavior of materials—the study of surface phenomena. In particular, friction, wear, adhesion, and lubrication of material surfaces will be discussed.

In choosing engineering materials for such applications as bearings, cams, sliding surfaces, shafts, turbine blades, helicopter rotors, crude oil pipelines, fountain pen points, and automobile tires, the surface properties of materials must be considered in conjunction with the bulk properties. For some applications, the surface characteristics may be the decisive factor in material selection. The importance of the subject matter can hardly be overemphasized. Yet, in spite of their significance, surface phenomena have not been studied as extensively as the bulk properties of materials. Consequently, many aspects of the subject matter are still controversial, in the sense that there are many differing views on the mechanisms involved in specific surface phenomena.

X-1-b Overview of the Friction and Wear of Materials

From experience with sliding surfaces, it is known that it takes a nonzero force to slide one solid surface over another surface. In other words, work is done between sliding surfaces. All this work is somehow consumed at the interface, while the bulk of the solid remains essentially unchanged. The question then is how the work is consumed at the surface. There are three major modes of energy consumption operating between the sliding surfaces. If any plastic and viscous deformations are involved at the interface, a part of the work done is transformed into thermal energy which raises the interface temperature. In many applications this mode of energy consumption is most important. The work done may also be consumed in creating new surfaces or interfaces. This occurs when wear particles and internal voids are generated and when the interface deforms. Finally, it may be stored in the form of residual elastic strain energy at the surface. The relative importance of these three energy consumption mechanisms depends on the process being considered, but only in unusual cases would the energy absorbed by creation of surface or strain energy exceed that dissipated as heat.

Interface phenomena are influenced by both environmental conditions and the physicochemical properties of the materials. Important environmental conditions are the nature and existence of the surrounding atmosphere and lubricant, as well as the magnitude of the applied load and displacement. The important material properties are such physical properties as microstructural characteristics and surface quality, and such chemical properties as chemical affinity, free energy, solubility, diffusivity, the stability of the oxide layer on the surface, and surface energy. The role of these properties must be evaluated in each case, since a sweeping generalization cannot as yet be made with the currently available knowledge.

X-2 SURFACE TOPOGRAPHY

The first step in understanding the friction and wear of metals is to consider the topography of solid surfaces. This aspect has been experimentally investigated using optical instruments, such as metallographs and electron microscopes, and mechanical tracing instruments, such as surface profilometers. According to these investigations, the surface of a metal is not smooth, but rather is made up of many asperities (tiny peaks). Therefore, when two metal surfaces are in contact, the real area of contact is much smaller than the apparent area of contact. The interaction of these asperities with adjoining surfaces governs the friction and wear behavior of solids.

The asperities are typically 10 to 300 μ in. high and 1,000 to 10,000 μ in. wide at their bases. The heights of the asperities depend on the mechanical process of generating the surface; the lapped and polished surface is the smoothest. The slope of the asperities is very shallow, being about 10^{-3}, and the distance between the major summits is approximately 2,000 μ in. or more (see Fig. X.1). The height of these asperities is greater than the resultant deformation of the surface under typical loading conditions. Therefore, as the applied load is increased, the real area of contact increases, mainly because the total number of asperities in contact increases. Because the displacement of the indented metal is restricted by the bulk material, the load required to deform these asperities is greater than the uniaxial yield stress of the metal. The loading condition at each

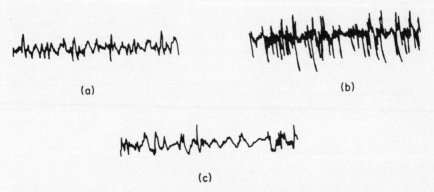

(a)   (b)

(c)

Fig. X.1 Profilograms of three metal surfaces. (a) Stainless steel 302, finish no. 4. (b) Nickel, nominal roughness 3 μ in. (c) Sample (b) examined in a perpendicular direction. Vertical magnification about 40,000; horizontal magnification about 3.3. (*Ref. 2.*)

asperity is similar to the case of the hardness indentation test (see Example X.1).

It has been found that, under normal load, as the highest asperities are flattened, in some cases the remainder of the surface (other than the deformed asperities) will rise. Depending on the shape of the surface profile, the number of asperities in contact and the projected area of contact change, so that the normal load L is equal to the product of the real area of contact, A_r, times the normal stress at each asperity. The lower limit for the real area of contact A_r occurs when the normal stress at each asperity is equal to the hardness of the softer metal. However, the analysis of actual surfaces indicates that the real area of contact is usually twice this lower limit. This *may* be due to the fact that the very surface layer is softer than the bulk, as discussed in Sec. X-4. The real area of contact A_r may then be expressed as

$$A_r = C \frac{L}{H} \tag{X.1}$$

where C may vary from 1 to 2.

EXAMPLE X.1—100% Real Area of Contact

A copper rod (1/2 in. in diameter and 1 in. long) is to be bonded end-to-end to a nickel rod with the same dimensions by diffusion bonding. It is absolutely necessary to have nearly 100% contact at the interface, i.e., the real area of contact must be nearly the same as the apparent area of contact. Describe how nearly 100% real area of contact can be obtained.

Solution:

Surfaces finished mechanically by lapping, etc., are not *perfectly* smooth and flat. The magnified surface invariably shows a large number of asperities, as shown in Fig. X.2(b). Each of the asperities will act as an indenter when the rods are brought together under the axial load. The stress required to push these asperities into the other surface is the same as the indentation hardness which is about three times the uniaxial yield stress of the softer metal, i.e.,

$$H_B = 3\sigma_Y \tag{a}$$

Therefore, the total axial force required to have 100% contact is

$$F = H_B A = 3\sigma_Y A \tag{b}$$

where F is the axial force, σ_Y is the yield stress of the softer metal, and A is the apparent area of contact.

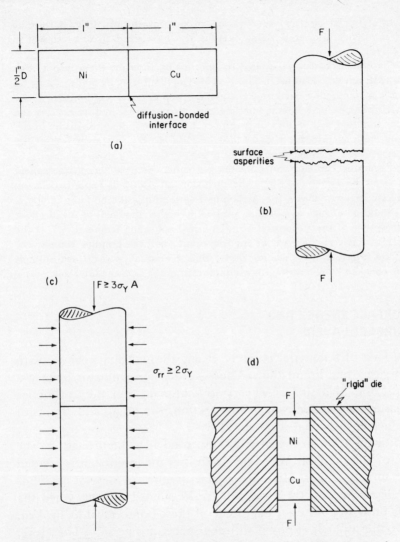

Fig. X.2 Diffusion bonding of copper and nickel rods for 100% real area of contact. (a) Final configuration desired; (b) exaggerated view of surface asperities; (c) lateral stress σ_{rr} greater than $2\sigma_Y$ is required for the application of the axial load $F \geq 3\sigma_Y A$; (d) a method of exerting the radial stress.

However, the maximum force the rods can withstand without undergoing general plastic deformation is only one-third of the force required for 100% contact, i.e.,

$$F_{max} = \sigma_Y A \tag{c}$$

Therefore, lateral stresses are required to prevent the rod from deforming plastically when the axial stress exceeds the bulk yield stress as shown in Fig. X.2(c).

The lateral stress can be applied to the rods by inserting them into a large block with a hole, as shown in Fig. X.2(d). The block may be made of a metal which is incompatible* with nickel and copper so as to prevent bonding at the lateral interfaces. The lateral bonding may also be prevented by coating the rods or the hole with oxides or graphite. If small amounts of diffusion of the coating material into the rods cannot be allowed, the lateral surface of the rods may have to be machined after bonding.

The bonding can best be done in a vacuum, or in an inert atmosphere such as argon or helium. Since the diffusion rate depends exponentially on temperature (i.e., follows the Arrhenius relationship), the bonding temperature should be as high as possible, preferably higher than half of the absolute melting point of the softer metal (i.e., normal annealing temperature). If the temperature is very close to the melting point, or if the bonding time is very long, the interface may become quite thick. The quality of diffusion bonds can be checked by measuring the contact resistance across the interface.

X-3 SURFACE ENERGY AND
SURFACE LAYERS

The surface of a material is rarely clean, always being covered with oxides, gases, and liquid films. These surface contaminants form as a result of the high surface energy of materials, and they influence various surface phenomena. In this section, the origin and effects of the surface energy are discussed.

Suppose an atom in the interior of a solid is to be brought to the surface of the material. When the atom is in the interior, it is bonded to the surrounding atoms, the actual number of nearest-neighbor atoms depending on the structure of the solid. However, when the atom is brought to the surface, some of these bonds must be broken, because the atom is exposed to open space on one side. Therefore, *work* must be done to bring an atom from the interior to the surface. The work to be done is approximately equal to the energy of the atomic bonds which must be broken. The atoms at the surface are then at a higher state of energy in comparison to those in the interior. The surface energy is equal to the work done in creating a unit area of surface. This state of higher energy is not confined to the

---

*Incompatible metals do not form solid solutions. In some books on friction and wear, incompatible metals are defined to be those which form solid solutions.

very top layer of atoms, but extends several atomic layers deep because of the disruption of the longer-range atomic interactions by the surface.

The surface energy of liquids can be obtained quite readily from measurements of the liquid surface configuration. However, the surface energy of a solid is difficult to measure. It is usually assumed to be the same as the surface energy of the material in the liquid phase at the melting point. The argument in the preceding paragraph can be used to estimate the surface energy from the heat of evaporation by noting that about 1/6 of the atomic bonds need to be broken to bring an atom to the surface, whereas for evaporation all bonds need to be broken. Approximating the surface energy as 1/6 of the heat of evaporation on an atomic basis seems to yield reasonable values.

The discussion given in the preceding two paragraphs is presented so as to emphasize the fact that the "unsaturated" state of the atoms near the surface make the atoms at the surface highly reactive, and, thus, the surface is never "clean." The surface energy of solids is lowered if gases or water molecules can be *physically adsorbed* to the solid surface by means of secondary-bond forces. Greater reduction in the surface energy can be made if chemical reactions such as oxidation take place at the surface through the establishment of primary bonds between the surface atoms and those that come into contact with the surface. (This type of reaction is called chemisorption.) Because of these reasons the surface of a metal readily becomes contaminated (see Fig. X.3). These contaminants are quite stable and cannot be readily removed from the surface. Therefore, in considering the friction and wear of solids, the role of these layers may not be neglected.

The higher state of energy also exists at internal surfaces such as voids and interfaces between the matrix, secondary-phase particles, inclusions, and grain boundaries. The basic principle behind chemical etching for grain boundary observation is based on the fact that chemical reactions occur more readily at higher energy sites. This higher energy state of interfaces even affects the decomposition rate of plastics and solid propellants when they are heated to high temperatures. Solid propellants decompose more rapidly and thus burn faster due to this "thermomechanical effect" of the presence of high-energy internal surfaces.

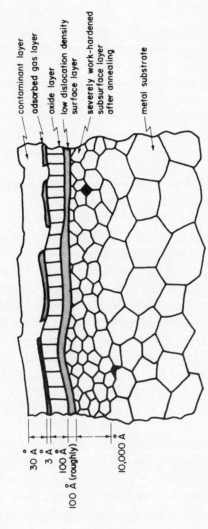

contaminant layer
adsorbed gas layer
oxide layer
low dislocation density surface layer
severely work-hardened subsurface layer after annealing

metal substrate

30 Å
3 Å
100 Å
100 Å (roughly)

10,000 Å

Fig. X.3 Schematic illustration of subsurface structure of an annealed metal with surface films. Dimensions are very approximate. (*Modified from Ref. 1.*)

X-4 MECHANICAL PROPERTIES OF MATERIALS
NEAR THE SURFACE

X-4-a Mechanical Properties of
Metal near the Surface

In addition to the chemical effects associated with surfaces, which were discussed in Sec. X-3, the mechanical properties of metals near the surface are also quite different from those of the bulk. The metal very near the surface strain hardens less than the subsurface layer and thus can undergo large plastic deformation when the surface material is deformed plastically. Since this layer affects the friction and wear properties of metals, it must be discussed.

In Chapter V it was shown that the dislocations very near and parallel to the surface experience image forces due to their proximity to the surface. When there is no continuous, coherent oxide layer adhering to the metal surface, the image force attracts the dislocations to the surface. When the image force is greater than the resisting drag force, commonly referred to as the dislocation friction stress,* these dislocations are attracted to the surface and disappear. Therefore, there tends to be a layer near the surface with low dislocation density. Some of the dislocations generated when the surface is deformed are located below this layer and thus are stable. Therefore, this subsurface layer accumulates dislocations and strain hardens more than the surface. The dislocation density in the low dislocation density zone may or may not be higher than the undeformed original bulk material. This means that the very surface layer can undergo larger plastic deformation than the subsurface layer. Qualitative experimental results show that this is so.

The thickness of the low dislocation density zone of the surface layer depends on the surface energy of the metal and the magnitude of the friction stress acting on the dislocations. The surface energy affects the thickness of this zone because the total surface area changes when dislocations emerge at the surface. The dislocation

---

*The dislocation friction stress is defined as the drag stress acting on dislocations which must be overcome for dislocations to move. This term should not be confused with the frictional forces acting on a sliding surface. This dislocation friction stress is caused by the Peierls stress, the internal stress created by other dislocations and substitutional and interstitial atoms and secondary-phase particles. The magnitude of the friction stress depends on the nodal spacing of dislocations, and the distribution and properties of any secondary-phase particles.

friction stress has been measured only for a limited number of metals. In commercial-grade metals, the dislocation friction stress is much smaller in f.c.c. metals than in b.c.c. metals.

A rough estimate of the image shear stress τ_i acting on a dislocation parallel to the surface is given by

$$\tau_i = \frac{Gb}{4\pi(1-\nu)h} \tag{X.2}$$

where G is the shear modulus, b the magnitude of the Burgers vector, ν Poisson's ratio, and h the distance from the surface. This force is opposed by the dislocation friction stress σ_f and the change in the surface energy as the dislocations emerge from the surface. Assuming that the dislocation friction stress dominates, Eq. (X.2) may be solved for h in terms of σ_f as

$$h = \frac{Gb}{4\pi(1-\nu)\sigma_f} \tag{X.3}$$

The dislocation friction stress in the low dislocation zone may be approximately expressed for solid solutions, precipitation hardened alloys, and overaged alloys as (Ref. 3)

$$\sigma_f = G\epsilon^2C \text{ (solid solution)}$$
$$\sigma_f = 2G\epsilon C \text{ (precipitation hardened alloy)} \tag{X.4}$$
$$\sigma_f = \frac{Gb}{\lambda} \text{ (overaged)}$$

where ϵ is the misfit strain, C the atomic concentration of substitutional atoms, and λ the spacing between hard particles. In the case of 3% silicon iron, $\sigma_f = 1{,}500$ kg/cm^2, $G = 3.9 \times 10^6$ kg/cm^2, $\nu = 0.3$, $b = 2.48$ Å, we find that h comes out to be 1/10 to 1/20 μm. For zone-refined pure copper, σ_f can be as small as 4 kg/cm^2, $G = 2.2 \times 10^6$ kg/cm^2, $\nu = 0.33$, h is equal to about 10 to 20 μm.

When a hard, continuous oxide layer adheres to the metal surface, the boundary condition at the oxide-metal interface demands that a dislocation existing in the metal experiences a force which repels it from the interface. This would also be the case if a hard slider were moving over a soft metal surface.

Even when a metal undergoes general plastic deformation such as in a tensile test, rather than by surface-traction-induced plastic deformation, as in wear, the material very near the surface work hardens less. This was shown experimentally by Fourie (Ref. 4). In this case, this soft layer extends deeper into the material than indicated by Eq. (X.3). Argon (Ref. 5) attributes this gradient in the flow stress of a plastically deformed metal to the relative inefficiency of the strain hardening process near the surface in comparison to that of the interior. It was argued that the material near the surface is subjected only to a unidirectional dislocation flux coming from the interior, while the dislocation flux of the interior comes from both sides. Therefore, interactions between dislocations which cause strain hardening are more probable in the interior than at the surface.

X-4-b Mechanical Properties of the Thermoplastic Surface

In certain crystalline thermoplastics the surface layer has different mechanical properties from the bulk. The existence of such a layer is attributed to the preferential nucleation of high molecular weight species during the crystallization process at certain nucleation sites (Ref. 6). According to this reasoning, low molecular weight species of the polymer are rejected to the molten plastic-air interface during the crystallization process giving a weak surface layer. At a molten polymer-metal interface, a region of high cohesive strength is produced in the plastic if the metal surface provides nucleation sites. This is caused by the rejection of the low molecular weight species from the interface into the bulk. Such a zone of high strength is shown in Fig. X.4 for the polyethylene/sulfochromated aluminum system. In the absence of nucleation sites, a weak plastic layer results at the plastic-metal interface.

X-5 FRICTION BETWEEN SLIDING METAL SURFACES

In this section, the frictional behavior of metals will be discussed. In particular, the difference between static and kinetic friction, the adhesion theory of friction, the role of surface energy, and selection of materials for sliding surfaces will be discussed. The frictional behavior of polymers is discussed separately in Sec. X-6, since it is quite different from that of metals.

Fig. X.4 Photomicrograph showing the region of high cohesive strength at the polyethylene/sulfochromated aluminum interface. (*Ref. 6.*)

It is now generally well recognized that sliding friction between solids is primarily controlled by the following two main mechanisms: the degree and quality of adhesion between the two sliding surfaces, and the plowing of the soft surface layers by hard asperities. The degree and quality of adhesion is influenced by the surface energy and the surrounding environment. The adhesion theory of friction assumes that complete adhesion exists between the contact surfaces of asperities as presented in this section. Although the "plowing" mechanism may be more important than the adhesion mechanism in many sliding situations, there is as yet no theoretical model for "plowing."

The existing theories are not descriptive enough to predict the frictional behavior of various metals under different sliding conditions, but they do provide qualitative insights into the problem.

X-5-a Phenomenological Aspects of Static and Kinetic Friction

Consider a slider block lying on a flat surface subjected to both normal load L and tangential load F. When the tangential force F, which is less than a critical value F_{max}, is applied to the block it will not move, except for the very small motion due to reversible elastic deflection at the interface. The static coefficient of friction is defined as the minimum tangential force required to initiate the tangential motion, F_{min}, divided by the normal load L acting on the interface, i.e.,

$$\mu_s = \frac{F_{min}}{L} \tag{X.5}$$

The coefficient μ_s is found to be independent of the apparent area of contact, the normal load, and the duration of loading. However, for very short duration of loading, less than 0.1 sec, μ_s is found to be a function of the loading time. The coefficient μ_s is often determined by placing the block on an inclined plane for which the inclination is adjustable, and determining the critical inclination at which the block starts sliding.

The tangential force required to sustain sliding is less than the force required to start it. This indicates that the kinetic coefficient of friction is less than the static coefficient of friction. The kinetic

TABLE X.1 Typical Friction Coefficients (*Ref.26*)

Conditions	Examples	Values
For clean surfaces		
General		
Clean, unlubricated surfaces	Steel on silver	0.5 to 0.3
	Leather on wood	
	Nylon on steel	
Exceptions to above		
Clean, similar metals other than	Copper on copper	1.5 to 0.8
those with close-packed hexagonal	Brass on brass	
structure	Chromium on chromium	
Clean, similar metals with close-packed	Titanium on titanium	0.65 to 0.45
hexagonal structure	Zinc on zinc	
Clean, duplex-structure alloys with	Copper-lead alloy on steel	0.3 to 0.15
a soft constituent, sliding against	Babbitt on steel	
either a hard metal or hard nonmetal		
Unusual nonmetals		
Rubber on other materials		0.9 to 0.6
Teflon on other materials		0.12 to 0.04
Graphite or carbon on other materials		0.16 to 0.08
For boundary-lubricated surfaces (covered by liquid lubricant)		
Ineffective lubricant	Water, gasoline, nonwetting liquid metals	Same as for clean surfaces
Fairly effective lubricant	Refined mineral oils, wetting liquid-metals, also metal surfaces nominally unlubricated but untreated to remove contaminants	0.3 to 0.15 or unlubricated value, whichever lower
Highly effective lubricant	Mineral oils with "lubricity" additives, fatty oils, good synthetic lubricants	
Metal on metal or	Steel on steel	0.10 to 0.05
metal on nonmetal	Nylon on steel	
Nonmetal on nonmetal	Nylon on nylon	0.20 to 0.10
For solid-film lubricated surfaces		
Hard metals covered by a thin layer of soft metal	Thin lead film on steel	0.20 to 0.08
Materials lubricated by a layer of graphite or molybdenum disulphide, either alone or compounded by a binder		0.12 to 0.06
For hydrodynamically lubricated surfaces		
A complex fluid film; produced by the sliding action, separates the surfaces. (This mode of lubrication generally applies only at speeds much in excess of 10 ft/min)		0.001 to 0.01

TABLE X.1 Typical Friction Coefficients (*Ref. 26*) (*Continued*)

Conditions	Examples	Values
For hydrostatically lubricated surfaces		
A complete fluid film, produced by external pressurization, separates the surfaces		0.001 to 0.000001 depending on the design parameters
For rolling-contact systems		
Pure rolling contact. Geometry carefully arranged so that pure rolling motion occurs over the contacting region	Cylinder rolling over a plane	0.001 to 0.00001
Normal rolling contact. Some shear occurs at the contacting region	Commercial ball bearings	0.01 to 0.001
Arbitrary geometry	Boulder rolling down a hillside	0.2 to 0.05
For naked surfaces		
Clean metals operated in a good vacuum (10^{-6} mm of mercury or better)		3.0, to adhering, lower for harder metals
Nonmetals (same conditions)		1.0 to 0.4

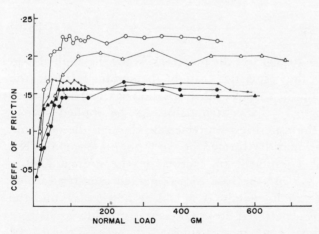

Fig. X.5 Coefficient of friction as a function of normal load at room temperature: ○ = tungsten-carbide tool sliding against lead at 0.292 ips; △ = tungsten-carbide tool sliding against aluminum at 0.334 ips; ° = tungsten-carbide tool sliding against copper at 0.261 ips; ▲ = tungsten-carbide tool sliding against copper at 0.318 ips; ● = tungsten-carbide tool sliding against stainless steel at 0.248 ips. (*Ref. 7.*)

coefficient of friction, F/L, is, in general, found to be independent of the apparent area of contact and the normal load (i.e., the friction force is proportional to the normal load); μ_k is also independent of the sliding velocity. When the normal load is so small that the surface contaminant layer cannot be penetrated by the asperities on the sliding surfaces, the kinetic coefficient of friction tends to be less than the "normal" coefficient. Over a wide roughness variation generated by typical machining processes, the coefficient of friction is found to be independent of roughness; but very rough and very smooth surfaces are found to have higher coefficients of friction. Typical coefficients of friction are given in Table X.1.

X-5-b   Adhesion Theory of Friction*

It was shown in Sec. X.2 that the real area of contact is smaller than the apparent area of contact. The adhesion theory of friction states that the junctions formed at the real area of contact adhere to each other, and that for sliding to occur these junctions must be sheared. When two surfaces slide over each other, the friction force (or the tangential force) must be equal to the product of the shear stress required to deform these junctions plastically and the area of these junction, i.e.,

$$F = \tau \cdot A_r \tag{X.6}$$

where τ is the shear strength of the junctions. When the applied load is so small that the asperities cannot penetrate through the contaminant layer on the surface, the magnitude of τ will be dictated by the shear strength of the contaminants. As the applied normal load is increased, the asperities can penetrate through the contaminants and junctions may be formed between the base metals. The dependence of the coefficient of friction on the normal load is illustrated in Fig. X.5 for the case of a cemented tungsten carbide tool sliding against various metals. It should be noted that until the contaminant layer is penetrated the coefficient of friction is small.

When metal-to-metal contact is established at these junctions, the critical shear strength τ may be assumed to be about equal to the strength to deform the weaker metal plastically. τ is normally

---

*This theory is the most often quoted theory on friction between sliding metallic surfaces. However, it does not fully explain all aspects of friction behavior.

assumed to be equal to the *bulk shear strength* of the weaker metal. Substituting Eq. (X.1) into Eq. (X.6), and denoting the critical shear stress by k, Eq. (X.6) may be written as

$$F = C \frac{kL}{H} \tag{X.7}$$

Eq. (X.7) may be rewritten for the coefficient of friction as

$$\mu = \frac{F}{L} = C \frac{k}{H} \tag{X.8}$$

In a typical indentation test for determining H, the plastically deformed zone extends into the bulk a distance which is about equal to the width of the asperity. Throughout this plastically deformed zone, including the junction, the shear strength k is assumed to be about the same.

In the absence of any shear stress acting on asperities, the indentation hardness at asperities is equal to $6k$ for all metals if the asperities are nearly flat. In this case, Eq. (X.8) reduces to

$$\mu = \frac{Ck}{6k} = \frac{C}{6} \tag{X.9}$$

Since C may vary from 1 to 2, the simple model predicts that the coefficient of friction may vary as

$$\frac{1}{6} \leq \mu \leq \frac{1}{3} \tag{X.10}$$

However, when there are shear stresses acting on the asperities in addition to the normal load, the normal stress and the shear stress can be less than $6k$ and k, respectively, for plastic deformation. For shallow asperities the upper-bound solution for a rigid plastic solid predicts the coefficient of friction to be about 0.4. (See Ref. 8 for details.)

The adhesion theory cannot explain all frictional behavior and cannot precisely predict the coefficient of friction. However, it explains some of the observed behavior. The existence of static friction is explained, since it takes the critical shear stress to plastically deform the junctions. According to this model, the friction force is proportional to the applied normal load, since the real area of contact increases to accommodate the applied load. The coefficient of friction is independent of the apparent area of contact, as seen in Eq. (X.8). The adhesion theory states that the coefficient of friction is a weak function of sliding velocity, since the critical shear stress of metals does not change appreciably with slight increases in temperature and shear strain rate. Furthermore, the effects of temperature and shear strain rate tend to compensate each other.

There are several experimental observations which cannot be explained based on the adhesion theory alone. For example, it has been observed that very clean surfaces have much higher coefficients of friction than the values predicted using the adhesion theory. Another major shortcoming of the adhesion theory is that it does not take into account the flow stress gradient existing near the surface which is discussed in Sec. X-4. It is reasonable to speculate that the existence of the soft surface layer will increase the actual area of contact by letting asperities sink into the opposite surface. In this case, a "plowing" type of action may occur at the sliding surfaces, increasing the coefficient of friction significantly. Further work needs to be done to clarify this question.

X-5-c Rabinowicz Theory of Friction

Among several reasons why the coefficient of friction between metals may normally be greater than that predicted by Eq. (X.10) is the influence of surface energy on the real area of contact, as considered by Rabinowicz (Ref. 1). The surface energy makes the real area of contact larger than that given by Eq. (X.1). If the overall surface energy change is denoted by W_{ab} (sometimes called the surface energy of adhesion), then

$$W_{ab} = \gamma_a + \gamma_b - \gamma_{ab} \qquad (X.11)$$

where γ_a and γ_b are the surface energies of the two contacting surfaces and γ_{ab} is the interface energy. The sum W_{ab} is always positive, i.e., the overall energy is decreased by bonding. Idealizing

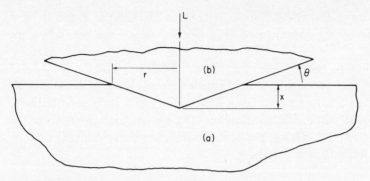

Fig. X.6 Penetration of a conical indenter into a flat surface.

the indentation of asperities as indentation by a conical indenter of material b penetrating into a half-space of material a, as shown in Fig. X.6, the work done by the normal load L during an infinitesimal indentation dx may be equated to the difference in the work done in deforming the material plastically and the surface energy change, i.e.,

$$L dx = \pi r^2 H dx - (2\pi r) \frac{dx}{\sin\theta} W_{ab} \qquad (X.12)$$

Equation (X.12) may be rewritten for the area in contact as

$$\pi r^2 = \frac{L}{H} + \frac{2\pi r}{\sin\theta} \frac{W_{ab}}{H} \qquad (X.13)$$

Equation (X.13) states that when the interfacial energy change is considered the projected contact area (πr^2) is larger than that given by Eq. (X.1) by an amount $(2\pi r/\sin\theta)(W_{ab}/H)$. Substituting Eq. (X.13) into the expression for the coefficient of friction gives

$$\mu = \frac{F}{L} = \frac{k}{H} \frac{1}{1 - 2W_{ab}/rH \sin\theta} = \frac{k}{H}\left(1 + K\frac{W_{ab}}{H}\right) \qquad (X.14)$$

where K is a geometric factor. Equation (X.14) predicts that the coefficient of friction can be high when the ratio of the surface energy of adhesion to hardness H is large and when the surface roughness angle θ is small. The coefficient of friction is found to be a linearly increasing function of W_{ab}/H for various metals sliding on like metals.

In addition to the surface energy contribution to the coefficient of friction, there are other contributions from such phenomena as plowing of the softer metal by a hard surface and surface roughness. The frictional force due to plowing may be larger than that due to the surface energy. This is particularly so when there is a soft surface layer as discussed in Sec. X-4.

X-5-d   Selection of Metals for Sliding Surface

Although there are usually exceptions to qualitative rules, the following may be used as general guides. A low coefficient of friction is likely if the two sliding surfaces are made of different metals which do not form a solid solution. A low coefficient of friction is also likely if the materials are hard, with a low ratio of surface energy of adhesion to hardness.

X-6   FRICTIONAL BEHAVIOR OF POLYMERS

X-6-a   Phenomenological Aspects of Frictional Behavior of Polymers

The frictional behavior of polymers differs from that of metals in the following respects:
1. The coefficient of friction is sensitive to the hydrostatic component of stress, and is thus dependent on the normal stress.
2. The coefficient of friction is very sensitive to the sliding speed, because of the large strain rate and temperature dependence of the flow stress of polymers.
3. The coefficient of friction of polymers is more sensitive to the ambient temperature which affects the flow stress.

The last two of these aspects are particularly applicable to thermoplastics.

The dependence of the coefficient of friction on the normal load is shown in Figs. X.7(a) and (b) for PTFE (polytetrafluoroethylene), phenolic, polyester, and epoxy. Note that the coefficient of friction can change substantially when the normal load is increased. The change is greater for thermoplastics than thermosetting plastics. The

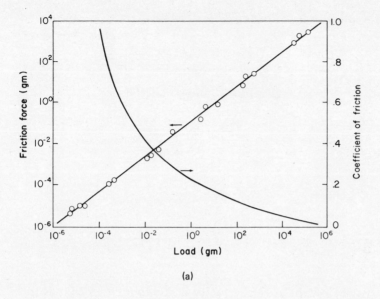

(a)

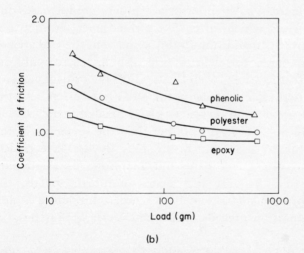

(b)

Fig. X.7 Effect of load on frictional properties of polymers sliding dry on steel. (a) Frictional force and coefficient of friction of PTFE vs. load; (b) coefficient of friction of reinforced thermosetting plastic vs. load. (*[a] from Ref. 9; [b] from Ref. 10.*)

dependence of the coefficient of friction on sliding speed is given in Fig. X.8 for polyethylene, polypropylene, and nylon. It should be noted that the coefficient of friction reaches a maximum value at a certain critical velocity. The increase in the coefficient of friction with

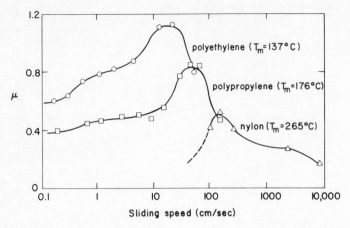

Fig. X.8 Friction-velocity dependence of linear crystalline polymers at 22° C, load = 100 gm. (*Ref. 11.*)

sliding velocity below the critical velocity is due to the fact that as the sliding velocity is increased, the interfacial temperature increases due to the mechanical work, which softens the surface layer and increases the real area of contact. In addition to this increase in the real area of contact, the shear force for deformation also increases with the sliding speed due to the strain rate effects. Until the critical velocity is reached, these two effects, namely, the strain rate effect and the increase in the real area of contact, play the dominant role in determining the frictional behavior, rather than the softening of the interfacial layer due to the temperature rise. Beyond the critical speed, the softening effect dominates and, therefore, the coefficient of friction decreases. At very high loads and high speeds, the interface temperature reaches the melting temperature and a hydro-dynamic sliding condition prevails. This point is clearly illustrated in Fig. X.8.

The coefficient of friction of polymers is found to rise with increasing temperature below the melting temperature. This is believed to arise from an increase in the area of contact resulting from lower hardness and increased wetability. This point is shown in Fig. X.9. The results shown in the figure are obtained by sliding a polyethylene block on a metal plate at a constant velocity at various temperatures. Note that the coefficient of friction increases until the melting point of polyethylene is reached and then it decreases.

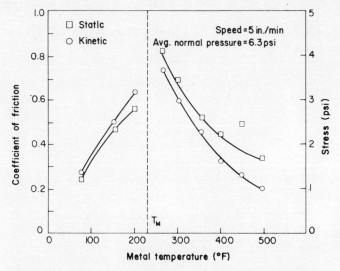

Fig. X.9 Friction coefficient and tangential stress on the sled of low-density polyethylene as a function of temperature. (*Ref. 12.*)

X-6-b Adhesion Theory for Frictional Behavior of Polymers

The friction force required to shear junctions formed between two sliding members at temperatures lower than the melting point may be written in the same form as for metals:

$$F = \tau \cdot A_r \tag{X.6}$$

The flow stress of polymers depends on the applied hydrostatic pressure, i.e.,

$$\tau = \alpha + \beta p \tag{X.15}$$

where α and β are material constants, and p is the hydrostatic pressure. The constant β denotes the dependence of minimum shear stress for flow on the hydrostatic pressure. Assuming that the hydrostatic pressure in the plastically deformed zone is nearly equal to the hardness, Eq. (X.15) may be written as

$$\tau = \alpha + \beta H \tag{X.16}$$

Substituting Eq. (X.16) into Eq. (X.6) and noting from Eq. (X.1) that $A_r = C L/H$, Eq. (X.6) may be written as

$$F = CL\left(\frac{\alpha}{H} + \beta\right)$$

$$\mu = \frac{F}{L} = C\left(\frac{\alpha}{H} + \beta\right)$$

$$\text{(X.17)}$$

Unlike the case of metals, A_r does not increase linearly with the applied normal load L, indicating that the hardness of plastics depends on the magnitude of the applied normal load. The results of indentation experiments show that the diameter of the indented area varies as a function of the applied normal load as

$$d = \left(\frac{L}{k_0}\right)^{1/n}$$

$$\text{(X.18)}$$

where k_0 and n are constants, and d is the indentation diameter. Pascoe and Tabor (Ref. 13) found n to be 2.7 for PMMA (Fig. X.10). By definition, d is related to the hardness [from Eq. (X.1)] by

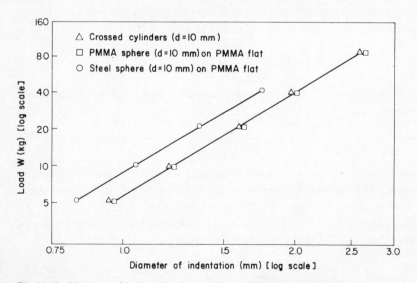

Fig. X.10 Diameter of indentation by a sphere on PMMA vs. applied load. (*Ref. 13.*)

$$d = \frac{2}{\sqrt{\pi}} \left(\frac{L}{H}\right)^{1/2} \tag{X.19}$$

Substituting Eqs. (X.18) and (X.19) into Eq. (X.17), one obtains

$$F = C\left(\frac{\pi\alpha}{4k_0^{2/n} L^{1-2/n}} + \beta\right)L = \mu L$$

$$\tag{X.20}$$

$$F = C\left(\frac{\pi}{4} \frac{\alpha d^2}{L} + \beta\right)L = \mu L$$

Equation (X.20) states that as L increases, μ decreases. In particular, if $n > 2$, μ approaches β as L is made very large. Conversely, it states that μ can never be less than β. Table X.2 gives the experimentally

TABLE X.2 Coefficients of Friction for Steel on Plastic [*]

Plastic	Rockwell hardness	½"-Diam. steel ball 1,000 g 4,000 g			α kg/mm²	β
		μ_s	μ_k	μ_k		
Polystyrene	38–40 M	0.43	0.37	0.36		
Phenol formaldehyde	89–90 M	0.51	0.44	0.37		
Polyvinylchloride	52 M	0.53	0.38	—		
Polypropylene	105–106 M	0.46	0.26	0.24	1.51	0.114
Polycarbonate	41–47 M	0.48	0.43	—		
Polymethylmethacrylate	88 M	0.64	0.50	0.49	5.13	0.204
Nylon 6.10	105 R	0.53	0.38	0.32	4.66	0.258
Nylon 6.6	110 R	0.53	0.38	0.36		
Nylon 6	91 R	0.54	0.37	0.38		
Polyoxymethylene	118 R	0.30	0.17	0.22		
Polyethylene (H.D.)	60 R	0.36	0.23	0.21	1.34	0.049
Polyethylene (L.D.)	25 R	0.48	0.28	0.26		
Polychlorotrifluoroethylene	112 R	0.45	0.27	0.26		
Polytetrafluoroethylene	5 R	0.37	0.09	0.10		
Polyimide	118 R	0.46	0.34	0.31		
A-B-S resin	105 R	0.40	0.27	0.29		
Polyphenylene oxide–styrene blend	118 R	0.60	0.46	0.41		

[*] μ_s = static coefficient of friction
μ_k = kinetic coefficient of friction
v (sliding velocity) = 0.001 cm/sec (*Frictional data from Ref. 15.*)

determined values of α, β, and μ for various plastics sliding against a steel sphere. The above derivation is only valid at low sliding speeds, where the temperature rise due to the mechanical work is small and does not affect the material parameters significantly. Equation (X.20) is found to predict the experimentally determined value of μ_k for PMMA closely when the values given in Table X.2 are used.

Grélan (Ref. 14) proposed an empirical equation for the friction force, which is of the form

$$F \propto L^m \tag{X.21}$$

where m varies from 0.7 to 0.95.

The combined state of loading at asperities should affect the value for indentation hardness and the shear stress for flow, as discussed for the case of metals. However, this effect has not been investigated.

X-6-c   Application of Time-Temperature Superposition to the Frictional Behavior of Polymers

In Chapter VI the equivalence of the time and temperature effects on the deformation of polymers was discussed. In this section the applicability of the time-temperature principle to the frictional behavior of polymers will be discussed. In particular, it will be shown that for rubber sliding at low sliding velocities of less than 1 cm/sec, the superposition principle can be applied to generalize the frictional behavior, and that at high sliding speeds and for other polymers, the superposition principle cannot be applied.

At very low sliding speeds, the temperature rise due to the heating of asperities during their deformation may be neglected and the time-temperature superposition may be applied to the frictional behavior under a set of restricted conditions. In order to establish the conditions under which the superposition principle may be applied, the deformation of the asperities may be idealized by assuming that a surface layer of thickness ΔS undergoes shear deformation due to the traction imposed by the other surface. The coefficient of friction of polymers due to the adhesion mechanism may be written as

$$\mu = \frac{F}{L} = \frac{\tau A}{L} \tag{X.22}$$

If it is assumed that the effect of hydrostatic pressure is negligible and that A/L remains nearly constant, Eq. (X.22) may be written as

$$\mu \propto \tau(\gamma, t) \tag{X.23}$$

where γ is strain and t is time.

The shear strain rate of the deformed surface layer at a reference temperature T_0 may be expressed as

$$\dot{\gamma} = \frac{V_0}{\Delta S} = \frac{\Delta\gamma}{\Delta t_0} \quad \text{at} \quad T = T_0 \tag{X.24}$$

where V_0 is the sliding velocity at $T = T_0$, ΔS the characteristic thickness of the deformed layer, and $\Delta\gamma$ the shear strain occurring in a time period Δt_0. Equation (X.24) may be expressed for Δt_0 as

$$\Delta t_0 = \frac{\Delta S}{V_0}\Delta\gamma \quad \text{at} \quad T = T_0 \tag{X.25}$$

If all other variables remain constant except the velocity and temperature, the time taken Δt_1 for shear deformation $\Delta\gamma$ at some other temperature T_1 may be expressed as

$$\Delta t_1 = \frac{\Delta S}{V_1}\Delta\gamma = a(T_1)\Delta t_0 \tag{X.26}$$

where $a(T_1)$ is the shift factor defined in Eq. (VI.15). The substitution of Eq. (X.25) into Eq. (X.26) yields

$$V_0 = a(T_1)V_1 \tag{X.27}$$

According to Eq. (X.27), if the asperity temperature is not affected by sliding and if the product of the thickness of the deformed layer and the strain increment, $\Delta S\Delta\gamma$, is not affected by either the temperature or velocity, the coefficient of friction may be correlated on a master curve as a function of $a(T)V$.

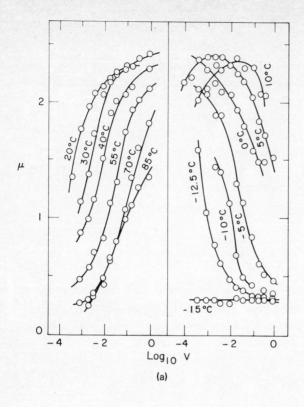

(a)

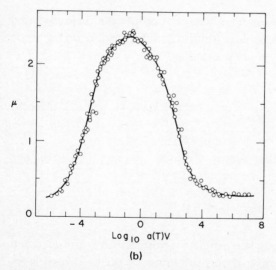

(b)

Fig. X.11 (a) Coefficient of friction as function of the sliding velocity at various temperatures of acrylonitrile-butadiene rubber on wavy glass. Curves are shown in two groups for clarity. (b) Master curve of the data shown in (a). Reference temperature is $20°$C.

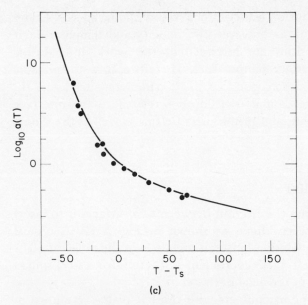

Fig. X.11 (*Continued*) (c) Shift factor $a(T)$ vs. $(T - T_s)$ for the rubber. The solid line represents the WLF equation given by $\log_{10}$ $a(T) = [-8.86(T - T_s)]/[101.5 + T - T_s]$, where $T_s = 50 + T_g(^\circ C)$ and $T_g = 21.4^\circ C$. (*Ref. 16.*)

Such a master curve was obtained for various rubbers sliding at speeds lower than 1 cm/sec by Grosch (Ref. 16) and Ludema and Tabor (Ref. 17). The experimental results of Grosch are shown in Fig. X.11. Figure X.11(a) shows the coefficient of friction of the acrylonitrile-butadiene rubber sliding against wavy glass. The master curve and the shift factor are shown in Figs. X.11(b) and X.11(c), respectively. Grosch also performed frictional tests with rubbers sliding against rough silicon carbide abrasive papers in order to induce large bulk deformation of the rubber as well as the local shear deformation of the surface layer. Based on these experiments, it was concluded that the frictional behavior of rubbers, which is attributable to the energy dissipation during the bulk deformation, is also viscoelastic and that the time-temperature superposition is applicable to this mode of frictional behavior.

The available experimental evidence on the frictional behavior of polymers other than rubber indicates that the superposition principle *may not* be applicable to these polymers, even at very low sliding velocities. However, this aspect of the frictional behavior needs further study.

The preceding discussion was confined to low sliding speeds. Under normal sliding conditions however, the velocity and temperature at the contacting asperities are interrelated; the work done during sliding raises the interface temperature. Therefore, it is not possible to relate the frictional behavior to a single parameter such as $a(T)V$. The temperature rise caused by sliding is related to the interface geometry, applied load, and physical properties as well as the sliding velocity.

X-7 ROLLING FRICTION

One means of reducing the friction between two surfaces is to insert a spherical ball between them, as shown in Fig. X.12. A typical coefficient of rolling friction between metals is in the range of 10^{-5} to 10^{-3}, whereas the typical coefficient of friction for a steel sphere rolling on a plastic is in the neighborhood of 10^{-2}.

If the ball makes point contacts with the flat plates, as shown in the figure, the force required to move the plate with a constant velocity will be zero. However, the ideal condition cannot be maintained for several reasons. The contacting bodies may deform both elastically and plastically. Once a nonzero contact area is established due to the deformation, slipping occurs between the ball and the plates. In addition to the friction force due to slipping and plastic deformation, the cyclic anelastic deformation induces hysteresis losses. Spinning of the ball about an axis perpendicular to the surfaces, which effectively increases the actual distance of sliding, is yet another source of frictional loss. Also, imperfections in roll geometry can contribute to the friction force.

In ball bearings, the fatigue fracture of the balls replaces wear as the major mode of failure. Since many balls are used in a bearing,

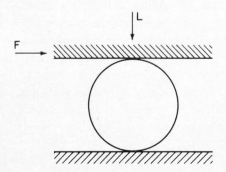

Fig. X.12 Rolling contact.

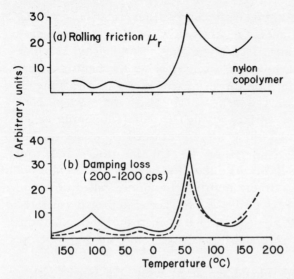

Fig. X.13 (a) Rolling friction of 3/16-in. steel ball over the surface of a nylon copolymer, as a function of temperature (load 1050 gm). (b) Low-frequency viscoelastic loss data for the same polymer as function of temperature. Solid line, damping loss or internal friction; dashed line, damping loss corrected for change in modulus $(-150° \ C < T < + 200°C)$. (*Ref. 17.*)

and fracture of a single ball constitutes failure of the entire bearing, the probability of a bearing failing is greatly increased over the probability of a ball failing.

When a metal sphere rolls on a plastic, a substantial part of the rolling friction is associated with the energy loss due to viscoelastic deformation of polymers. In Fig. X.13 the rolling friction of a 3/16-in. steel ball over the surface of a nylon copolymer is compared with the low-frequency (i.e., 1 $\sec^{-1}$) viscoelastic loss data for the same polymer as a function of temperature. The similarity of the curves should be noted. Since the typical coefficient of friction in sliding between steel and plastics is around 0.2 to 0.6, which is about an order of magnitude larger than the rolling friction, it can be concluded that the sliding friction is primarily caused by the plastic shearing of the surface rather than by the viscoelastic deformation of the polymer.

X-8 WEAR OF METALS

A phenomenon which is intimately associated with friction is wear. Wear of metals can be caused by a number of different mechanisms.

One of the most important wear mechanisms is that occurring between two sliding surfaces at relatively low speed. It is the primary cause of wear for cams, gears, bearings, and many machine components. Metals can, however, wear by other mechanisms. Wear occurs when a hard surface or abrasive particles dig into soft metals (commonly called abrasive wear). It can also occur when hard particles impinge on the surface at high speeds (commonly called impingement erosion). All these aforementioned processes do not involve any substantial temperature rise. On the other hand, tool wear at high cutting speeds occurs due to the heating of the chip-tool interface and weakening of the matrix (commonly called diffusion wear). Many of these wear processes are accelerated or decelerated by chemical reaction of the surface, such as corrosion and oxidation. In many applications, several of these wear mechanisms operate simultaneously.

The discussion of the phenomenon of wear will begin with a description of the experimental observations made in wear experiments and then proceed to a discussion of the theories which attempt to explain these observations. Some theories which are not wholly consistent with the experimental observations will be included in the discussion because they have historical significance and are widely quoted in the literature on wear.

X-8-a Experimental Observations of the Wear Process for Metals at Low Sliding Speeds

Wear tests are commonly done using a pin-on-disk or a cylinder-on-cylinder type of apparatus (see Fig. X.14). In the pin-on-disk apparatus, a plate of one material is rotated beneath a pin of the other material. The harder material is usually used as the pin and is called the slider. The normal load is applied by weights on the pin. In a typical set of experiments, the weight lost from the sample is measured as a function of the sliding distance at a variety of normal loads and sliding speeds. These experiments show that over a wide range of loads, and at sliding speeds in a range where heating of the interface is small, the weight lost from the softer material is proportional to the distance slid S and to the normal load L. The weight loss is independent of the sliding speed and the temperature. At very low normal loads, this proportional behavior is no longer observed as there appears to be a threshold load below which wear does not occur.

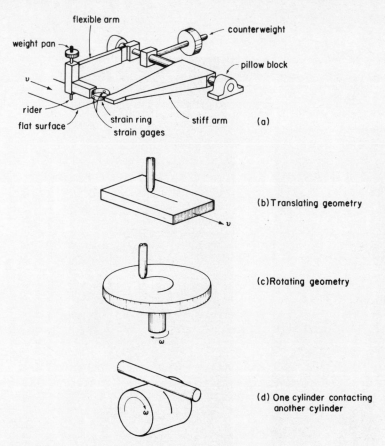

Fig. X.14   (a) Friction force measurement with a dynamometer. The friction force compresses the strain ring. (b) Translational motion of the disk relative to the slider pin. (c) Rotational motion of the disk relative to the slider pin. (d) Two contacting cylinders. (*Ref. 1.*)

In addition to measuring the rate of material loss, one can also make microscopic observations of the wear surface and of the particles produced during wear. By sectioning the wear sample through the wear track, the subsurface damage to the sample can be observed. Such observations help to provide an understanding of the relationship between the microstructure of the test materials and their wear rates.

i) *Subsurface Shear Deformation*

As shown in Figs. X.15–X.18, which are sectional views of the material beneath the wear tracks on a number of samples, the sliding of the rider over the surface produces a layer of sheared material

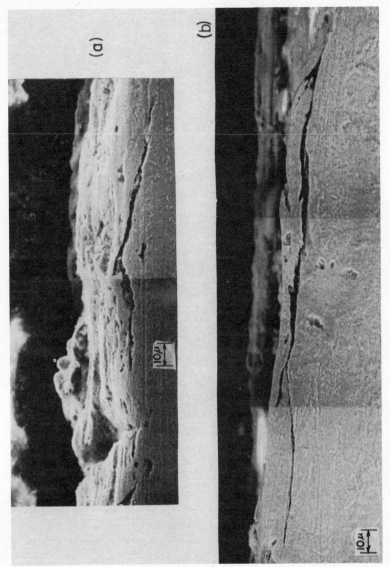

Fig. X.15 Crack formation underneath the wear track of annealed, commercially pure copper: (a) perpendicular to the wear track; (b) parallel to the wear track. (Grain size = 15μm; 52100 bearing steel slider pin; test temperature = 120°C; normal load = 1,816 gm; sliding speed = 0.5 cm/sec; argon was flushed over the sliding surface; H_B = 40 kg/mm^2). (*Ref. 20.*)

beneath the wear track. The strain in the material is maximum at the surface and decreases with increasing depth below the wear track. The shear strain at the surface is usually quite large, exceeding 16 in the case of steel and over 100 in the case of copper. The thickness of the sheared layer depends on the material being worn, but it is usually between 25 and 80 μm. In some cases, the boundary between the deformed and undeformed material is quite sharp, as, for example, in Fig. X.17.

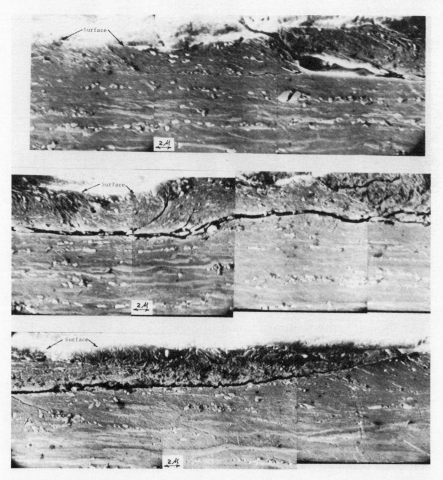

Fig. X.16 Three segments of a subsurface crack in cold-worked AISI steel sliding against 52100 steel. (a) and (c) show the ends of the crack, while (b) shows the midsection of the crack. Note the crack propagation along carbide particles. The crack is 200 μm long. (H_B = 184 kg/mm^2; normal load = 1.81 kg; sliding velocity = 76 cm/min; wear factor = 1.5 x 10^{-8} cm^2/kg; argon was flushed over the surface.) (*Ref. 21.*)

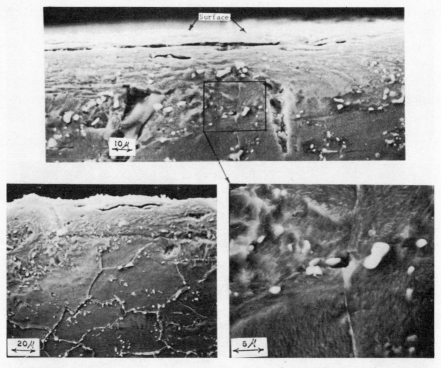

Fig. X.17 Subsurface cracks and deformation of annealed AISI 1020 steel sliding against 52100 steel. Note the voids around inclusions and the sharp demarcation line between deformed and undeformed zones in (b). (Grain size = 35 μm; H_B = 80 kg/mm^2; normal load = 1.81 kg; sliding velocity = 76 cm/min; wear factor = 2.3 x 10^{-8} cm^2/kg; the sliding surface was flushed with argon.) (*Ref. 21.*)

ii) *Subsurface Cracks*

As shown in the sectional views in Figs. X.15–X.18, one result of the shear deformation in the layer adjacent to the surface is the production of subsurface cracks. Figure X.15 shows a sectional view of a subsurface layer beneath the wear track on an annealed copper specimen. Note that the cracks run both perpendicular to and parallel to the wear track, but are longer in the direction parallel to the track. Figure X.16 shows a similar long crack about 200 μm in length in a sample of cold-rolled AISI 1020 steel. In this case, the crack has propagated along a line of hard carbide particles which are strung out along the rolling direction of the steel. The spacing between the lines of carbide particles in this case determines the spacing of the cracks and the thickness of the sheets which will be

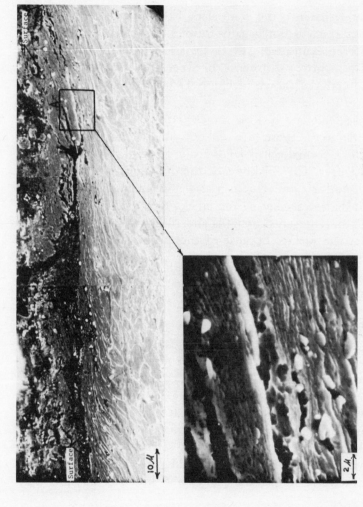

Fig. X.18 Subsurface deformation and void formation in annealed AISI 1020 steel sliding against 52100 steel. Note the severe plastic deformation at the surface, the void formation around inclusions, and the cracks along inclusion in the grain boundaries of the undeformed region. One of the major differences between this specimen and that of Fig. X.17 is the grain size; chemical compositions of both steels were nearly indentical. (Grain size = 6 μm; H_B = 102 kg/mm^2; wear factor = 2.5 × 10^{-8} cm^2/kg; argon was flushed over the sliding surface.) *(Ref. 21.)*

formed during the wear process. The spacing in this material is about 2 to 5 μm. A careful examination of the photomicrograph also shows that at other points, voids have formed around the hard carbide particles.

The process of crack formation and growth is affected by the grain size of the material. When the grain size of the metal is very small, a long continuous crack is not usually formed. Figures X.17 and X.18 show two different annealed AISI 1020 steels sectioned along the wear track. The primary difference between these two steels is the grain size. The material shown in Fig. X.17 has a grain size of 35 μm while that in the material shown in Fig. X.18 is 6 μm. In the coarse-grained specimen, a long crack is present, while in the fine-grained material, there are many small cracks developing concurrently. The formation of voids around carbide particles is also seen in these figures. The thickness of the deformed region in these specimens is about 20 μm. One can also note from Fig. X.18, that the shear deformation is causing the cracks to elongate in the shear direction, and that the crack lengths and the porosity of the metal are greater near the surface because of the increased shear deformation.

iii) *Wear Particles*

Further insights into the wear mechanism can be obtained by examining the shapes and sizes of the wear particles. The process of subsurface crack formation described above would be expected to produce flat sheetlike wear particles, because the length of the subsurface cracks is much greater than their distance below the surface. Such a wear sheet about to leave the surface is shown in Fig. X.19. The sliding direction in the figure is from right to left. Wear particles of shapes other than flat sheets can also be produced by subsurface cracking. The shape of wear particle is changed by cold work or further fracture after it separates from the bulk of the material. Equiaxed particles could also be formed when cracks are formed on nonparallel planes and intersect at short distances.

Wear particles for examination can be collected from used lubricating oil using the Ferrograph technique* (Ref. 18) and examined using a bichromatic microscope[†] or using a scanning

---

*Developed by Trans-Sonics, Inc., Burlington, Mass.

[†]A bichromatic microscope is a microscope which employs colored transmitted and reflected light simultaneously to illuminate the sample (Ref. 18).

Fig. X.19 Wear sheet formation in iron (with 8,000 ppm tungsten) viewed from the top at 30.5°. The sliding direction of the slider is from right to left. (Normal load = 2.25 kg; friction coefficient = 0.5; distance slid = 90 m; sliding velocity = 1.8 m/min; apparent wear = 1.6 mg; wear factor = 1.0×10^{-8} cm²/kg.) (*From S. Jahanmir, N. P. Suh, and E. P. Abrahamson, II, Microscopic Observations of the Wear Sheet Formation by Delamination,* Wear, *vol. 28, p. 235, 1974.*)

Fig. X.20 A stereo picture of typical wear particles collected from the lubricating oil used in gun mounts. Although there are a variety of shapes, most of the wear particles are flat. (*Ref. 20.*)

electron microscope (SEM). The microscopic examinations of steel and bronze wear particles collected from oil used to lubricate gun mounts reveal a number of features. The size of the wear particles depends on the material. The bronze particles are typically thicker and larger than the steel particles. The average thickness of the bronze particles is about 5 μm, while that of the steel particles is approximately 1 to 2 μm. However, there is a wide spread in the particle size and many particles which have a submicron thickness are also present.

Figure X.20 shows a number of wear particles which are typical of those examined in a stereo pair. Many of the particles shown are in the form of thin sheets as would be expected to result from the crack formation process shown in Figs. X.15 to X.18. Most of the wear particles in Fig. X.20 are steel, but some of the larger particles are bronze. Note that there are many different shaped particles, including a spherical one near the left-hand bottom corner of the figure. However, the particles tend not to be equiaxed, but more like flat sheets.

X-8-b Archard's Theory of the Wear of Metals at Low Sliding Speeds

Archard (Ref. 19) proposed a model which he called "adhesive" wear theory for the wear of metals at low sliding speeds. This theory presents a very simplified picture of the wear phenomenon which is not very consistent with the microscopic observations described in the previous section. However, since this theory is widely referred to in the literature on wear and as the first quantitative theory on wear which explained many of the observed phenomenon, it has historical significance. The development of the adhesive wear theory is closely related to the adhesion theory for friction. The theory postulates that when asperities come into contact and adhere strongly to each other, the subsequent separation occurs in the bulk of the weaker asperity. This creates a particle from one surface which is then attached to the other material. When these transferred particles become free, then loose wear particles are formed, and wear, observed as weight loss, occurs.

According to Archard's model, when the adhered junctions are established during sliding, hemispherical wear particles of the junction diameter d form, as shown in Fig. X.21. If all the junctions

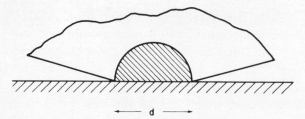

Fig. X.21 Archard's idealization of a wear fragment.

have the same diameter, the number of such junctions, n, is derived as follows:

$$A_r = \frac{L}{H} = n\left(\frac{\pi}{4}d^2\right)$$

$$n = \frac{4L}{\pi d^2 H} \tag{X.28}$$

Assuming that the junctions remain in contact during a sliding distance d, the total number of junctions formed per unit distance slid, N_t, is

$$N_t = n\left(\frac{1}{d}\right) = \frac{4L}{\pi d^3 H} \tag{X.29}$$

The total volume of the material removed during sliding through a distance S is given by

$$V = ZN_t S\left(\frac{\pi}{12}d^3\right) = Z\left(\frac{LS}{3H}\right) \tag{X.30}$$

where Z, a constant which gives the probability that a strong junction is formed, is known as the wear constant.

Equation (X.30) is called the Archard wear equation and states that the volume worn away is proportional to the normal load and the distance slid and is inversely proportional to the hardness of the

TABLE X.3 Wear Constants of Various Sliding Surfaces

Test condition	Metal combinations	Wear constant (Z)	Hardness (kg/mm^2) of the softer metal
Air and Room Temp. (Ref. 1)	zinc on zinc	160×10^{-3}	30
	cooper on copper	32×10^{-3}	—
	stainless steel on stainless steel	21×10^{-3}	250
	copper on low-carbon steel	1.5×10^{-3}	—
	low carbon steel on copper	0.5×10^{-3}	—
Argon and Room Temp. (Ref. 21)	5400 steel on cold-rolled 1020 steel	0.28×10^{-3}	184
	5400 steel on annealed 1020 steel (grain size 35 μm)	0.18×10^{-3}	80
	52100 steel on inclusion- and void-free iron	0.014×10^{-3}	35

material. Wear constants for several sliding surfaces are given in Table X.3 (Refs. 1 and 21). It should be noted that the wear constants indicate the amount of material lost by the weak surface, but not necessarily the number of loose particles formed. The actual number of loose particles formed is lower than the number of wear particles transferred from one surface to the other by a factor of 1/3 to 1/1,000, according to Rabinowicz (Ref. 1). This is presumed to be because particles can be transferred between the surfaces many times before they become free wear particles.

Archard's adhesion theory of wear has been widely accepted since the phenomenological relationship between wear volume, sliding distance, normal load, and hardness is reasonably consistent with experimentally observed results. However, the theory is weak in that it does not agree with many observed phenomena, and it does not take into account any of the physics or physical metallurgy of metal deformation. Furthermore, the theory does not provide insight into what metals or combination of metals will have good wear resistance, because the constant Z cannot be quantitatively related to any other material parameters.

The Archard theory emphasizes the role of hardness in the wear process. If Z is a material constant, anything which increases the hardness of a given material should increase its wear resistance. This is often found not to be the case. For example, the wear rate for high-purity iron, which is very soft compared to steel, would be expected from Archard's theory to be quite high relative to steel, since they are both primarily iron. In fact, an experiment which is described in the next section shows that the wear rate for the soft high-purity iron is quite low. The Archard theory would also seem to imply that copper sliding on copper would wear much faster than steel sliding on steel, since the latter is much harder than the former. This is not necessarily the case. It is found that the wear rates for these two combinations are about the same.

Archard's theory presents a picture of particles being pulled from the surface of the softer material, on a single pass of the slider, by the creation of a fracture in the weaker asperity. From this picture one would expect to find the material below the wear surface of the softer body to be relatively undeformed at a depth below the height of the asperities. This is in complete disagreement with the deep sheared zones observed in wear samples as described in the previous section. Since the failure in the Archard model always occurs in the weaker asperity, the harder material should remain undeformed and should not wear. This disagrees with the fact that, in a wear test on combinations of dissimilar metals, shear deformation is observed below the surface of the harder metal as well as in the softer metal. It is also at odds with the observation that even hard metals wear slowly when sliding against rubber.

Finally, Archard's model predicts that wear particles will be nearly equiaxed in shape and have a size comparable to the diameter of asperity contacts. As shown in the previous section, wear particles are often sheetlike and can be quite large compared to asperity dimensions.

X-8-c Delamination Theory of Wear for Metals Sliding at Low Speeds

Recently a new theory, called the delamination theory of wear, has been advanced to explain the wear of metals at low sliding speeds (Refs. 20 and 21). This theory is based on a wear mechanism which is consistent with the observations described in Sec. X-8-a and takes into account the effects of physical metallurgy on deformation and fracture processes in metals.

The delamination theory of wear assumes that the effect of the sliding contact is to cause plastic shear deformation of the material at the surface. The shearing of the surface may be accomplished by the adhesion between asperities, as described in the adhesion theory of friction, or it may result from the plowing of the surfaces by asperities on the harder body. This plowing action may be enhanced if a soft layer exists on the material as described in Sec. X-4. When a soft layer is present, the asperities of the harder material can sink farther into the soft material, and the shear stress transmitted to the subsurface layer will be increased.

The eventual result of the shearing of the material at the surface will be the creation of cracks and voids in the sheared region. The cracks can be nucleated by a number of mechanisms. Most commercial materials contain numerous hard particles which can contribute to crack formation. Cracks and holes can be nucleated at the surface of the particles by the stresses created by dislocation pile-ups (Refs. 22–24) or by fracture of the particles under the influence of the dislocation pile-up. Decohesion of the particle-matrix interface can occur as a result of the concentration of plastic strain around the undeformable particle (Ref. 25). In the absence of hard inclusions, cracks can be nucleated by a number of mechanisms involving direct interaction of dislocations (Ref. 26). However, in the absence of inclusions, much more plastic strain will probably be required to cause a crack to be nucleated.

The location of the cracks beneath the surface can be determined in a number of ways. When the material is inhomogeneous, such as in the case of the steel containing strings of carbide particles (Fig. X.16), the structure of the material will determine the spacing of the cracks. If work hardening is ineffective near the surface, giving rise to a soft surface layer, as described in Sec. X-4, crack nucleation will probably occur below the soft layer. In other cases, the crack spacing will probably be determined by the statistics of the distribution of fracture sites and the shape of the strain gradient below the surface. It will also depend on the stress distribution below the surface, because crack formation and subsequent propagation will be favored in regions where the hydrostatic component of stress is least compressive.

After cracks and voids are nucleated beneath the wear surface, they must grow to a critical size before they will link up to the surface and form a wear particle. These cracks and voids grow and link together by three different mechanisms: growth of voids (Ref. 27), crack propagation, and intensified plastic shear deformation between the voids. After the original creation of the defect and before the cracks become long enough to allow the creation of a free particle, considerable additional plastic deformation will have to take place. At the time that the cracks become large enough to cause formation of a free particle, they should be much longer than the diameter of asperity contacts.

The processes of crack formation, elongation, and linkup are clearly shown in Fig. X.22. Figures X.22(b) and X.22(c) graphically illustrate the start of a crack and its linkup with other cracks. It is on the way to becoming of critical length to cause delamination of the top layer of material to form a free wear particle. Figure X.22(a) shows that only the inclusions and voids near the surface show this type of crack propagation. This indicates the large amounts of shear deformation which are required to form a delaminated wear particle.

According to this picture of the wear process, the wear particles would be expected to be thin sheets. However, the process of wear particle formation is not necessarily over with the creation of a free particle. When a loose particle is formed it is initially trapped between the sliding surfaces. The deformation of the particle which occurs while it is between the sliding surfaces will also contribute to its final shape. A number of things can happen to the particle. If it is soft and ductile, it may roll up, forming a large but more equiaxed particle. If it is hard and relatively brittle, it may be fractured into a number of smaller and more equiaxial particles. In these cases, the shape of the wear particles which are recovered after they have escaped from the wear surfaces may not reflect the way that the particles were created.

This qualitative picture of the wear process is in good agreement with most of the observed phenomena associated with wear. It predicts that free particles will be formed only after considerable plastic deformation has occurred at the surface and that, therefore, the material below a wear surface will be severely deformed. Plastic deformation and wear can even occur in the hard material. When the flow stress of the hard metal is less than three times that of the soft metal, the hard metal can undergo shear deformation due to the stress

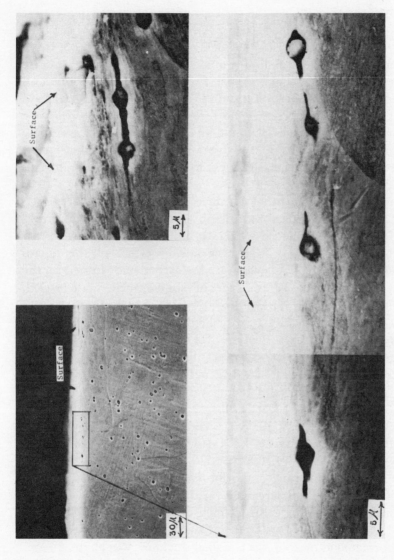

Fig. X.22 Void formation around inclusions and crack propagation from these voids near the surface of iron with 1.3% molybdenum. (Grain size ~ 1,000 μm; $H_B = 116$ kg/mm²; normal load = 2.25 kg; sliding velocity = 180 cm/min; wear factor = 2.6×10^{-8} cm²/kg; cylinder test; argon was flushed over the surface.) *(Ref. 21.)*

concentrations at the contacts. However, since the shear stress created at the interface is primarily limited by the soft material, the deformation of the hard material occurs much more slowly than that of the soft material. When the flow stress of the harder material is much larger than three times that of the soft metal, wear of the hard material can still occur through crack nucleation and propagation. Local microplastic deformation can occur even when the average stress is less than the yield stress of the metal. The dislocations generated during the microplastic deformation can pile up at hard obstacles and nucleate cracks. The foregoing would imply that the wear rate of the hard material would be much less than the wear rate of the soft material and that the depth of the deformation zone in the hard material would be much less than in the soft material. This agrees well with the experiments.

The theory also predicts that the microstructure of the material, in terms of the concentration and distribution of phases, will have a strong effect on the observed wear rate, and therefore that wear rate cannot be simply related to hardness. This effect of the microstructural dependence of the wear rate is clearly demonstrated by the following experiment on the wear of zone-refined, zone-leveled, high-purity iron which was nearly free of voids or inclusions.

When this metal is tested for wear, the number of subsurface cracks and voids visible at a magnification of 3,000 was very much smaller than in metals with inclusions. In fact, there were many regions free of any cracks. The wear rate for the metal was correspondingly low, in spite of the fact that this metal was extremely soft. The wear factor κ, which is defined later in this section, of this "inclusion-and void-free" iron was only 0.4×10^{-8} cm^2/kg (wear constant = Z = 0.14×10^{-4}), whereas it was 1.5×10^{-8} cm^2/kg (wear constant = $Z = 2.8 \times 10^{-4}$) for cold-rolled AISI 1020 steel, although the hardness was 35 kg/mm^2 for iron and 184 kg/mm^2 for the steel. The thickness of the plastically deformed region near the surface continuously increased in this specimen until the specimen had been wear tested for 50 min at a constant surface speed of 180 cm/min under a normal load of 2.25 kg. The final thickness was about 20 μm.

On the basis of the phenomenological picture of the wear process discussed above, it is possible to create simplified models which lead to an equation for the wear rate. Two such models which are based on slightly different assumptions are discussed in the appendix to the chapter (Appendix X.A). Both models lead to a wear equation of the form

$$W = \kappa \cdot LS \qquad \qquad (X.31)$$

where W is the volume of material worn away, L is the normal load, S is the distance slid, and κ is called the wear factor. The wear factor κ depends on a number of material parameters. The exact form of κ depends on the model chosen. The major difference between this relationship and Archard's adhesive wear formula is that the material parameters which appear in κ do not indicate a simple relationship between the wear rate and the hardness of the material.

X-8-d General Rules for Reducing Wear

The following general rules might be used in selecting metals for wear applications:

1) For a given metal with a given microstructure, the harder the surface, the greater will in general be its wear resistance. Therefore, increasing the strength of the surface by solution hardening, as in the case of oxygen or nitrogen dissolved in titanium, should increase wear resistance.

2) Hardening by formation of solid solutions (e.g., brass and bronze), which does not form large second phases, is desirable, since the shear deformation of the surface layer can be minimized by increasing the flow strength without introducing sites for crack nucleation. Other methods of minimizing the shear deformation of the surface layer and, thus, the wear rate are to lubricate the surface well with low-strength fluids and/or solids, choose a pair of metals which do not form solid solutions, and to make the grain size small. Lubrication primarily reduces the tractions applied to the surface. Lubrication can decrease the wear rate by several orders of magnitude, while decreasing the coefficient of friction by an order of magnitude.

3) In choosing a pair of metals, if one surface can be allowed to wear without jeopardizing the functional requirements of the system, the other metal should be as hard as possible and incompatible with the first. Some of the hard materials are oxides, carbides, nitrides, borides, and diamond.

4) If the oxides of the sliding metals are hard, a means should be provided for eliminating loose wear particles from the sliding

surfaces. These hard particles can cause severe abrasive wear. Hard foreign particles are similarly detrimental.

X-8-e Abrasive Wear

Abrasive wear occurs when an abrasive particle is entrapped between two sliding surfaces, plowing material from both surfaces. Abrasive wear can also occur when a hard rough surface is sliding against a softer one, plowing the soft one. The former type of abrasive wear is sometimes called three-body abrasive wear, while the latter is known as two-body abrasive wear. The rate of abrasive wear, like the adhesive wear, is proportional to the normal load and the distance slid and inversely proportional to the hardness of the wearing material. However, it may also be affected by microstructure and other metallurgical variables as in the case of sliding wear.

X-8-f Erosion of Surfaces by Impingement of Solid Particles

Erosion may be defined as gradual material removal that occurs when a solid surface is impinged on by solid or fluid particles, or when a surface undergoes chemical reactions. The erosion by solid-particle impingement (impingement erosion), by fluid particle impingement and by fluid cavitation (cavitation erosion) will be discussed here. In recent years, detrimental impingement erosion has become a serious problem in such applications as aircraft propellers, gas and steam turbine blades and vanes, pipes for crude oil transportation, and chemical reaction vessels. Erosion by solid particles of a bend in a transport pipe is illustrated in Fig. X.23.

Figure X.24 illustrates a solid particle impinging on a surface. Experimentally it has been found that the erosion rate depends on the impingement angle α, the particle velocity, and the size and density of the particles. Erosion also depends very sensitively on the nature of the material being eroded away. As shown in Fig. X.25, maximum erosion of ductile materials, such as aluminum, occurs at an impingement angle of somewhat less than 30°, whereas the maximum erosion of brittle materials such as glass occurs near 90°.

A simple theoretical model for impingement erosion may be derived based on a few experimental observations.* Figure X.26

---

*The analysis is similar to that given by Neilson and Gilchrist (Ref. 30).

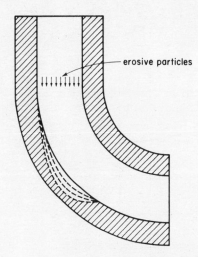

Fig. X.23 Attack of a bend in a transport line by erosive particles. (*Ref. 28.*)

shows the subsurface damage done to an annealed copper specimen by silicon carbide particles (254 μm in average diameter) impinging on the copper surface at 350 ft/min at an impingement angle of 55°. There are subsurface cracks parallel to the surface, the cracks extending to a depth of 10 μm. Other photomicrographs show evidence of craters forming on the surface, with the displaced metal forming mountains next to the crater. Some of these mountains are either flattened under impact or fractured off. At an impingement angle of 90° (normal impact), the crack patterns are almost random, the cracks extending as deep as 20 μm. Similar cracks are also observed in steel. These micrographs indicate that impingement erosion of metals occurs by a mechanism very similar to that which occurs under sliding conditions, as explained by the delamination theory.

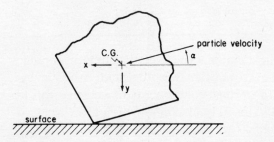

Fig. X.24 Idealized picture of an abrasive grain striking a surface. (*Ref. 29.*)

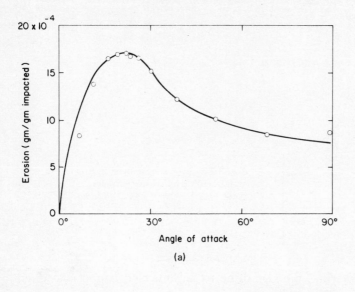

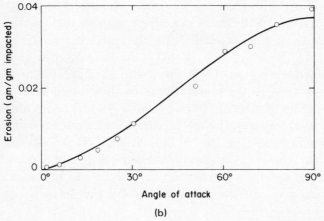

Fig. X.25 Erosion vs. angle of attack. (a) Aluminum eroded by 210-μm alumina particles at 494 ft/sec; (b) glass plates eroded by 210-μm alumina particles at 354 ft/sec. (*Ref. 30.*)

For the purpose of deriving a mathematical relationship for impingement erosion, a simplifying assumption may be made: the damage done by the tangential and normal components of velocity will be assumed to be uncoupled. The normal component of the velocity will be assumed to create cracks which cause fatigue failure or brittle fracture of the surface, while the tangential component of the velocity will be assumed to cause failure by plastic deformation.

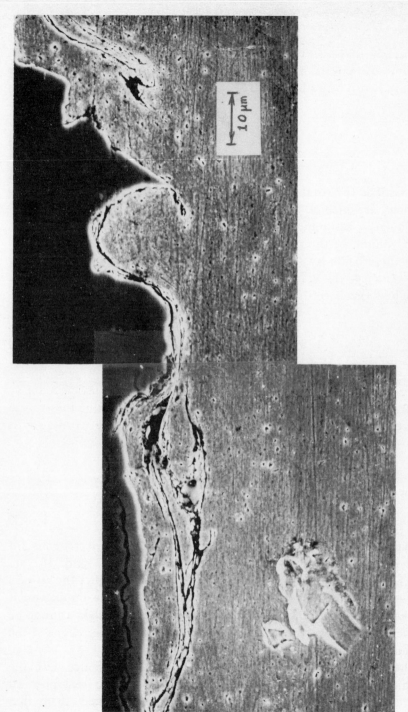

Fig. X.26 Subsurface damage of annealed copper due to SiC (254 μm) impingement at a speed of 350 ft/sec and at an impingement angle of 55°.

Assuming that a fixed amount of energy E_f is needed to erode a unit volume of material by the fatigue mechanism, the data show that the volume worn away by the fatigue mechanism W_f may be written as

$$W_f = \frac{1/2\, M(V \sin\alpha - V_l)^2}{E_f} \qquad (X.32)$$

where M is the total mass of abrasive particles, and V_l is the limiting velocity at which no fatigue fracture takes place.

Similarly, assuming that the specific energy needed in removing a unit volume of the surface material by shear deformation is E_s and that the erosion by shear is caused by the horizontal component of the velocity, the volume worn away by the shearing mechanism W_s may be written as

$$W_s = \frac{1/2\, M(V^2 \cos^2\alpha - V_p^2)}{E_s} \qquad (X.33)$$

where V_p is the residual horizontal component of particle velocity after impact. If the particle is completely stopped during the erosion process, V_p is then equal to zero. The total volume worn away is then given as

$$W = W_f + W_s = \frac{1/2\, M(V \sin\alpha - V_l)^2}{E_f} + \frac{1/2\, M(V^2 \cos^2\alpha - V_p^2)}{E_s} \qquad (X.34)$$

Equation (X.34) is used primarily as a correlation scheme for experimental results. The constants E_f, E_s, V_l, and V_p are determined from a set of experimental results to predict the erosion curves as a function of the incident angle α and the impingement velocity V. The extrapolation of the experimental data cannot be made too extensively, because E_f and E_s are actually functions of the impingement velocity and angle.

The experimental results are in reasonable accordance with the prediction of the theoretical model. However, the real physical phenomenon seems to be much more complicated than that indicated by the equation.

As shown in Fig. X.27, the erosion resistance of metals increases with increasing melting point and, thus, with increasing modulus.

EXAMPLE X.2—Prevention of Impingement
Erosion

A sandblasting nozzle is made of low-carbon steel. This nozzle wore rapidly by particle impingement. The air inlet tube just opposite the inlet of the abrasives also wore heavily from the impingement of particles perpendicular to the air inlet (see Fig. X.28). After a week's service, a hole was formed on the air inlet tube. Suggest a means of reducing the wear.

Solution:
The abrasives are impinging on the air inlet tube almost perpendicularly. In order to reduce the fatigue wear of the inlet tube, a rubber tube was pushed around the tube in order to increase V_l and reduce the wear rate according to Eq. (X.34). The tube could then be used for more than half a year.

The wear of the nozzle cannot be prevented by this technique, because the erosion of the nozzle is mainly by the shear deformation and cutting mechanism. In this case the solution is to choose a very hard material. When the nozzle was made of tungsten carbide, it lasted several years without failure.

Cavitation erosion is a major problem when cavitation in liquids occurs near solid surfaces. The pressure distribution in turbines and pipes and around ship propellers may be such that a portion of the liquid may be under tension or at very low pressures and consequently boils, forming a bubble. The bubbles formed in this manner may then collapse suddenly producing a shock wave or a liquid jet,

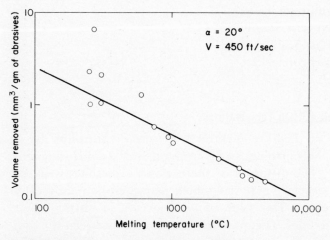

Fig. X.27 Erosion vs. melting temperature. (*Ref. 31.*)

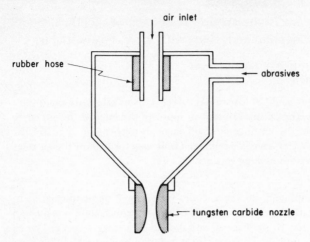

Fig. X.28 Sandblasting nozzle. (Ref. 28.)

which may impinge on the surface. Sometimes cavitation erosion may impose the upper limit to the turbine speed and thus on the maximum efficiency of a given size turbine.

The following list ranks materials in the order of cavitation erosion resistance in turbines:

1. Stellite
2. 17-7 Cr–Ni Stainless Steel (Weld)
3. 18-8 Cr–Ni Stainless Steel (Weld)
4. Ampco No. 10 weld material
5. Cast Ampco No. 18 bronze
6. Nickel-aluminum bronze
7. 18-8 Cr–Ni Cast Stainless Steel
8. 13% Cr, Cast Stainless Steel
9. Manganese bronze, cast
10. Cast Steel

X-8-g Wear of Metals at High Temperatures

Wear of metals at high temperatures is often controlled by a diffusion process. This involves atoms from one surface diffusing into the other surface. The diffusion of these atoms sometimes weakens the matrix of the surface material, which may then be removed by other wear processes. This type of wear occurs in cutting tools at high cutting velocities. The interface temperature between the tool and chip may reach as high as $1100°C$ under certain conditions.

Since the diffusion of atoms is a thermally activated process, the wear equation may be written in terms of the Arrhenius equation as

$$W = \kappa \cdot LS = \kappa_0 \cdot \exp\left(\frac{-\Delta E}{kT}\right) \cdot LS \quad * \quad (X.35)$$

where ΔE is the activation energy for the diffusion process.

EXAMPLE X.3—Prediction of Tool Life

A cemented tungsten carbide tool is used to cut AISI 4340 steel which has a Brinell hardness $H_B = 250$ kg/mm^2. When the steel is cut at a velocity of 600 ft/min, a depth of cut of 0.05 in., and a feed of 0.0052 in. per revolution, the tool life is found to be 54 min. Predict the tool life when the cutting speed is changed to 1,150 ft/min at the same depth of cut and feed rate. Assume that the activation energy is 50,000 cal/mole.[†] The cutting velocity V and the depth of cut, t, which for turning corresponds to the feed, are related to the temperature at the chip-tool interface by

$$\frac{T}{T_{ad}} = 0.4\left(\frac{Vt}{\alpha}\right)^{1/3} \quad (a)$$

where T_{ad} = $H/\rho c$ and T is in $°$F
     H = Hardness = 250 kg/mm^2 = 355,000 psi
     ρc = volumetric heat capacity = 202 in.-lb/in.^{3} $°$F
     α = thermal diffusivity = 1.84 in.^{2}/sec
     R = 1.1 cal/mole $°$R
     t = .0052 in.

Solution:
First calculate the temperatures T_1 and T_2 for the cutting speeds $V_1 = 600$ ft/min and $V_2 = 1,150$ ft/min using Eq. (a). $T_{ad} = H/\rho c = 1757°$F.

$$T_1 = T_{ad}(0.4)\left(\frac{V_1 t}{\alpha}\right)^{1/3} = 1,757(0.4)\left[\frac{600(12)(0.0052)}{60(1.84)}\right]^{1/3} °\text{F}$$
$$= 490°\text{F}$$
$$= 950°\text{R}$$

(b)

*Activation energies for diffusion are often given in units of cal/mole. When this type of activation energy is used, Eq. (X.35) should be written $W = \kappa \exp(-\Delta E/RT)\cdot LS$, where R is the gas constant, which has the value of $R = 1.1$ cal/mole $°$R.

[†]At low cutting speeds (when the interface temperature is below 1350$°$F), the activation energy is about 20,000 cal/mole.

$$T_2 = 1{,}757\,(0.4)\left[\frac{1{,}150\,(12)(0.0052)}{60\,(1.84)}\right]^{1/3}$$

$$= 607°\text{F}$$

$$= 1067°\text{R}$$

(b)
(Cont.)

If tool failure occurs at the same value of W in all cases, then, since L, H, and κ_0 are constants, Eq. (X.38) may be written as

$$\frac{W}{\kappa_0 L} = V\tau \exp\left(\frac{-\Delta E}{RT}\right) = \text{const} \tag{c}$$

where τ is the tool life. Therefore,

$$\tau_2 = \frac{V_1 \tau_1}{V_2}\exp\left(\frac{-\Delta E}{RT_1} + \frac{\Delta E}{RT_2}\right) \tag{d}$$

Substituting the temperatures calculated above, together with the given values of V_1, V_2, τ_1, and ΔE, gives

$$\tau_2 = \frac{600}{1{,}150}\,(54_{\text{min}})\exp\left[\frac{50{,}000}{1.1\,(1067)} - \frac{50{,}000}{1.1\,(950)}\right]$$

$$\tau_2 = 28.2\exp(42.6 - 47.8)$$

$$= 28\exp(-5.2)$$

$$= 0.15\ \text{min}$$

(e)

The experimental results obtained under the conditions specified give a tool life of 2.5 min. Since tool failure does not always occur at the same amount of wear this difference is not unreasonable.

X-8-h Other Wear Mechanisms

Wear can also occur by *corrosive* attack of the surface. The wear of materials by corrosive attack has little dependence on sliding. However, when the corrosion rate is small, sliding affects the wear rate. When the corrosion product is hard and brittle, the layer can break off, exposing new surface for corrosion. In this case, the corrosion rate is accelerated by sliding, and, thus, a greater wear of the surface results. On the other hand, when the corrosion product is soft or has lubricating effects, it can lower the wear rate under the sliding conditions.

Wear by *fretting* occurs when contact surfaces, such as between shafts and pressed-on inner races of ball bearings undergo small oscillatory motions between the contacting bodies. Under this condition, small wear particles form. Fretting wear follows the wear equation, Eq. (X.31), reasonably well, since the early stage of fretting wear is caused by the delamination process. The wear factor is a constant equal to that under sliding conditions, except at very small sliding amplitudes. When the amplitude is very small, the wear factor decreases with amplitude. This is due to the fact that the shear deformation at the surface is less than a critical value necessary to link the subsurface cracks. This type of wear can be minimized using boundary lubricants, or by completely eliminating the small oscillatory motions. The latter is difficult to accomplish in practical situations.

X-9 WEAR OF POLYMERS

X-9-a Wear Mechanisms for Polymers

The basic causes of wear in polymers are believed to be similar to those for metals. Additional wear mechanisms that play a role are associated with melting and extrusion of thermoplastics at the interface and thermo-oxidative wear of thermosetting plastics. The dominant wear mechanism of thermoplastics at high speeds is associated with melting. When the temperature is sufficiently high, thermosetting plastics tend to wear by charring through decomposition of the polymeric structures. In thermo-oxidative wear, broken primary bonds of thermosets react with molecules such as oxygen.

The maximum allowable load P at a given sliding velocity V (or conversely the maximum sliding speed V at a given load P) is attained when the interface temperature reaches the melting or the decomposition temperature of plastics. When this condition is reached, the wear rate increases suddenly, resulting in catastrophic failure of the sliding surfaces. The failure condition is specified in terms of the product PV at a given sliding condition. This concept, known as the PV limit, is used to define the limiting operating conditions for plastics. Although the product PV is proportional to the surface temperature rise at low sliding speeds, according to Cook and Bhushan (Ref. 32), the experimental results show that the value of the limiting PV has to be specified at a given loading condition.

The wear equations, Eqs. (X.30) and (X.31), are found to describe the wear rate of plastics reasonably well, although in this case the wear constant is found to depend on the magnitude of the normal load. This effect of the normal load on the wear rate may be due to the dependence of the flow stress of plastics on the hydrostatic component of stress. Table X.4 gives the wear constant for several polymers. Lewis (Ref. 35) proposed a somewhat simpler correlation scheme for wear of polymers based on an empirical relation, which may be written as

$$W = \kappa PVt \ (\text{in}^3) \tag{X.36}$$

where κ is the wear factor in units of $(\text{in}.^3 \ \text{min/ft-lb-hr})$, V is the sliding velocity in (ft/min), t is time (hr), and P is load (lb).

X-9-b Polymers as Bearing Materials

The major advantage of using plastics as bearing materials is their low cost, coupled with excellent corrosion resistance, dry sliding without excessive wear, and their quiet and clean operation. The major disadvantages are their low melting temperatures, low thermal conductivity, and low mechanical strength. However, in many applications, plastic bearings and cams are more than adequate for the service requirements, if they are made thin.

Thermoplastics are used as bushings, small bearings, cams, pump seals and piston rings. Thermoplastic bearings are most often used unfilled, in dry conditions (although Teflon bearings are sometimes impregnated with metallic fillers for increased heat transfer, which is

TABLE X.4   Wear Rates of Pins of Various Polymers at a Load of 400 gm. Speed = 180 cm/sec. Polymer Pins Were Slid against Hardened Tool Steel. (*From Ref. 33.*)

Combination of materials	Wear rate (10^{-10} cm^3/cm)	Hardness (10^6 gm/cm^2)	Z
Teflon	200	0.5	2.5×10^{-5}
Perspex (PMMA)	14.5	2.0	7×10^{-6}
Molded Bakelite x 5073	12.0	2.5	7.5×10^{-6}
Laminated Bakelite 292/16	1.8	3.3	1.5×10^{-6}
Molded Bakelite 11085/1	1.0	3.0	7.5×10^{-7}
Laminated Bakelite 547/1	0.4	2.9	3×10^{-7}
Polyethylene	0.3	0.17	1.3×10^{-7}
Nylon (from Ref. 34)			1.5×10^{-6}

not, however, very effective), or in conjunction with such solid lubricants as MoS_2 (molybdenum disulfide) and graphite.

Thermosetting plastics are rarely used without reinforcement. Thermosets are reinforced with fabrics or fibers to increase their resistance to mechanical loads, especially shock loads. For these bearings, water is often used as the lubricant.

General performance data for self-lubricating plastic bearing materials are given in Table X.5.

X-10 LUBRICATION

In many applications, the friction and wear of sliding surfaces are undesirable, if not detrimental. As a means of reducing the wear and friction of materials, lubricants are introduced between the sliding surfaces to eliminate or minimize the metal-to-metal contact. The lubrication methods may be broadly categorized as follows:

1. *Fluid-Film Lubrication (Hydrodynamic Lubrication)*—The metal-to-metal contact is completely eliminated by supporting the applied load by a pressurized fluid film. Wear of the metals is not a major problem in this case. The journal bearing is a good example of a bearing which uses hydrodynamic lubrication.
2. *Boundary Lubrication*—The adhesion between the sliding surfaces is partially eliminated by introducing polar molecules which are physically adsorbed onto the metal surfaces. The thickness of the layer can range from a monolayer to several hundred angstroms.
3. *Solid Film Lubrication*—The metal-to-metal contact is completely eliminated by introducing either solid lubricants, such as graphite, MoS_2 and Teflon, or additives which form thick oxides and sulfides (400 to 1,000 Å thick) upon reaction with the metal surfaces.*

Hydrodynamic lubrication theory is beyond the scope of this book, since the problems involve fluid mechanics. The discussion here will mainly concentrate on boundary lubrication and solid-film lubrication. In recent years, there have been many investigations made into elastohydrodynamic lubrication. Elastohydrodynamic

*Some authors discuss solid-film lubrication as a subclass of boundary lubrication.

TABLE X.5 Performance Data for Self-Lubricating Plastic Bearing Materials (*Ref. 34*)

Performance characteristic or property	Unmodified polymers					Modified polymers				
	Nylon	Acetal	Fluorocarbon	Polyimide	Phenolic	Nylon, graphite filled	Acetal, TFE fiber filled	Fluorocarbon, wide range of fillers	Polyimide, graphite filled	Phenolic TFE filled
Maximum load/ projected area (zero speed), psi	4,900	5,200	1,000	10,000	4,000	1,000	1,800	2,000	10,000	4,000
Speed, continuous operation (5-lb load), max ft/min	200–400	500	100	1,000	1,000	200–400	800	1,000	1,000	1,000
PV for continuous service, 0.005-in. wear in 1000 hr	1,000	1,000	200	300	100	1,000	2,500	2,500 50,000*	3,000	5,000
Limiting PV at 100 ft/min	4,000	3,000	1,800	100,000	5,000	4,000	5,500	30,000	100,000	40,000
Coefficient of friction	0.20– 0.40	0.15– 0.30	0.04– 0.13	0.1– 0.3	0.90– 1.1	0.1– 0.25	0.05– 0.15	0.04– 0.25	0.1– 0.3	0.05– 0.45
Wear factor, $\kappa \times 10^{-10}$, in.^{3}/min/ ft-lb-hr	50	50	2,500	150	250– 2,000	50	20	1–20	15	10
Elastic modulus, bending, 10^6 psi	0.3	0.4	0.08	0.45	5	0.4	0.4	0.4	0.63	5
Critical temperature at bearing surface, °F	400	300	500	600	300–400	400	300	500	600	300–400
Resistance to: humidity chemicals	Fair Good	Good Good	Excellent Excellent	Good Good	Good Good	Fair Good	Good Good	Excellent Excellent	Good Good	Good Good
Density, gm/cm^3	1.2	1.43	2.15–2.20	1.42	1.4	1.2	1.54	2.15–2.25	1.49	1.4
Cost index for base material	1.4	1	5	15	—	1.5	6	5	15	—

*Exceeds limiting PV.

lubrication involves thin fluid films and deformation of the surfaces. The subject and even the definition of boundary lubrication are still in a state of flux.

The purpose of boundary lubrication is to separate the sliding surfaces using long-chain organic molecules with permanent dipoles at one of their ends [e.g., chlorotrifluorohydrocarbons, of the composition $CF_3(CF_2 - CFCl_nCF_3)$]. Each molecule is physically adsorbed at its active end to the metal surface. An ideal boundary lubrication exists when these molecules line up so that the long axis of each molecule is perpendicular to the metal surface, forming a complete film. Wear is minimized by boundary lubrication since the shear stress at the plane of easy shear between the film surfaces is very small, thus eliminating the plastic shear deformation of the surface layer.

As these surface layers slide over each other, they wear down and therefore means of replenishment must be provided. This can be done by flooding the interface with boundary lubricant dissolved in a solvent such as mineral oil. The wear rate is extremely low at room temperature. However, as the temperature increases, these molecules become increasingly randomly oriented, and their wear rate increases. At temperatures higher than the melting point of the lubricant, the breakdown of the lubricant occurs, and metal-to-metal contacts develop with correspondingly increasing metal wear rate.

The role of the solid lubricant is to separate the contacting metal surfaces. The solid lubricant may be soft materials which easily shear or hard oxides and sulfides formed on the metal surface to minimize the plastic deformation of the metal surface.

These lubricants may lower the frictional force by an order of magnitude, while decreasing the wear rate by several orders of magnitude. The frictional force in a well-lubricated case is that associated with the stress required to shear the lubricant.

X-11 ADHESION

X-11-a Engineering Applications of Adhesive Bonding

One of the increasingly important processes in engineering related to surface phenomena is the bonding of structural and machine parts using adhesives. In addition to the more traditional uses of adhesives, it is an important method of constructing rigid structural panels, such as the honeycomb panels used in aircraft. Adhesive bonding has advantages over other methods of joining when the parts are either thin or are small articles. The advantages are that when proper adhesives are used, the load *can* be distributed over a large area, the strength-to-weight ratio can be increased by cross bonding such as in plywood panels, dissimilar materials can be joined, the adhesive joint may be used as a moisture barrier and electric insulator, and above all, adhesive joints can be made cheaper and faster. The disadvantages are that an adhesive joint cannot be easily taken apart, bonds cannot be easily inspected, very careful surface preparation and process control are essential, some adhesives take a long time to cure, and the service temperature may be limited.

There are many different kinds of adhesives: natural, semi-synthetic, and synthetic. Most thermosetting plastics, such as epoxies, polyurethanes, phenol-formaldehydes, urea-formaldehydes, melamine-formaldehydes, and polyesters, make good adhesives. Many thermoplastics, such as PMMA, polyvinyl acetate, polyvinyl alcohol, and cellulose nitrate (nitrocellulose) are also used as adhesives. Such elastomers as natural rubber, neoprene, and buta-diene-acrylonitrile are also used as adhesives. Natural materials such as starch, dextrin, and asphalt make good adhesives for wood, paper, and tiles, when dissolved in a solvent, although asphalt is also applied hot. Thermosetting plastics are normally applied after mixing two reacting components for curing in-place, whereas thermoplastics are either dissolved in a solvent for application or applied in the molten state. The cost of these adhesives ranges from a few cents per pound to several dollars per ounce. (For a detailed discussion of various adhesives[see Refs. 36 and 37].)

X-11-b Physicochemical Considerations of Adhesion

The conditions required for formation of a good adhesive junction will be considered here.* These conditions may be divided into those which affect the *bond strength* between the adhesive and adherend, and those which affect the overall *joint strength*. The bond strength, which is the tensile stress required to pull the adhesive from the

---

*There are many theories for adhesion, all with limited ranges of applicability. Among the better known theories are the chemical bond theory, surface energy theory, diffusion theory, polar theory, and electrostatic theory.

adherend, is affected by chemical properties, such as the free energies of the surfaces and interfaces, and also by physical characteristics, such as the strength of the surface layer. The joint strength, which is the force per unit area required to break the joint apart, is influenced by shrinkage of resins, mechanical properties of the adhesive, and the conditions of the adherend surface.

Different adhesive mechanisms are responsible for bonding under different situations. For example, in bonding two pieces of porous sheets such as paper, the bonding may be accomplished by diffusion of adhesives into the paper, physical adsorption and absorption, and mechanical bonding. On the other hand, metal-to-metal bonding by an adhesive may only involve physical adsorption and chemisorption, whereas the bonding of thermoplastics to metal by a solvent-type adhesive may involve diffusion of solvents and adhesives into the plastic, followed by physical adsorption, involving secondary-bond forces.

In all these cases, good wetting of the surface by the adhesive is one of the most important requirements. In Sec. X-3 the fact that the surface of materials has a higher free energy than the bulk and that chemisorption and physical adsorption at the surface act to lower the free energy was discussed. Chemisorption involves the formation of primary bonds, often by the formation of oxide layers. In physical adsorption moisture and gas molecules are attracted to the surface and form secondary bonds. For good adhesion, the adhesive must wet the surface of the adherend thoroughly without entrapping air bubbles. In order to accomplish good wetting, the total free-energy change of the surface and interface must be less than the free energy of the original surface. Therefore, the surface must either be free of surface contaminants, or the adhesive must displace the contaminants to achieve good wetting.

When the adherend can be dissolved by the adhesive (such as when PMMA is bonded with Duco Cement*), intermolecular diffusion can aid the bonding. In this case, the free-energy change must be such that

$$\Delta F = \Delta H - T\Delta S < 0 \qquad (X.37)$$

where ΔH is the heat of mixing, T is the temperature, and ΔS is the entropy of mixing. When Eq. (X.37) is satisfied, wetting is easily

---

*Tradename of duPont de Nemours Co.

accomplished. Many nonpolar and moderately polar substances are such that ΔF is greater than zero. The wetting characteristics of adhesives on a nonsoluble substrate are often specified in terms of the surface energy γ and the wetting angle θ as*

$$\cos \theta = \frac{\gamma_{sv} - \gamma_{sL}}{\gamma_{Lv}} \qquad (X.38)$$

where the subscripts sv, sL, and Lv denote the solid-vapor, solid-liquid, and liquid-vapor interfaces. When θ is less than $90°$, the adhesive spreads over the surface.

The adhesive bond strength is derived primarily from the secondary-bond forces, i.e., hydrogen bond, van der Waals, etc. (see Chapter VI). In some limited, exceptional cases, primary bonds, such as covalent and ionic bonds, play a role. The joint strength cannot exceed the bond strength, but the work required to break the joint can greatly exceed the bond energy. The work required to break the joint is used in propagating cracks and in deforming visco-elastic-plastic adhesives. A crack in the adhesive joint does not normally propagate directly across the adhesive, but rather zigzags between the adherends, sometimes propagating along the adhesive-adherend interface. Before the final failure occurs, thousands of cracks may develop. Thus, because of the large crack area generated, and because of the viscous and plastic work dissipated in forming the surfaces, the work required to break an adhesive joint will be greater than the surface energy of the plane area.

The zigzagging pattern can be caused either by the residual stress existing in the adhesive joint due to the shrinkage of the adhesive during curing, or by inhomogeneities in the material. If the adhesive is viscoelastic, the failure mechanism depends on the speed of testing. At very high speeds, brittle fracture of the adhesive may occur, or the adhesive-adherend interface may separate, if the cohesive strength is greater than the bond strength. At low speeds both viscous and plastic deformation occur, sometimes accompanied by the formation of voids in the adhesive.

*The free surface energy per unit area (sometimes called the specific free energy) is equal to the surface tension in a liquid.

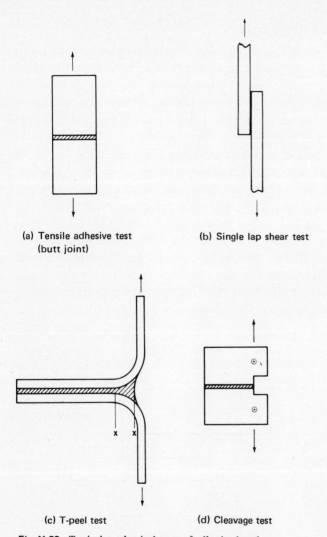

(a) Tensile adhesive test
(butt joint)

(b) Single lap shear test

(c) T-peel test

(d) Cleavage test

Fig. X.29   Typical mechanical tests of adhesive bonds.

A strong adhesive joint to a plastic cannot be made if a weak surface layer is present on the surface of the plastic, as discussed in Sec. X.4. The joint strength can be increased considerably by bombarding the surface with ionized gas to cross link the weak molecules. For example, the joint strength between polyethylene/ epoxy adhesive/aluminum can be increased substantially by exposing the polyethylene to ionized gas for about 10 sec (Ref. 6).

X-11-c Mechanics of Adhesive Joints

The lap joint shown in Fig. X.29 is the most commonly used joint. Butt joints, which are subjected to tensile loading across the thickness of the adhesive are usually weak, since peeling may start at localized points and propagate. However, when testing adhesives, tensile tests of butt joints are also used, as well as shear tests of lap joints, T-peel tests, and the cleavage tests, as shown in the figure.

The stress distribution in a lap joint has been analyzed (Refs. 38 and 39), the derivation of which is beyond the scope of this book. The results of the analysis for the stress distribution along the shear plane in joints with a relatively inflexible cement are plotted in Fig. X.30. The expression σ_{22} is the tearing stress caused by the bending moment. When the adhesive is ductile, the stress concentration near the leading edge of the lap joint is less severe. In this case, the joint fails when the yield zones in the adhesive spread over the entire bonded area. Therefore, the load-carrying capacity is proportional to the total bonded area. When the adhesive is a glassy brittle type, the

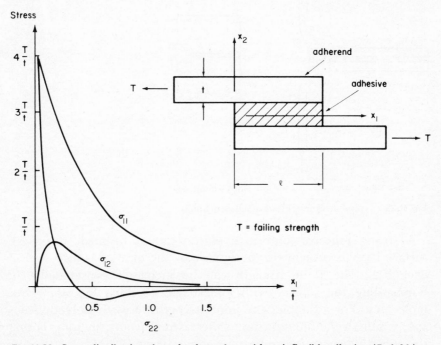

Fig. X.30 Stress distribution along the shear plane with an inflexible adhesive. (*Ref. 39.*)

effect of the stress concentration is important, and the bonded area is not very important. In this case, the length of the overlap is not critical, since the crack propagation will be controlled by the maximum tensile stress near the bond edge (Ref. 40).

X-8-d   Practical Rules of Thumb

To obtain a good adhesive joint the following rules of thumb may be used:

1) The thickness of the adhesive should be as small as possible, consistent with the smoothness of the adherend surface.

2) The adhesive should not shrink too much during curing (e.g., use epoxies rather than polyester).

3) The adhesive should be less rigid than the adherend.

4) When plastics with weak surface layers are joined, the weak surface layer should be removed or treated. In the case of metals the surface should be degreased or roughened.

5) Fillers should be incorporated to reduce shrinkage.

6) Solvents should be evaporated from adhesives before joining impervious substrates.

REFERENCES

1. Rabinowicz, E.: "Friction and Wear of Materials," John Wiley, New York, 1966.
2. Ku, P. M. (ed.): Interdisciplinary Approach to Friction and Wear, NASA SP-181, 1967.
3. Cottrell, A. H.: "Dislocation and Plastic Flow in Crystals," Oxford (Clarendon), New York, 1953.
4. Fourie, J. T.: The Plastic Deformation of Thin Copper Single Crystals, *Canad. J. Phys.*, vol. 45, pp. 777–786, 1967.
5. Argon, A. S.: Effects of Surfaces on Fatigue Crack Initiation, *Corrosion Fatigue*, NACE-2, pp. 176–182, 1972.
6. Schonhorn, H: Adhesion to Low Energy Polymers, "Adhesion," Gordon and Breach Science Publishers, New York, 1969.
7. Suh, N. P., and B. J. Sanghvi: Frictional Characteristics of Oxide-Treated and Untreated Tungsten-Carbide Tools, *J. Eng. for Ind.*, vol. 93, pp. 455–460, 1971.
8. Green, A. P.: Friction between Unlubricated Metals: A Theoretical Analysis of the Junction Model, *Proc. Roy. Soc. (London)*, vol. 228, p. 191, 1955.
9. Allan, A. J. G.: Plastics as Solid Lubricants and Bearings, *Lubric. Eng.*, vol. 14, p. 211, 1958.
10. Pinchibeck, P. H.: A Review of Plastic Bearings, *Wear*, vol. 5, pp. 85–113, 1962.
11. McLaren, K. G., and D. Tabor: Friction of Polymers at Engineering Speeds: Influence of Speed, Temperature, and Lubricants, *Instn. of Mech. Engrs. Paper No. 18*, Lubrication and Wear Convention, Bournemouth, May, 1963.
12. Chung, C. I.: New Ideas about Solids Conveying in Screw-Extruders, *SPE J.*, vol. 26, no. 5, pp. 32–44, 1970.

13. Pascoe, M. W., and D. Tabor: The Friction and Deformation of Polymers, *Proc. Roy. Soc. (London)*, vol. A235, pp. 210-224, 1956.

14. Grélan, N.: Friction Between Single Fibers, *Proc. Roy. Soc. (London)*, vol. A212, p. 491, 1952.

15. Steijn, R. P.: Friction and Wear of Plastics, *Met. Eng. Quart., ASM*, May, 1967.

16. Grosh, K. A.: Relation between Friction and Visco-Elastic Properties of Rubber, *Proc. Roy. Soc. (London)*, vol. A274, pp. 21-39, 1963.

17. Ludema, K. C., and D. Tabor: The Friction and Viscoelastic Properties of Polymeric Solids, *Wear*, vol. 9, pp. 329-348, 1966.

18. Seifert, W. W., and V. C. Westcott: A Method for the Study of Wear Particles in Lubricating Oil, *Wear*, vol. 21, pp. 27-42, 1972.

19. Archard, J. F.: Contact and Rubbing of Flat Surfaces, *J. Appl. Phys.*, vol. 24, pp. 981-988, 1953.

20. Suh, N. P.: The Delamination Theory of Wear, *Wear*, vol. 25, pp. 111-124, 1973.

21. Suh, N. P., S. Jahanmir, E. P. Abrahamson, II, and A. P. L. Turner: Further Investigation of the Delamination Theory of Wear, *J. Lubr. Tech.*, in press.

22. Zener, C.: Micro Mechanism of Fracture, *Fracturing of Metals, ASM*, pp. 3-31, Novelty, Ohio, 1949.

23. Stroh, A. N.: The Formation of Cracks as a Result of Plastic Flow, *Proc. Roy. Soc. (London)*, vol. A223, pp. 404-414, 1954.

24. Stroh, A. N.: The Formation of Cracks in Plastic Flow II, *Proc. Roy. Soc. (London)*, vol. A232, pp. 548-560, 1955.

25. McClintock, F. A.: On the Mechanics of Fracture from Inclusions, *Ductility, ASM*, pp. 255-277, Metals Park, Ohio, 1968.

26. McClintock, F. A., and A. S. Argon: "Mechanical Behavior of Materials," p. 549, Addison-Wesley, Reading, Mass., 1966.

27. McClintock, F. A., S. M. Kaplan, and C. A. Berg: Ductile Fracture by Hole Growth in Shear Bands, *Intern. J. Fract. Mech.*, vol. 2, pp. 614-627, 1966.

28. Bitter, J. G. A.: A Study of Erosion Phenomena, Parts I and II, *Wear*, vol. 6, pp. 5-21, 169-190, 1963.

29. Finnie, I.: The Mechanism of Erosion of Ductile Metals, *Proc. 3d U.S. Nat. Cong. Appl. Mech.*, 1958.

30. Neilson, J. H., and A. Gilchrist: Erosion by a Stream of Solid Particles, *Wear*, vol. 11, pp. 111-122, 1967.

31. Smeltzer, C. E., M. E. Gulden, and W. A. Compton: Mechanisms of Metal Removal by Impacting Dust Particles, *ASME Paper 69-WA-Met-8*, 1969.

32. Cook, N. H., and B. Bhushan: Sliding Surface Interface Temperatures, *ASME Paper No. 72-Lub-34*, 1972.

33. Archard, J. F., and W. Hirst: The Wear of Metals under Unlubricated Conditions, *Proc. Roy. Soc. (London)*, vol. A236, p. 397, 1956.

34. Steijn, R. P.: Friction and Wear of Plastics, *Met. Eng. Quart., ASM*, May, 1967.

35. Lewis, R. B.: Predicting Wear of Sliding Plastic Surfaces, *Mech. Eng.*, October, 1964.

36. Skeist, I. (ed.): "Handbook of Adhesives," Reinhold Publishing Corp., New York, 1962.

37. Shields, J.: "Adhesive Handbook," CRC Press, Cleveland, 1970.

38. Eley, D. D. (ed.): "Adhesion," Oxford University Press, London, 1961.

39. Alner, D. J.: "Aspects of Adhesion 5," CRC Press, Cleveland, 1969.

40. Wang, T. T., F. W. Ryan, and H. Schonhorn: Effect of Bonding Defects on Shear Strength in Tension of Lap Joints Having Brittle Adhesives, *J. Appl. Polymer Sci.*, vol. 16, pp. 1901-1909, 1972.

41. Gupta, P. K., and N. H. Cook: Junction Deformation Models for Asperities in Sliding Interaction, *Wear*, vol. 20, pp. 73-87, 1972.

PROBLEMS

X.1 In order to determine if the compressor blades for jet engines can be made of graphite fiber-reinforced composite materials with epoxy matrix, erosion tests were performed by impinging the composite surface with alumina powder (Al_2O_3). The following data were obtained at the impingement velocity of 300 ft/sec:

Impingement angle α	Erosion weight loss (gm/gm of Al_2O_3 impacted)
10°	50×10^{-4}
20°	75×10^{-4}
25°	170×10^{-4}
40°	130×10^{-4}

Under actual operating conditions, the maximum tolerable erosion rate is 800×10^{-4} gm/gm of Al_2O_3 impacted. Under normal operating conditions of jet engines, the impingement velocity is about 750 ft/sec and the average impingement rate of abrasive particles is 500 gm/hr.

a) Determine whether the fiber-reinforced composite will meet the above requirement, and determine the hourly erosion rate under the stated conditions.

b) Discuss what you might do to the surface to increase the blade life.

X.2 Given Teflon, nylon, polyimide and polyethylene, which material would you use to make a bearing for use at a sliding speed of 100 cm/sec and normal pressure of 100 gm/cm^2? Why?

X.3 It is found that using a lubricant between two sliding surfaces reduces the coefficient of friction by a factor of 10. It is also found that the lubricant reduces the wear rate by a factor of 200. Why?

X.4 A drive shaft connecting two machines suffers a bending deflection because the two machines are misaligned by an angle θ. The bending moment is applied to the ends of the shaft by a pair of journal bearings which can be assumed to apply the moment in the form of a force couple, as shown in the figure. The deflection of the shaft can be assumed to be negligible compared with the bending moment. The bending of the shaft is subjecting the shaft to fatigue and also causing excessive wear on the bearings. Using the dimensions and material properties given, answer the following questions:

a) What is the fatigue life of the shaft in terms of revolutions?

b) What is the size of the fatigue crack when final failure occurs? (i.e., what is the critical crack size?)

c) If the bearings are assumed to be worn out when 0.03 in.[3] of material has been worn away, what is the life of the bearings in terms of shaft revolutions?

d) Estimate the order of magnitude of the increase in bearing life which will be achieved by lubrication.

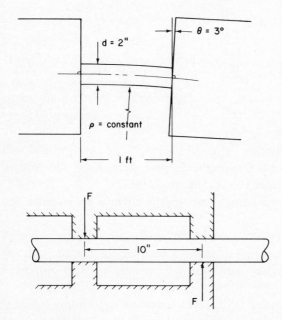

Shaft: 2-in. diameter
    1 ft between machines
    4340 steel
    σ_u = 230,000 psi
    Fatigue limit = 115,000 psi
    k_{1c} = 38,000 psi in.$^{1/2}$

Bearings: Free cutting
    Phosphor bronze
    σ_u = 50,000 psi
    Wear coefficient on
        4340 steel = 5×10^{-4}

X.5   The machine component shown in the figure below is supported by pins at either end and is loaded by a force applied at the center which varies with time as $F = F_m + F_a \sin \omega t$.

The oscillating load may cause fatigue of the bar. In addition, the rotation of the ends about the pins due to the deflection causes fretting wear at the pins. The bar is made from 2024-T3 aluminum and the pins from hardened 4340 steel. The wear coefficient for these materials is $Z = 2 \times 10^{-3}$ and you may assume that fretting obeys the usual wear equation. The bar must be replaced when the wear reaches a value of $V = 5 \times 10^{-3}$ in.^{3}. If $F_m = 2,000$ lb, $F_a = 200$ lb, and the bar has dimensions shown below, will the part fail by fatigue or by wear?

Assume that the fatigue does not occur at the pins and that there is no significant stress concentration at the point of application of the load. The angle of rotation of the end of a simply supported beam loaded at the center is $\theta = FL^2/16EI$. The fatigue strength of the aluminum at 10^7 cycles is 20 ksi, and the fatigue life at the tensile strength ($\sigma_u = 90$ ksi) is 10^3 cycles. Assume that the hardness H is approximately equal to three times the tensile strength for 2024-T3 aluminum.

If this part is to last for 10^9 cycles of loading, what recommendations would you make? (Qualitative answers only.)

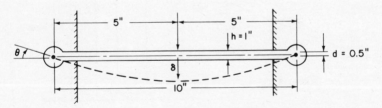

[$E = 10^7$ psi; thickness of the bar = 0.4"]

X.6   One of the dangers of a space flight is that the rocket fuel may ignite prematurely, before the space vehicle is ready to take off. This will surely lead to the incineration of the astronauts in the space module at the tip of the rocket. In order to allow an escape route, a stainless steel rope is provided, as per diagram. In case of fire, the astronauts can slide down the wire to safety by holding on to a contrivance which slides along the wire.

What are the qualitative requirements for the friction and the wear between the device and the wire? (High, low, constant, fluctuating, etc.?) Should the combination's frictional properties be affected by the presence of moisture or contaminants?

In some preliminary tests, the block that rides on the rope was made of a stainless steel coated with a lead film .001″ thick. This was fine for the first few feet of travel, but then the lead wore off and unsatisfactory sliding occurred.

Discuss the following suggested remedies:

a) Make the block out of pure lead.
b) Make the block out of a free machining steel (lead content .35%).
c) Make the block out of titanium.
d) Make the block out of a tin bronze.
e) Make the block out of rubber.
f) Do you have a candidate material?

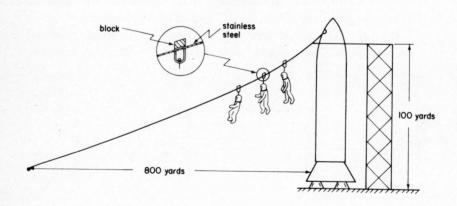

X.7 Indicate which of the following adhesive joints are good joints:

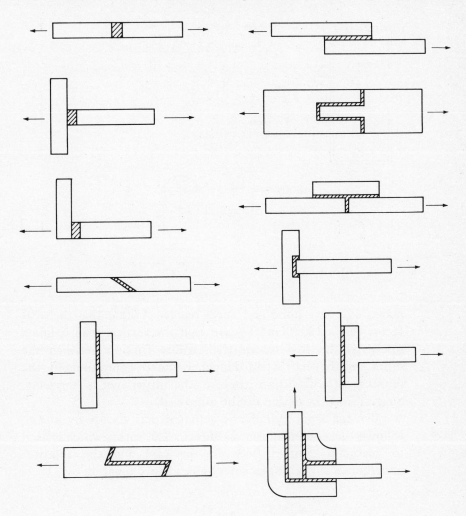

X.8 The rotating drum of a commercial clothes dryer is attached to its shaft by two screws as shown. Due to the eccentric loading of the drum and the clearance around the screws, there is some relative motion between the screws and the drum. The initial clearance between the screw shank and the hole is 0.005 in. When this clearance becomes larger than 0.020 in., the dryer

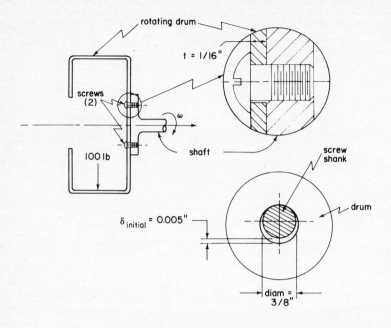

makes excessive noise and needs repairs. Assume that the wear factor is 1×10^{-9} cm^2/kg and that the vertical load is taken solely by the two screws (i.e., ignore friction between the drum and the shaft's end flange). Estimate approximately the number of revolutions before the dryer needs servicing. Suggest how the design can be improved.

X.9 A valve for a gas cylinder is actuated periodically by a cam-follower-lever-mechanism, as shown. The spring which is used to maintain contact between the cam and the follower undergoes a cyclic displacement given approximately by

$$d = d_0 + d_{max} \sin 2\pi\omega t$$

where ω is the rotational velocity of the cam. If it is not within 0.005 in. of the original closed design position, the valve will not seal properly.

Hardened steel (assume zero wear) is used for the cam-follower and brass for the cam itself. Using the data below, calculate the useful life of the cam, with and without lubrication.

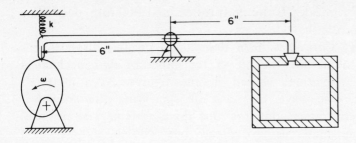

Data: d_0 = 1/2 in.

d = 1/4 in.

k = spring constant = 200 lb/in.

ω = 6.28 radians/sec

κ = wear factor without lubricant = 4×10^{-7} cm²/kg

κ_l = wear factor with lubricant = 4×10^{-10} cm²/kg

t = thickness of cam = 1/4 in.; r_{max} = max radius of cam = 1.5 in.

X.10 A rotating shaft in an engine is pressed on a ball bearing as shown. The end of the shaft carries a vertical load F. The stress between the shaft and the bearing is primarily caused by the externally applied load, the stress due to the press-fit being much smaller. During the operation there is an oscillating relative slip of 0.002 in. along the circumferential direction between the shaft and the bearing. The shaft is replaced when the weight of worn metal (determined by collecting worn metal particles carried by lubricating oil circulated through the system) is 20 gm. The shaft is made of AISI 4340 steel, and the inner race of the ball bearing is 52100 steel. The wear factor for the fretting wear under the operating condition is

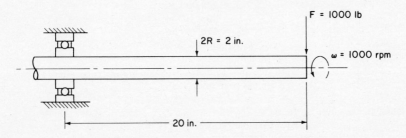

4×10^{-10} cm^2/kg. If the shaft rotates at 1,000 rpm, determine the life of the shaft.

X.11 Tootherty Corp. has been designing a positive displacement system to provide the power for either of two mechanical forming operations shown in the figures. As a materials engineer, you have the task of specifying the *minimum* component dimensions for this drive system. The physical layout and materials have already been chosen, and are listed with appropriate physical properties.

The two forming processes are as follows:

1) Sideways extrusion of a compacted, sintered powder-metallurgy component. The work material may be modeled as exhibiting a shear flow stress of 2,500 psi. Neglect friction on the facing plates, but consider that a sticking condition prevails everywhere else.

2) A pressure-forming operation of a 5/16 in. thick brass sphere from a radius of 2 to 2 1/4 in. The piston acts on a fluid which pressurizes the sphere. A pressure relief valve will be preset to a 10% greater pressure than typically required to set one of the spheres.

The power device will provide a piston stroke of 2 in. in each case. Use the maximum force necessary as the load requirement. Neglect eccentric loading due to angular changes of the device except where motion is a necessary parameter (i.e., in general, the loads can be modeled as occurring in the shown configuration).

Material Properties

Sideways extrusion:	k = 2.500 psi
Spherical forming:	70–30 leaded brass
	$\bar{\sigma} = 100,000 \, (\bar{\varepsilon})^{0.5}$ psi
Moving arms:	4340 steel
	σ_Y = 230,000 psi
	E = 30 × 10^6 psi
	σ_u = 250,000 psi
	k_{1C} = 85,000 psi in.$^{1/2}$
	K_f(notch) = 1.5
Pins and bearings:	4340 steel pins (3/8″ diameter) on brass bushings (well lubricated)
	$Z = 2 \times 10^{-6}$
Plastic tubing (air supply):	Delrin
	Operating temperature = 110°F

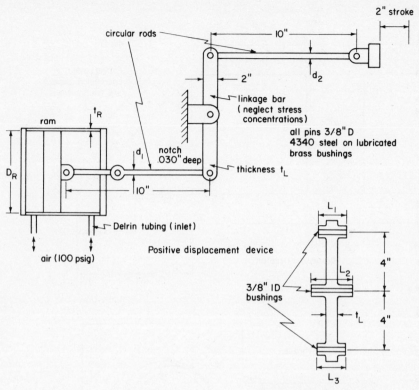

Positive displacement device

Bearing surfaces of pivot arm

The linkage components are made of 4340 steel for which data is given. The parts are to be designed for a life of 10 years or 10^6 cycles, whichever is relevant. The bushings for the three bearing pins are made of brass, and the surfaces are well lubricated. Dimensional tolerances of the bearing surfaces must be held within 0.005 in. in the stated lifetime.

The pneumatic ram is also made of 4340 steel and the air supply can provide a controlled pressure of 100 psi. To prevent leakage, the maximum gap opening between the piston and cylinder must be less than +.005 in. greater than its manufactured size. You may neglect wear of the piston and cylinder due to adequate lubrication.

A 0.030 in. deep notch is to be cut into the tension arm so that a linear velocity transducer can be attached for control

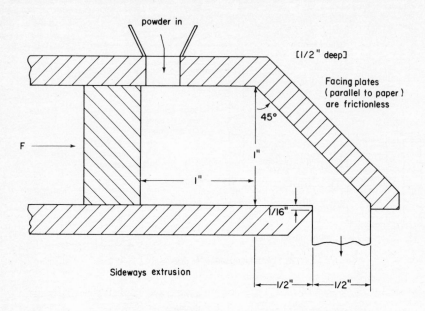

Sideways extrusion

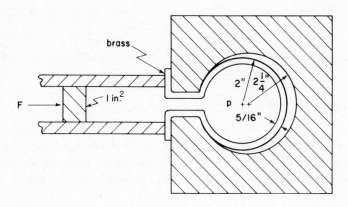

Spherical forming

purposes. The notch provides a fatigue strength reduction factor of $K_f = 1.5$, among other possible effects.

The strain in the inlet tubes is to be kept under 5% during the 10-year operating life of the device. The tubes are made of Delrin, and the operating temperature of the machine is anticipated to be 110°F. Specify a maximum value of r/t for these tubes.

Consider all possible failure modes for this device and specify minimum dimensions for all components: bearing lengths, rod diameters, linkage thickness, wall thickness (all unspecified dimensions in the drawing). Be explicit as to the reasoning behind your judgments and assumptions.

APPENDIX X-A

Models for Delamination Wear

There are two possible sets of assumptions which can be made to build a model of wear by the delamination process. In one model, it is assumed that at some fraction of the asperity contacts, a strong sliding junction is formed which causes the material to be sheared until a wear sheet is formed. A wear sheet is created by interaction of one set of asperities. In the other model, it is assumed that the creation of a wear sheet is a cumulative process which results from the material being sheared a small amount by each passing asperity, but that creation of a wear sheet will occur only after a large number of asperities have passed each point on the surface.

1 Single-contact Wear Sheet Formation

It may be assumed that a wear particle is formed by interaction with a single asperity. The number of wear particles formed per unit distance slid as estimated from the wear rate and the wear particle size is often of the same order of magnitude as the number of asperity contacts expected in this distance. It is also assumed that wear particles can be formed by a single pass of a slider over a surface during a wear experiment. When this is assumed to be the nature of the process, the following assumptions can be made about the wear process.

a) It m..y be assumed that void and crack nucleation and the critical amount of shear deformation are produced when the two contacting asperities undergo a critical sliding displacement S_0. Since the state of stress and strain under the contacting asperities is independent of the normal load, S_0 should be a constant for a given material.

b) The number of wear sheets, N, formed in a sliding distance S_0, is proportional to the number of asperities in contact at any instant in time. The proportionality constant need not be 1, since contacting asperities may not produce wear sheets.

c) The area of the wear sheet, A, will be proportional to the area of the asperity contact.

d) The thickness of a wear sheet, h, depends on the structure and the mechanical properties of the wearing material and is independent of the normal load and sliding speed.

Assumptions a) and b) imply that if S is the total distance slid, NS/S_0 will be the total number of wear sheets formed in a sliding distance S. Assumptions b) and c) imply that the actual worn area in a wear experiment is proportional to the number of asperities in contact, which depends on the load. This is consistent with the results of pin-on-disk wear experiments, as illustrated in Fig. A.1. Normally when the slider moves over the disk, many wear tracks are formed. The width of each wear track is much smaller than the width of the apparent area of contact between the slider and the disk. When the normal load is changed, there is a change in the width of the wear tracks and the number of wear tracks formed, so that the actual worn area is in fact a function of the applied normal load.

Using the above assumptions, the volume of material worn from the contacting surfaces can be written as

$$W = N_1 \left(\frac{S}{S_{o_1}} \right) A_1 h_1 + N_2 \left(\frac{S}{S_{o_1}} \right) A_2 h_2 \qquad \text{(A1)}$$

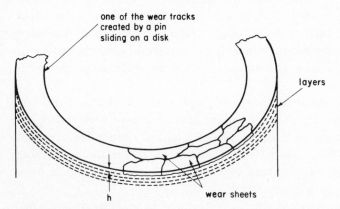

Fig. A-1 Schematic illustration of the delaminated wear sheets under an isolated track. Note that there are many similar tracks under a slider pin. (*Ref. 20.*)

where A is the average area of the delaminated sheet, h is the thickness of the delaminated sheet, and S is the distance slid. The subscripts 1 and 2 refer to the metals sliding against each other. The term NS/S_0 is equal to the number of wear sheets removed. The distance S_0 is likely to take on smaller values when compatible materials are slid over each other, as a result of the greater shear stress established at the interface, and, also, when inclusions are present in the material, because inclusions accelerate crack nucleation and growth.

The average surface area of each delaminated sheet is assumed to be proportional to the real area of contact per asperity, a_r, as follows:

$$A = C a_r \tag{A2}$$

where C is a proportionality constant. Gupta and Cook (Ref. 41) showed that the real area of contact per asperity, a_r, and the number of asperities in contact, n, are both proportional to the applied load L by

$$\begin{aligned} a_r &\propto L^{0.09} \\ n &\propto L^{0.91} \end{aligned} \tag{A3}$$

By assumption b) above, n is proportional to N. The substitution of Eqs. (A2) and (A3) into Eq. (A1) yields

$$W = \left(\frac{K_1 h_1}{S_{o_1}} + \frac{K_2 h_2}{S_{o_2}}\right) LS = \kappa \cdot LS \tag{A4}$$

where κ is a wear factor, and K_1 and K_2 are constants which depend primarily on the surface topography. This equation corresponds to Eq. (X.31).

2 Multiple-contact Wear Sheet Formation

It may be assumed that each asperity which passes a given point on a wearing surface produces only a small increment of plastic deformation in the surface layer, and that a loose wear particle is produced only after a large number of asperities have passed. In this case the critical sliding distance for the production of a layer of wear particles will not be a constant but will depend on the number and distribution of contacting asperities. In this model, it is assumed that a loose wear layer is formed when a critical plastic displacement d_c has occurred at the surface. It will be assumed that the plastic displacement d_p which has occurred at a point on the surface is proportional to the total distance slid and to the percentage of time that any point on the surface has been in actual contact. If A_r is the total real area of contact at any time, and A_T is the actual area of the wear track on the wearing surface, each point on the wear track should be in actual contact with the mating surface for a fraction of the total time equal to the ratio A_r/A_T. Therefore, the plastic displacement is given by

$$d_p = \frac{BSA_r}{A_T} \tag{A5}$$

where B is a proportionality constant, and S is the distance slid. When d_p reaches its critical value d_c, a layer of loose wear particles will be formed. The total area of the layer will be equal to the wear track area A_T. The thickness of the layer, h, is assumed to be determined by the structure and mechanical properties of the material and to be independent of the normal load L. The critical sliding distance for producing a layer of loose particles is found by setting $d_p = d_c$ and solving Eq. (A5) for S, giving

$$S_c = \frac{d_c A_T}{BA_r} \tag{A6}$$

where the subscript c denotes the critical values. When the total sliding distance S is much greater than S_c, the ratio $S/S_c = N$ equals the total number of layers removed.

The volume of material worn away, W, is then given by

$$
\begin{aligned}
W &= N_1 h_1 A_{T_1} + N_2 h_2 A_{T_2} \\
&= \frac{S}{S_{c_1}} h_1 A_{T_1} + \frac{S}{S_{c_2}} h_2 A_{T_2}
\end{aligned}
\tag{A7}
$$

where the subscripts 1 and 2 refer to the two materials of the contacting pair. Substituting for S_c, using Eq. (A6), gives

$$W = \frac{B_1 h_1 A_r S}{d_{c_1}} + \frac{B_2 h_2 A_r S}{d_{c_2}} \qquad \text{(A8)}$$

Since the real area of contact is proportional to the normal load L, according to $A_r = CL/H$, this can be written

$$W = \frac{B'_1 h_1}{d_{c_1} H} LS + \frac{B'_2 h_2}{d_{c_2} H} LS \qquad \text{(A9)}$$

$$= \kappa \cdot LS$$

where, $B' = BC$ is a new constant, and H is the hardness of the softer material.

This formula depends on a number of material parameters in addition to the hardness H. The way in which the thickness h of the removed layer depends on the material structure was discussed in Sec. X.8. The critical plastic displacement to form loose particles, d_c, should be qualitatively related to the ductility of the material in shear. Increasing hardness is usually accompanied by decreasing ductility, especially when the hardness is achieved by introducing second-phase particles. Thus, the two constants in the denominator often have compensating behavior.

Appendix

Conversion Factors

	English System		Metric System
Length:	Inch	25.4	Millimeters (mm)
	Foot	0.3048	Meters (m)
	Yard	0.9144	Meters
	Mile	1.609	Kilometers (km)
Area:	Inch^2	645.2	Millimeters^2 (mm^2)
		6.45	Centimeters^2 (cm^2)
	Foot^2	0.0929	Meters^2 (m^2)
	Yard^2	0.8361	Meters^2 (m^2)
Volume:	Inch^3	16,387.	Millimeters^3 (mm^3)
		16.387	Centimeters^3 (cm^3)
		0.0164	Liters (l)
	Quart	0.9464	Liters
	Gallon	3.7854	Liters
	Yard^3	0.7646	Meters^3 (m^3)
Mass:	Pound mass (lbm)	0.4536	Kilograms (kg)
		453.6	Grams (g)
	Ounce mass (ozm)	28.35	Grams (g)
	Ton mass (tonm)	907.18	Kilograms (kg)
		0.907	Ton (t)

Conversion Factors (*Continued*)

	English System		Metric System
Force:	Ounce force (ozf)	0.2780	Newtons (N)
	Pound force (lbf)	4.448	Newtons
Energy:	Btu	1055.	Joules (J=W-sec)
		252.2	Calories
		6.585×10^{21}	Electron Volts
	Foot pound (ft-lbf)	1.3558	Joules
		0.1383	kgf-m
	Kilowatt-hr	3.6×10^6	Joules
Power:	Horsepower (hp)	0.746	Kilowatts (kw)
Pressure or Stress:	Pounds/square inch (psi)	0.07031	kgf/cm^2
		6.895	Kilopascals (kPa)
	Atmosphere (Atm)	1.033	kgf/cm^2
	Inches of H_2O	0.00254	kgf/cm^2
		0.2491	Kilopascals
	dynes/square cm	1.45×10^{-5}	psi
Velocity:	Feet/second	30.48	cm/sec
		1.097	km/hr
	Miles/Hour	1.6093	km/hr
		44.70	cm/sec
Temperature:	Degrees fahrenheit ($^\circ F$)	($^\circ F - 32$)/1.8	degrees Celsius ($^\circ C$)
	$^\circ R (= ^\circ F + 459.7)$	.556	$^\circ K (= ^\circ C + 273.2)$

Useful Constants

Stefan-Boltzmann number	$= 0.171 \times 10^{-8}$ Btu/ft^2 hr $^\circ R^4$
Avogadro's number	$= 6.022 \times 10^{23}$ molecules/gm-mole
Gravitational constant (G)	$= 6.670 \times 10^{-11}$ m^3/kg-sec^2
Gravitational acceleration	$= 9.813$ m/sec^2
	$= 32.17$ ft/sec^2
Universal gas constant (R)	$= 8.31434 \times 10^7$ erg (gm-mole)$^{-1}$ $^\circ K^{-1}$
	$= 1.986$ Btu (lbm-mole)$^{-1}$ $^\circ R^{-1}$
Boltzmann's constant (k)	$= 1.38 \times 10^{-16}$ erg/atom $^\circ C$
	$= 6.79 \times 10^{-23}$ in.-lb/atom $^\circ R$

Tables of Physical Properties

Steels

Material	Condition	Density lb/in³	Melting range °F	Coeff. of them. exp. 10^{-6} °F⁻¹	Young's modulus (10^6 psi)	Shear modulus (10^6 psi)	0.2% yield strength (10^3 psi)	Tensile strength (10^3 psi)	Elongation %	Brinell hardness	Comments and uses
Plain Carbon											
AISI C1020	Cold drawn	0.283	2750–2775	8.4	30	11.6	51	61	15	12	Crankshafts, tie rods, bolts, easily welded
C1040	Cold drawn	0.283	2700–2750	8.3	30	11.6	71	85	12	170	Gears, crankshafts, camshafts
C1095	Hot rolled	0.283	–	8.1	30	11.6	66	120	10	248	Heavy machine parts, difficult to weld
B1112	Cold drawn	0.283	–	8.4	30	11.6	57	68	10	137	Case hardened for surface finish and wear, nuts, sleeves, handles
Low Alloy											
AISI 1340	1550° Quench 600° Temper	0.283	–	7.9	30	11.6	206	227	11	450	Axles, shafts, gears, bolts
2340	1425° Quench 600° Temper	0.283	2600–2620	8.0	30	11.6	205	222	11	435	Gears, cams
3140	1500° Quench 600° Temper	0.283	–	–	30	11.6	209	228	11	450	Axles, shafts, pump liners
4340	1525° Quench 600° Temper	0.283	2740–2750	8.1	30	11.6	230	250	9	485	Heavy duty gears, aircraft tubing
5140	1500° Quench 600° Temper	0.283	2750–2760	7.4	30	11.6	211	232	11	450	Transmission gears, springs
8640	1525° Quench 600° Temper	0.283	2745–2755	8.2	30	11.6	220	240	10	470	Gears, cams, hand tools
Tool											
AISI W1	Working condition	–	–	–	30	11.6	–	–	–	587–680	Shallow hardened, cutlery, taps, woodworking tools
H20	Working condition	–	–	–	30	11.6	–	–	–	426–547	Mandrels, metal extrusion dies
T1	Working condition	–	–	–	30	11.6	–	–	–	656–680	High-load, high-temperature applications, cutting tools, broaches
F1	Working condition	–	–	–	30	11.6	–	–	–	560–680	Good wear resistance at low temperatures, wire drawing dies, forming tools
Stainless											
AISI 304	Austenitic Annealed	0.283	2550–2650	10	28	10.7	35	85	55	150	General purpose, weldable
316	Austenitic Annealed	0.283	2500–2550	9	28	10.7	35	85	55	150	For highly corrosive, high-temp. applic.
410	1800° Quench, 1000° Temper, Martensitic	0.283	2700–2790	6	29	11.1	115	145	20	300	General purpose, heat treatable
440C	1900° Quench, 600° Temper, Martensitic	0.283	2500–2750	5.5	29	11.1	275	285	2	580	High hardness, bearings, races, nozzles

Aluminums

Material	Condition	Density lb/in³	Melting range °F	Coeff. of them. exp. 10^{-6} °F^{-1}	Young's modulus (10^6 psi)	Shear modulus (10^6 psi)	.2% yield strength (10^3 psi)	Tensile strength (10^3 psi)	Elongation %	Brinell hardness	Comments and uses
1100	Annealed Strain hardened H18	0.098	1190–1215	14.2	9.5–10.0 9.5–10.0	3.8 —	5 22	13 24	45 15	19 35	Weak, but used where ease of forming is required
2024	T4	0.100	935–1180	13.7	10.6	4.0	47	68	20	120	High strength, moderate toughness, used in aircraft structural components, hardware, rivets
3003	O (Annealed) H18	0.099	1190–1210	13.9	10.0 10.0	3.8 3.8	6 27	16 29	40 10	55 28	Good formability, stronger than 1100-aluminum, used in cookware, storage tanks, piping
5154	H32	0.096	1100–1190	14.4	10–10.3	3.8	30	39	15	67	Exc. corrosion resist. in comb. w/ weldability and strength, marine applic., pressure vessels, trucks + trailers
6061	T6	0.098	1080–1200	14.1	10.0	3.8	40	45	17	95	Corrsoion resist. and strength, boats appliances, furnitures, railings, piping
7075	T6	0.101	890–1180	14.4	10.4	3.9	73	83	11	150	High strength, good corrosion resist., used in aircraft structural components

Tables of Physical Properties (*Continued*)

Heat Resisting Alloys

Material	Composition	Yield strength* 75°F	Yield strength* 1000°F	Tensile strength* 75°F	Tensile strength* 1000°F	Temper temperature °F	Stress* 10^2-hr life	Stress* 10^3-hr life	Stress* 10^4-hr life	Stress* 10^5-hr life	Stress for 1% creep 10^4 hr	Stress for 1% creep 10^5 hr
17-4 PH	Precipitation hardened stainless 10.5%CR 4.25Ni 0.55i 3.6Cu 0.4Mn 0.04C	190	80	200	100	600	160	155	140	—	—	100
						800	135	115	50	—	—	—
Hastelloy-X	Nickel base 20%Fe 22Cr 9Mo 0.15C	55	45	115	95	1200	40	30	20	—	—	—
						1800	1	—	—	—	—	—
Inconel X-550	Nickel base 15%Cr 7Fe 2.3Ti 1.0Nb 1.2Al 0.4Si 0.7Mn 0.04C	90	85	160	140	1500	30	18	—	—	—	—
						1800	3	—	—	—	—	—
Waspaloy	Nickel base 19%Cr 14Co 4Mo 3Ti 1.3Al 1Fe 0.1C	120	110	190	175	1200	110	85	—	—	—	—
						1500	40	25	—	—	—	—
Udimet-500	Nickel base 19%Cr 19Co 4Mo 4Fe 3Ti 2.9Al 0.1C	110	110	175	175	1500	45	30	—	—	18	—
						1800	9	6	—	—	—	—
310 (Stainless)	Austenitic stainless steel 25%Cr 20Ni 2Mn 1.5Si 0.25C	50	35	85	70	1000	40	30	25	25	30	15
						1500	10	5	3	1	3	1

*All units in 1,000 psi.

Common Metals

	Material	Condition	Density lb/in³	Melting range °F	Coeff. of therm. exp. $10^{-6}\ °F^{-1}$	Young's modulus (10^6 psi)	Shear modulus (10^6 psi)	0.2% yield strength (10^3 psi)	Tensile strength (10^3 psi)	Elongation %	Brinell hardness	Comments and uses
Copper and Copper Alloys	ASTM B133 Deoxidized Copper	Soft	0.323	1981	9.8	17.0	6.4	10	32	45	–	High electrical and thermal conductivity, radiators, electrical conductors, hardware, plumbing
		Hard	0.323	1981	9.8	17.0	6.4	45	50	10	83–92	
	Commercial Bronze	1/4 Hard	0.318	1870–1910	10.2	17.0	6.4	35	45	25	76	Excellent cold-working properties, hardware and munitions
	Red Brass	1/4 Hard	0.316	1810–1880	10.4	16.0	6.0	39	50	25	89	Copper replacement, stronger and more ductile
	Cartridge Brass	1/4 Hard	0.308	1680–1750	11.1	16.0	6.0	40	54	43	89	Best combination of ductility and strength of any brass, radiators, munitions
	Manganese Bronze	Hard	0.302	1590–1630	11.8	15.0	5.6	50	75	30	130	High strength and wear resistance, clutch disks, shafts
	Free Cutting Phosphor Bronze	1/2 Hard	0.321	1700–1830	9.6	16.0	6.0	40	48	25	107	Easily machined, bearing material due to lead content
	Nickel Silver	1/4 Hard	0.314	1830–1900	9.0	18.0	6.8	45	65	23	110	Exc. corrosion resistance, optical goods
	High Silicon Bronze	Hard	0.308	1780–1880	10.0	16.0	6.0	55	92	22	157	Corrosion resist. with strength of mild steel, piston rings, propellers
Other Metals	Monel	70% Ni, 30% Cu	0.319	–	–	26.0	–	40	85	35	185	High strength and corrosion resistance
	Inconel	72% Ni, 17% Cr, 10% Fe	0.308	2540–2600	–	31.0	–	50	100	35	170	Excellent resistance to high-temperature oxidation
	Hastelloy X	47% Ni, 22% Cr, 18% Fe	0.297	2350	7.7	–	–	52.2	114	43	–	Jet engine tailpipes, turbine blades, and afterburners
	Magnesium M1A	1.2% Mn, Hard	0.063	1198–1200	14.5	–	–	26	35	7	54	Moderate mechanical properties, excellent weldability
	Titanium	Commercially Pure	0.164	3029	5.4	–	–	70	80	20	–	Chemical processing equipment
	Ti-Mn Alloy	9.0% Mn	0.171	–	6.0	–	–	140	150	18	–	Moderate high-temperature strength, aircraft skin
	Corroding Lead	Sand Cast	0.410	–	16.0	2.0	–	0.8	1.8	30	3.2–4.5	For cable sheathing, ammunition, solder, bearings and seals
	Arsenical Lead	12–15% Sb, "Babitt"	0.365	595–667	–	4.2	–	–	10.3	2	20	High-load, high-speed bearings
	Zamak-3 (Zinc)	4.3% Al, 0.25% Cu	0.240	717–728	15.2	–	–	–	41	10	82	Die casting of automotive parts, hardware and toys
	Hard Tin	Annealed 395°F	–	441–446	–	–	–	–	3.3	–	–	Collapsible tubes and foil
	Pewter	91% Sn, 7% Sb, 2% Cu Cold Rolled	0.263	471–563	–	–	–	–	7.6	50	8	Ornamental household items
	Platinum	70°F	0.775	3215	5.6	21	–	–	24	35	–	
	Tungsten	70°F	0.697	6170	2.56	50	–	–	20–600	0–3	–	

Tables of Physical Properties (*Continued*)

Polymers

	Plastic type	Condition	Density lb/in³	Coefficient of therm. exp. 10^{-5} °F^{-1}	Optical characteristics	Tensile modulus (10^5 psi)	Tensile strength (10^3 psi)	Compress. strength (10^3 psi)	Tensile elong. %	Izod impact. en. (70°F)	Compression molding Temperature °F	Compression molding Pressure (ksi)	Injection molding Temperature °F	Injection molding Pressure (ksi)
Thermoplastics	Acetal	Polymer	0.051	4.5	Translucent	4	10	18	15-75	1-2.5	350-400	1-5	350-450	10-20
	ABS	Molded	0.038	5.0	Avail. in colors	-	4-7	-	-	2-10	300-375	1-8	400-500	6-30
	Polymethylmethacrylate		0.042	2.7-5.0	Opaque	4-5	7-11	12-18	2-10	0.4	300-425	2-10	325-500	10-20
	Polytetrafluoroethylene	PTFE (Teflon)	0.078	5.0	Opaque	0.6	2-4.5	1.7	200-400	3.0	-	2-5	-	-
	Nylon	Type 6	0.041	4.0	Translucent or opaque	1.5-4	7-12	7-13	25-300	1-3	-	-	450-600	10-25
	Polycarbonate	Unfilled	0.043	4.0	Transparent to opaque	3.5	8-9	12	60-100	15	300-400	1	400-500	10-20
	Polyethylene	High density	0.034	6.0	Translucent to opaque	0.6-1.5	3-5	3	15-100	1-20	300-450	0.5-0.8	300-600	10-20
	Polyethylene	Low density	0.033	9.0	Translucent to opaque	0.2-0.4	1-2	-	90-800	No break	275-350	0.1-0.8	300-700	8-30
	Polypropylene	Polymer	0.033	3.0-5.0	Translucent	1.5-2	4-4.5	6-8	200-700	1-6	350-450	0.3-3	400-600	10-30
	Polystyrene	Polymer	0.040	3.0-4.0	Transparent (89%)	4-5	5-9	11-16	1-2	0.3	250-400	1-10	325-600	10-30
	ABS	ABS	0.039	3.0-7.0	Translucent to opaque	1-4	2-9	2-11	10-140	1-10	300-400	1-8	350-550	6-30
Thermosetting plastics	Polyester	65% Glass cloth	0.055-0.078	-	Opaque	-	30-70	25-50	-	5-30	70-250	0-0.3	-	-
	Epoxy	70% Glass cloth	0.072	-	Opaque	-	20-60	50-70	-	11-26	70-250	0-0.3	-	-
	Phenolic	60% Glass cloth	0.069	-	Opaque	-	40-60	35-40	-	10-35	-	-	-	-

Ceramics

Material	Density lb/in³	Melting point °F	Coefficient of therm. exp. 10^{-6} °F^{-1}	Young's modulus (10^6 psi)	Compress. strength (1,000 psi)	Tensile strength (1,000 psi)	Elongation %	Knoop hardness	Thermal shock resistance	Comments and uses
SiC Silicon Carbide	0.112	4940°F	2.7	13	15	-	-	2,500	Excellent	Cutting tools
WC Tungsten Carbide	0.550	5025	3.0	60-95	520-800	130	-	2,500	Excellent	Tube sockets, coil forms, values and seats
ZrO₂·SiO₂ Zircon	-	-	-	21	60-100	5-12	-	-	-	-

Index

Interpolate LCF by log N
(PSWC chart)

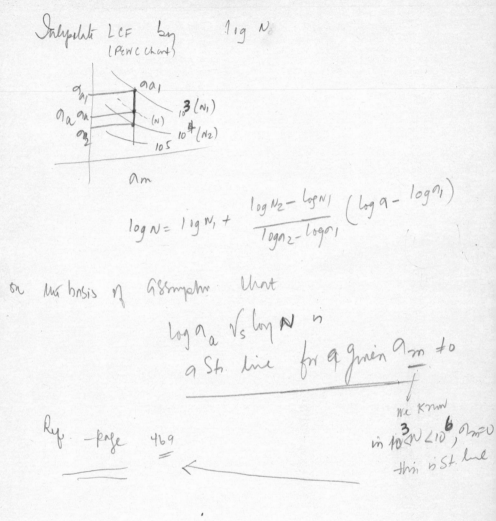

$$\log N = \log N_1 + \frac{\log N_2 - \log N_1}{\log \sigma_2 - \log \sigma_1}(\log \sigma - \log \sigma_1)$$

on the basis of assumption that

log σ_a Vs log N is

a St. line for a given $\sigma_m \neq 0$

Ref. - page 469

we know
in $10^3 < N < 10^6$, $\sigma_m = 0$
this is St. line